Inverse Heat Transfer

Heat Transfer
A Series of Reference Books and Textbooks

Afshin J. Ghajar
Regents Professor, School of Mechanical and Aerospace Engineering,
Oklahoma State University

Engineering Heat Transfer, Third Edition
William S. Janna

Conjugate Problems in Convective Heat Transfer
Abram S. Dorfman

Thermal Measurements and Inverse Techniques
Helcio R.B. Orlande, Olivier Fudym, Denis Maillet, Renato M. Cotta

Introduction to Thermal and Fluid Engineering
Allan D. Kraus, James R. Welty, Abdul Aziz

Advances in Industrial Heat Transfer
Alina Adriana Minea, Editor

Introduction to Compressible Fluid Flow, Second Edition
Patrick H. Oosthuizen and William E. Carscallen

District Cooling: Theory and Practice
Alaa A. Olama

Advances in New Heat Transfer Fluids: From Numerical to
Experimental Techniques
Alina Adriana Minea, Editor

Finite Difference Methods in Heat Transfer
M. Necati Özişik, Helcio R.B. Orlande, Marcelo J. Colaço, Renato M. Cotta

Convective Heat and Mass Transfer, Second Edition
S. Mostafa Ghiaasiaan

The Art of Measuring in Thermal Sciences
Josua Meyer and Michel De Paepe

Inverse Heat Transfer
Fundamentals and Applications

Second Edition

M. Necati Özisik
Helcio R. B. Orlande

CRC Press
Taylor & Francis Group
Boca Raton London New York

CRC Press is an imprint of the
Taylor & Francis Group, an **informa** business

Second edition published 2021
by CRC Press
6000 Broken Sound Parkway NW, Suite 300, Boca Raton, FL 33487-2742

and by CRC Press
2 Park Square, Milton Park, Abingdon, Oxon, OX14 4RN

© 2021 Helcio R. B. Orlande

First edition published by CRC Press 2000

CRC Press is an imprint of Taylor & Francis Group, LLC

ISBN: 978-0-367-82067-1 (hbk)
ISBN: 978-0-367-72526-6 (pbk)
ISBN: 978-1-003-15515-7 (ebk)

Typeset in Times
by codeMantra

To Teresa, Fernanda and Arthur José

Contents

PART I *Introduction and Parameter Estimation*

PART II   Function Estimation

PART III *State Estimation*

Preface

The second edition of *Inverse Heat Transfer: Fundamentals and Applications* is published 21 years after the publication of the original book. During this period, inverse problems have evolved beyond academic works and are now an important tool in engineering practice. Also, methods commonly used within the statistics community have become popular for the solution of inverse problems. In heat transfer, applications have significantly progressed from the estimation of a *cause* by knowing the *effects*, like the identification of a boundary heat flux from temperature measurements in heat conduction. Nowadays, inverse heat transfer problems commonly deal with quite complicated situations, involving coupled phenomena of different natures, measurements of diverse quantities and several unknowns that vary in space and time. The evolution of inverse heat transfer problems within the last 21 years, in terms of solution methodologies and applications, is reflected in this second edition of the book.

The first edition of *Inverse Heat Transfer: Fundamentals and Applications* was divided into two parts, which defined the book subtitle. The first part (*Fundamentals*) included two chapters with basic concepts and four solution techniques, while the second part (*Applications*) was composed of four chapters with applications of inverse problems in conduction, convection and radiation. The material presented in this second edition was rearranged in three parts, in order to accommodate three additional solution techniques, as well as new applications. An effort was made to keep the text with a didactical structure, by presenting concepts, methods and examples with an increasing complexity level.

The three parts of this book are:

Part I: Introduction and Parameter Estimation;
Part II: Function Estimation;
Part III: State Estimation.

Inverse problems are solved in Parts I and II with a whole-domain approach, where all the available measurements are simultaneously used for the estimation of parameters and/or functions. Part I is focused on the estimation of constant parameters, while Part II is aimed at the estimation of functions, which appear in the mathematical formulation of heat transfer problems. In Part III, measurements are sequentially assimilated to estimate dynamic variables that uniquely define the state of a heat transfer problem.

Within each part of the book, chapters were organized in terms of the solution methodology and of the information considered available for the solution of the inverse problem. Examples involving different heat transfer modes and coupled phenomena were included in each chapter, rather than in separate chapters as in the first edition of the book.

The solution methodologies covered in this book are:

Technique I: Levenberg-Marquardt Method
Technique II: Conjugate Gradient Method
Technique III: Conjugate Gradient Method with Adjoint Problem for Parameter Estimation
Technique IV: Markov Chain Monte Carlo (MCMC) Method
Technique V: Conjugate Gradient Method with Adjoint Problem for Function Estimation
Technique VI: Kalman Filter
Technique VII: Particle Filter

Techniques I, II and III are applied in Part I to the solution of parameter estimation problems by considering the minimization of objective functions. In this book, the definition of an objective function is based on statistical hypotheses regarding the measurement errors, as well as on any *prior* information available for the unknown quantities before the solution of the inverse problem. Techniques I, II and III are presented in Chapter 2 for cases where there is no *prior* information about the unknown parameters. Chapter 3 deals with parameter estimation by also solving a minimization problem, but where the objective functions are obtained from the *posterior probability distribution*. While in Chapter 2 the objective function aims at the maximization of the *likelihood function* (statistical model of the measurement errors), the objective function in Chapter 3 aims at the maximization of the *posterior distribution*, which takes into account the prior information and the measurement error statistical models. The prior information is formally combined with the information provided by the measurements by applying Bayes' theorem. Therefore, the solutions of inverse parameter estimation problems considered in Chapter 3 fall within the Bayesian framework of statistics. Part I also includes Chapter 4, in which the solution of the parameter estimation problem is obtained with stochastic simulation of the posterior probability distribution. In this case, inference on the posterior distribution is obtained through inference on its samples, which are generated with the Markov Chain Monte Carlo (MCMC) method (Technique IV).

Part II deals with the solution of function estimation problems. In such cases, the ill-posed character of the inverse problem is unquestionably apparent and the solution needs to be obtained in terms of an approximate well-posed problem, by utilizing regularization techniques. In Chapter 5, Technique V is applied for cases involving the estimation of functions without any prior information regarding their functional forms. Technique V was advanced by the group led by Dean O. M. Alifanov of the Moscow Aviation Institute and belongs to the class of *Iterative Regularization* methods. In Chapter 6, inverse problems of function estimation are solved within the Bayesian framework of statistics, either by the minimization of an objective function or by stochastic simulation.

Part III is focused on the solution of *State Estimation* problems, where the available measured data are used together with prior knowledge about the phenomena of interest and the measurement devices, in order to sequentially produce estimates of the desired dynamic variables. The Kalman filter (Technique VI), which is the optimal solution for linear problems with additive Gaussian noises, is presented in Chapter 7. Chapter 8 presents a sequential Monte Carlo technique, commonly denoted as Particle Filter (Technique VII), which can be applied to nonlinear and/or non-Gaussian problems.

Newcomers are recommended to sequentially read the book by following the order of the chapters. The material in Part I can serve for the solution of many inverse problems of practical interest and the reader might not immediately need to go over Parts II and III. More experienced readers might be particularly interested in the subject of a specific chapter. A list of symbols is not presented and the general nomenclature is established along the book. On the other hand, examples that have their own particular nomenclature have the associated symbols defined in the corresponding sections. References are given at the end of the book. The literature on inverse problems is vast and about 700 papers containing the keyword *Inverse Heat Transfer* have been published per year in the last 3 years. Therefore, references were cited with the only objective of supporting the text and examples, that is, a literature review on inverse problems was neither intended nor performed.

The first edition of this book was limited to the presentation and application of Techniques I, II, III and V (Technique V was denoted as Technique IV in the first edition). Techniques IV, VI and VII rely on the formalism of the Bayesian framework of statistics, that is, the solution of the inverse problem is obtained through statistical inference on the posterior probability distribution of the unknowns. The Bayesian techniques included in this second edition reflect my works after the publication of the first edition of the book. In fact, in 2005, Professor Jari Kaipio and Professor Ville Kolehmainen visited COPPE/UFRJ and gave a very comprehensive course based on the book *Statistical and Computational Inverse Problems* by J. P. Kaipio and E. Somersalo.[1]

---

[1] Kaipio, J., Somersalo, E., *Statistical and computational inverse problems*, Applied Mathematical Sciences 160, Springer-Verlag, New York, 2004.

After this course, I have been mainly using solution methodologies within the Bayesian framework of statistics.

Texts and examples that I have previously prepared for METTI Advanced Schools were very helpful for the preparation of the second edition of this book. These schools are periodically organized by the METTI Group, which is a division of the French Heat Transfer Society - SFT. METTI is the acronym for *Thermal Measurements and Inverse Techniques* (in French). The participation in METTI Schools is strongly encouraged for those interested on inverse heat transfer problems. METTI Schools usually cover basic and advanced subjects in the form of lectures and hands-on tutorials.

Just after obtaining my Ph.D. in 1994 under the direction of Professor Ozisik at North Carolina State University, I was surprised with his invitation to join him in the preparation of the first edition of this book. Professor Emeritus M. N. Ozisik dedicated his life to education and research in heat transfer, being an iconic worldwide leader in thermal sciences. He published more than 300 research papers in international journals and conferences. He was the author of 11 books, most of them bestsellers that were re-edited several times and published in different languages. I now have the privilege to write this second edition of *Inverse Heat Transfer*: *Fundamentals and Applications* and to honor Professor Ozisik by keeping his name as the first author. I consider completely inappropriate the situation where authors of a new book edition, who were not authors in previous editions (and, in some cases, not even known by the original authors), have their names appearing ahead of deceased authors of the original work.

This second edition of *Inverse Heat Transfer*: *Fundamentals and Applications* should be ready by December 2019. However, I had the opportunity to work with Professor George S. Dulikravich, at Florida International University in Miami, from January to June 2020. The time needed to take care of my own personal matters in preparation for this trip has delayed the book to 2020. My wife and I were then caught in Florida by the COVID-19 pandemic and we had to return to Rio de Janeiro in March 2020. It was challenging to finish the book during this difficult period, while hearing everyday bad news about the huge number of people dying due to COVID-19 around the world and especially in Brazil. Our precautions against contamination have confined my whole family at home. Being together with my wife, daughter and son, everybody in good health and trying to keep high spirits, has undoubtedly motivated me to conclude this work. With love, I express my deepest gratitude to Teresa, Fernanda and Arthur José, for their support, motivation and for bringing me joy, not only during these strange and unimaginable days, but at all times.

This book would not be possible without the support of Professor Afshin J. Ghajar, who is the *Heat Transfer* Series Editor for CRC Press/Taylor & Francis. For several years, Professor Ghajar has been encouraging me to write this second edition of *Inverse Heat Transfer*: *Fundamentals and Applications*, which finally becomes a reality. I am indebted to Professor Ghajar, as well as to Mr. Jonathan W. Plant, who used to be the Executive Editor for Mechanical Engineering & Applied Mechanics of the Taylor & Francis/CRC Press Group, and retired in November 2019. The publication of this book has been afterwards managed by Mrs. Kyra Lindholm, the Editor for Thermal-Fluids, Aerospace/Aviation, Energy, Nuclear Engineering of the Taylor & Francis/CRC Press Group. Mrs. Lindholm's diligent work, taking care of all edition and publication details, is sincerely appreciated. Also, I would like to thank all the CRC Press staff involved in the edition and publication of the book.

Throughout the years, my research has been continuously supported by agencies of the Brazilian government (CNPq—National Council for Scientific and Technological Development; CAPES—Brazilian Foundation for the Advancement of Graduate Studies) and of the local government of Rio de Janeiro (FAPERJ—Foundation "Carlos Chagas Filho" for Research Support of the State of Rio de Janeiro). Without citing names, I would like to thank my current and former students, as well as all my collaborators. Beyond the joint work on publications and research projects, they have significantly influenced my professional life and many became very good friends.

Helcio R. B. Orlande

Preface of the First Edition

Inverse Heat Transfer Problems (IHTP) rely on temperature and/or heat flux measurements for the estimation of unknown quantities appearing in the analysis of physical problems in thermal engineering. As an example, inverse problems dealing with heat conduction have been generally associated with the estimation of an unknown boundary heat flux, by using temperature measurements taken below the boundary surface. Therefore, while in the classical direct heat conduction problem the cause (boundary heat flux) is given and the effect (temperature field in the body) is determined, the inverse problem involves the estimation of the cause from the knowledge of the effect. An advantage of IHTP is that it enables a much closer collaboration between experimental and theoretical researchers, in order to obtain the maximum of information regarding the physical problem under study.

Difficulties encountered in the solution of IHTP should be recognized. IHTP are mathematically classified as ill-posed in a general sense, because their solutions may become unstable, as a result of the errors inherent to the measurements used in the analysis. Inverse problems were initially taken as not of physical interest. due to their ill-posedness. However, some heuristic methods of solution for inverse problems, which were based more on pure intuition than on mathematical formality, were developed in the 50's. Later in the 60's and 70's, most of the methods, which are in common use nowadays, were formalized in terms of their capabilities to treat ill-posed unstable problems. The basis of such formal methods resides on the idea of reformulating the inverse problem in terms of an approximate well-posed problem, by utilizing some kind of regularization (stabilization) technique. In this sense, it is recognized here the pioneering works of scientists who found different fonns of overcoming the instabilities of inverse problems, including A. N. Tikhonov, O. M. Alifanov and J. V. Beck.

The field of inverse heat transfer is wide open and diversified. Therefore, an orderly and systematic presentation of the scientific material is essential for the understanding of the subject. This principle has been the basic guideline in the preparation of this book.

This book is intended for graduate and advanced undergraduate levels of teaching, as well as to become a reference for scientists and practicing engineers. We have been motivated by the desire to make an application-oriented book, in order to address the needs of readers seeking solutions of IHTP, without going through detailed mathematical proofs.

The main objectives of the book can be summarized as follows:

- Introduce the fundamental concepts regarding IHTP;
- Present in detail the basic steps of four techniques of solution of IHTP, as a parameter estimation approach and as a function estimation approach;
- Present the application of such techniques to the solution of IHTP of practical engineering interest, involving conduction, convection and radiation; and
- Introduce a formulation based on generalized coordinates for the solution of inverse heat conduction problems in two-dimensional regions.

The book consists of six chapters.

Chapter 1 introduces the reader to the basic concepts of IHTP.
Chapter 2 is concerned with the description of four techniques of solution for inverse problems. The four techniques considered in this book include:

Technique I: The Levenberg-Marquardt Method for Parameter Estimation
Technique II: The Conjugate Gradient Method for Parameter Estimation
Technique III: The Conjugate Gradient Method with Adjoint Problem for Parameter Estimation
Technique IV: The Conjugate Gradient Method with Adjoint Problem for Function Estimation

These techniques were chosen for use in this book because, based on the authors' experience, they are sufficiently general, versatile, straightforward and powerful to overcome the difficulties associated with the solution of IHTP.

In **Chapter 2** the four techniques are introduced to the reader in a systematic manner, as applied to the solution of a simple, but illustrative, one-dimensional inverse test-problem, involving the estimation of the transient strength of a plane heat-source in a slab. The basic steps of each technique, including the iterative procedure, stopping criterion and computational algorithm, are described in detail in this chapter. Results obtained by using simulated measurements, as applied to the solution of the test-problem, are discussed. The mathematical and physical significances of sensitivity coefficients are also discussed in Chapter 2 and three different methods are presented for their computation. Therefore, in Chapter 2 the reader is exposed to a full inverse analysis involving a simple test-problem, by using the four techniques referred to above, which will be applied later in the book to more involved physical situations, including Conduction Heat Transfer in **Chapter 3**, Convection Heat Transfer in **Chapter 4** and Radiation Heat Transfer in **Chapter 5**.

Chapter 6 is concerned with the solution of inverse heat conduction problems of estimating the transient heat flux applied on part of the boundary of irregular two-dimensional regions, by using Technique IV. The irregular region in the physical domain (x,y) is transformed into a rectangle in the computational domain (ξ,η). Different quantities required for the solution are formulated in terms of the generalized coordinates (ξ,η). Therefore, the present formulation is general and can be applied to the solution of boundary inverse heat conduction problems over any region that can be mapped into a rectangle. The present approach is illustrated with an inverse problem of practical engineering interest, involving the cooling of electronic components.

The pertinent References and sets of Problems are included at the end of each chapter. The proposed problems expose the reader to practical situations in a gradual level of increasing complexity, so that he(she) can put into practice the general concepts introduced in the book.

We would like to acknowledge the financial support provided by CNPq, CAPES and FAPERJ, agencies for science promotion of the Brazilian and Rio de Janeiro State governments, as well as by NSF-USA, for the visits of M. N. Özisik to the Federal University of Rio de Janeiro (UFRJ) and of H. R. B. Orlande to the North Carolina State University (NCSU). The hospitality of the Mechanical Engineering Departments at both institutions is greatly appreciated. This text was mainly typed by M. M. Barreto, who has demonstrated extreme dedication to the work and patience in understanding our handwriting in the original manuscript The works of collaborators of the authors, acknowledged throughout the text, were essential for transforming an idea for a book into a reality. We would like to thank Prof. M. D. Mikhailov for invaluable suggestions regarding the contents of Chapter 2 and Prof. R M. Cotta for introducing us to the editorial vice president of Taylor & Francis. We are indebted to several students from the Department of Mechanical Engineering of the Federal University of Rio de Janeiro, who helped us at different points during the preparation of the book. They include E. N. Macedo, R.N. Carvalho, M. J. Colaço, M. M. Mejias, L. M. Pereira, L. B. Dantas, H. A. Machado, L. F. Saker, L. A. Sphaier, L. S. B. Alves, L. R. S. Vieira and C. F. T. Matt. H. R. B. Orlande is thankful for the kind hospitality of several friends during his visits to Raleigh, who certainly made the preparation of this book more pleasant and joyful. They include the Ferreiras, the Gonzalezes and the Özisiks. Finally, we would like to express our deep appreciation for the love, prayers and support of our families.

M. Necati Özisik
Helcio R. B. Orlande

Authors

M. Necati Özisik (1923–2008) retired in 1998 as Professor Emeritus of North Carolina State University's Mechanical and Aerospace Engineering Department, where he spent most of his academic career. Professor Özisik dedicated his life to education and research in heat transfer. His outstanding contributions earned him numerous awards, including the Outstanding Engineering Educator Award from the American Society for Engineering Education in 1992. He authored eleven books, many of which were published in different languages.

Helcio R. B. Orlande earned a BS in mechanical engineering at the Federal University of Rio de Janeiro (UFRJ) in 1987 and an MS in mechanical engineering from the same university in 1989. After earning a Ph.D. in mechanical engineering in 1993 at North Carolina State University, he joined the Department of Mechanical Engineering of UFRJ, where he was the Department Head from 2006 to 2007. His research areas of interest include the solution of inverse heat and mass transfer problems as well as the use of numerical, analytical and hybrid numerical-analytical methods of solution of direct heat and mass transfer problems. He is a member of the Scientific Council of the International Centre for Heat and Mass Transfer and a Delegate in the Assembly for International Heat Transfer Conferences. He serves as an Associate Editor for the journals *Heat Transfer Engineering, Inverse Problems in Science and Engineering, High Temperatures – High Pressures* and *International Journal of Thermal Sciences.*

Part I

Introduction and Parameter Estimation

1 Basic Concepts

For more than 50 years, the theory and application of inverse problems has drawn the interest of researchers in almost every branch of science and engineering. Mechanical, aerospace, chemical and nuclear engineers, mathematicians, astrophysicists and statisticians are all interested in this subject, each group with different applications in mind.

The space program has played a significant role in the advancement of solution techniques for inverse heat transfer problems (IHTPs) in late 1950s and early 1960s. For example, aerodynamic heating of space vehicles is so high during reentry in the atmosphere that the surface temperature of the thermal shield cannot be measured directly with temperature sensors. Therefore, temperature sensors are placed beneath the hot surface of the shield and the surface temperature is recovered by inverse analysis. Inverse analysis can also be used for the estimation of the thermophysical properties of the heat shield during operating conditions at such high temperatures.

Recent technological advancements often require the use of complicated experiments and indirect measurements, within the research paradigm of inverse problems. With the advent of modern complex materials having thermophysical properties strongly varying with temperature and position, the use of conventional methods for determining thermophysical properties has become unsatisfactory. Similarly, the operation of modern industrial concerns is becoming more and more sophisticated, and an accurate *in situ* estimation of thermophysical properties under actual operating conditions is many times necessary. The IHTP approach can provide satisfactory answers for such situations because it enables to conduct experiments as close to the real conditions as possible.

Nowadays, inverse analyses are encountered in single- and multi-mode heat transfer problems, dealing with multi-scale phenomena. Applications range from the estimation of constant heat transfer parameters to the mapping of spatially and timely varying functions, such as heat sources, fluxes and thermophysical properties. In a more general sense, inverse problems involving other physical phenomena are constantly present in our daily life, such as in medical imaging with x-ray tomography or magnetic resonance. Another common everyday application of inverse problems is the estimation of dynamic states, like the weather forecast, with evolution and measurement stochastic models.

Difficulties associated with the solution of IHTPs should be recognized. Inverse problems are mathematically classified as *ill-posed*, whereas standard heat transfer problems are *well-posed* [1–22]. The concept of well-posedness, originally introduced by Hadamard [1], requires that the solution of a mathematical problem must satisfy the following three conditions:

- Existence;
- Uniqueness;
- Stability under small changes in the input data.

For several cases, the existence of a solution for an IHTP may be assured by physical reasoning. On the other hand, even for these cases, proofs of the uniqueness of the inverse problem solution are in general challenging [2–22]. For example, consider a transient heat conduction problem in a slab with thermally insulated surfaces, initially at a uniform temperature. A sensor at some position inside the slab is used to measure the transient variation of the local temperature. If there is a change in the measured temperature larger than the measurement uncertainties, then there exists a causal characteristic, say, an internal heat source that promotes this temperature variation. However, questions immediately come to our minds, such as: Is the information provided by the temperature measurements sufficient to distinguish if the heat source is constant or transient? Is the heat source uniform or does it vary spatially in the slab?

The questions of existence and uniqueness are even more complicated for inverse problems that are not related to the estimation of causal characteristics, such as the heat source in the above example. Consider that heat flux and temperature measurements are taken at the surface of a body for the identification of the internal spatial variation of its thermophysical properties. This inverse problem is actually a thermal tomography [23–25], which can be used to detect internal structures of the body. Several questions thus arise: Do multiple internal structures exist? What are the locations and shape of multiple internal structures, if they really exist? Are the surface measurements capable of identifying internal structures?

Despite possible mathematical proofs of solution existence and uniqueness, inverse problems are very sensitive to errors in the measured input data. Thus, the solution of an inverse problem requires special stabilization techniques, commonly referred to as *regularization* techniques. In fact, the ill-posed inverse problem is not really solved, but an approximate solution is obtained based on an approximate well-posed *regularized* problem.

For a long time it was thought that, if any of the conditions required for well-posedness were violated, the problem would be unsolvable or the results obtained from such a solution would be meaningless, hence would have no practical importance. As a result, interest waned by mathematicians, physicists and engineers on the solution of inverse problems [5]. With the development of *Tikhonov's regularization technique* [3,26–28], *Alifanov's iterative regularization technique* [2,6,29–43] and *Beck's sequential function specification technique* [5,44], the interest on the solution of IHTPs was revitalized. Tikhonov's regularization procedure modifies the least squares norm by adding smoothing terms in order to reduce the unstable effects of the measurement errors. In Alifanov's iterative regularization method, the stopping criterion is chosen so that the final solution is stabilized with respect to errors in the input data. In Beck's sequential technique, regularization is obtained from the averaging properties of least squares and from the measurements taken at future time steps. More recently, stochastic simulation techniques are becoming very popular for the solution of inverse problems. These simulation techniques generate samples of a statistical distribution and inference on the distribution is obtained through inference on the samples. In particular, the solution of an inverse problem within the *Bayesian framework* of statistics is recast in the form of statistical inference on the so-called *posterior probability density* [17–20].

One of the earliest discussions of thermal inverse problems is due to Giedt [45], who examined the heat transfer at the inner surface of a gun barrel. Stolz [46] presented a procedure for estimating surface temperature and heat flux from the temperature measurements taken within a body being quenched. It is also important to mention the pioneer work of Calderón [47], who dealt with solution uniqueness for the inverse problem of estimating the spatially variation of conductivity within a medium, from the knowledge of the potential and of the flux over the surface.

The literature on IHTPs is abundant. By the time of publication of this book, a search with the keyword *inverse heat transfer* in a scientific database resulted in almost 10,000 documents, with ~700 documents being published per year in the last 3 years. Around 55% of the inverse heat transfer publications are classified in the subject area of *Engineering*. Therefore, the use of inverse analysis in heat transfer is indeed a practical tool in thermal engineering [48,49].

In this chapter, we present a general discussion of IHTPs including basic concepts, classification, an overview of various solution techniques and difficulties involved in such solutions.

1.1 INVERSE HEAT TRANSFER PROBLEM CONCEPT

Inverse heat transfer problems deal with the estimation of unknown quantities appearing in the mathematical formulation of thermal sciences' processes, by using measurements of temperature, heat flux, radiation intensities, etc.

The significance of this concept is better envisioned by referring to the following one-dimensional transient heat conduction problem, in a slab of thickness L. The temperature distribution in the slab is initially $F(x)$. For times $t>0$, a transient heat flux $q(t)$ is applied on the boundary $x = 0$, while the boundary $x = L$ is maintained at the constant temperature T_L. The mathematical formulation of this problem is given by:

$$\frac{\partial}{\partial x}\left(k\frac{\partial T}{\partial x} \right) = \rho c_p \frac{\partial T}{\partial t} \text{ in } 0 < x < L, \text{ for } t > 0 \tag{1.1.1a}$$

$$-k\frac{\partial T}{\partial x} = q(t) \text{ at } x = 0, \text{ for } t > 0 \tag{1.1.1b}$$

$$T = T_L \text{ at } x = L, \text{ for } t > 0 \tag{1.1.1c}$$

$$T = F(x) \text{ for } t = 0, \text{ in } 0 < x < L \tag{1.1.1d}$$

For the case where the boundary conditions $q(t)$ and T_L, the initial condition $F(x)$ and the thermophysical properties ρ, c_p and k are all known, the problem given by equations (1.1.1a–d) is concerned with the determination of the temperature distribution $T(x, t)$ in the interior region of the solid, as a function of time and position. This is a *direct heat transfer problem*.

We now consider a problem similar to that given by equation (1.1.1a–d), but with the boundary condition function $q(t)$ at the surface $x = 0$ unknown, while all the other quantities appearing in these equations, such as T_L, $F(x)$, k, ρ and c_p, are known. We then wish to estimate the unknown boundary heat flux $q(t)$. To compensate for the lack of information on this boundary condition, measured temperatures $Y(x_{meas}, t_i) \equiv Y_i$ are given at an interior point x_{meas} ($0 < x_{means} < L$) at different times t_i ($i = 1, 2,..., I$), over a specified time interval $0 < t \leq t_f$, where t_f is the final time. This is an *inverse problem* because it is concerned with the *estimation* of the unknown surface condition $q(t)$.

Here, the terminology *estimation* is used in place of determination. The reason is that the measured temperature data used in the inverse analysis contain measurement errors. As a result, the quantity recovered by the inverse analysis (i.e., the boundary condition $q(t)$ in the example above) is not exact and should be reported with the associated uncertainties.

Then, the mathematical formulation of this *inverse problem* can be written as:

$$\frac{\partial}{\partial x}\left(k\frac{\partial T}{\partial x} \right) = \rho c_p \frac{\partial T}{\partial t} \text{ in } 0 < x < L, \text{ for } 0 < t \leq t_f \tag{1.1.2a}$$

$$-k\frac{\partial T}{\partial x} = q(t) = ? \,(\text{unknown}) \text{ at } x = 0, \text{ for } 0 < t \leq t_f \tag{1.1.2b}$$

$$T = T_L \text{ at } x = L, \text{ for } 0 < t \leq t_f \tag{1.1.2c}$$

$$T = F(x) \text{ for } t = 0 \text{ in } 0 < x < L \tag{1.1.2d}$$

with temperature measurements taken at an interior location x_{meas}, at different times t_i, given by:

$$Y\left(x_{meas}, t_i \right) \equiv Y_i \text{ at } x = x_{meas}, \text{ for } t = t_i \, (i = 1,2,...,I) \tag{1.1.3}$$

Originally, IHTPs have been associated with the estimation of an unknown boundary heat flux, by using temperature measurements taken below the boundary surface of a heat conducting medium [5], such as in the example above. This type of inverse problem is referred to as a *boundary inverse heat transfer problem*. Analogously, one envisions IHTPs of unknown initial condition, energy generation, thermophysical properties and so on.

The methodology for the *verification and validation* (V&V) of computational codes in fluid dynamics and heat transfer has been established in a standard published by ASME [50]. *Validation* is defined as the process of determining the degree to which a model is an accurate representation of the real world from the perspective of the intended uses of the model. Hence, validation necessarily requires experimental data, which is quantitatively compared to the results of the computational simulation. Uncertainties in the experimental data and in the computational results should be appropriately taken into account in the process of validation. Also, as established in the ASME V&V Code, the experimental data can be indirectly measured, that is, obtained from the solution of an inverse problem. Validation must be preceded by *code verification* and *solution verification*. *Code verification* establishes if the code accurately solves the incorporated mathematical model, that is, if the code is free of mistakes for the simulations of interest. On the other hand, *solution verification* estimates the numerical accuracy of a particular calculation through grid refinement [50].

As demonstrated by Beck [51], the use of inverse problems represents a research paradigm that is in fact complementary to that established in the ASME V&V Code [50]. In the inverse analysis paradigm, the results obtained from numerical simulations and from experiments are not compared *a posteriori*, but a close synergism exists between experimental and theoretical researchers during the course of the study, in order to obtain the maximum of information regarding the problem under analysis.

The solution of an inverse problem relies on the computational solution of the direct (forward) problem, which is used, together with the available experimental data, for the estimation of parameters and/or functions appearing in the mathematical formulation of the phenomena being studied. Therefore, the solution of inverse problems requires code verification and solution verification. On the other hand, the validation of the code comes out as part of the inverse problem solution, for example, through the analysis of the residuals. The residuals are given by the differences between the measurements and the dependent variables that are obtained from the solution of the direct problem. The residuals are expected to be small (of the order of magnitude of the measurement errors) and uncorrelated (without any deterministic behavior), if the mathematical formulation and the solution of the direct problem appropriately model the phenomena under analysis, with the estimated parameters and/or functions [4]. Another important characteristic of the inverse problem paradigm is that uncertainties on the judged *known* parameters or functions can be taken into account in the solution procedure, by using techniques within the Bayesian framework of statistics [17–20].

1.2 CLASSIFICATION OF IHTPs

So far, we considered an IHTP of conduction. Similarly, we can have inverse problems of convection, body or surface radiation, mixed modes of heat transfer and numerous others. Therefore, IHTPs can be classified in accordance with the nature of the heat transfer process, such as:

IHTP of conduction;
IHTP of convection (forced or natural);
IHTP of surface radiation;
IHTP of radiation in participating medium;
IHTP of phase change (melting or solidification);
IHTP of coupled modes of heat transfer.

Another classification can be one based on the type of characteristic to be estimated. For example,

IHTP of boundary conditions;
IHTP of thermophysical properties;
IHTP of initial condition;
IHTP of source term;
IHTP of geometric characteristics of a heated body.

IHTPs can be one-, two- or three-dimensional. Also, inverse problems can be linear or nonlinear. The factors affecting the linearity of an inverse problem will be apparent in the following chapters.

The *time domain* over which measurements are used in the inverse analysis may be another way to classify the methods of solution [5]. Consider the estimation of the boundary heat flux $q(t)$ in the time domain $0 < t \le t_f$ that was discussed in Section 1.1, given by equations (1.1.2 and 1.1.3). Three different possible time domains for the measurements used in the estimation of the heat flux component $q(t_i)$ at time $t_i < t_f$ include:

a. up to time $t_i < t_f$;
b. up to time $t_i < t_f$ plus few time steps;
c. the whole time domain $0 < t \le t_f$.

Methods based on the time domains (a) and (b) are *sequential* in nature. Although apparently attractive, they have the disadvantage that the solution algorithms are extremely sensitive to measurement errors. The use of measurements up to time t_i plus few time steps, such as in Beck's function specification technique [5,44], improves the stability of the sequential algorithms. Beck's approach is based on the fact that the temperature response is lagged with respect to a boundary heat flux excitation, as discussed below in Section 1.3. In the class of inverse problems designated as *state estimation problems*, solutions are obtained sequentially based on time domains (a) or (b). In such kind of problems, the available measured data are used together with prior knowledge about the phenomena under analysis and the measuring devices, in order to sequentially produce estimates of the desired dynamic variables. This is accomplished in such a manner that the error is minimized statistically [17,52–55], as will be apparent later in the book. We note, however, that sequential methods based on the time domains (a) and (b) generally become unstable when small time steps are used in the inverse analysis [5].

The *whole time domain* approach (c) is very powerful because very small time steps can be taken for the solution. This is quite important in order to estimate, with good resolution, time-dependent unknown functions, such as the boundary heat flux of the example above. However, methods based on the whole time domain are not as computationally efficient as the sequential ones.

Inverse problems can be solved either as a *parameter estimation* or as a *function estimation* approach. If some information is available on the functional form of the unknown quantity, the inverse problem is reduced to the estimation of few unknown parameters. Let us consider the boundary inverse problem given by equations (1.1.2 and 1.1.3) and assume that the unknown function $q(t)$ can be represented as a polynomial in time in the form:

$$q(t) = P_1 + P_2 t + P_3 t^2 + \cdots + P_N t^{N-1} \tag{1.2.1}$$

or in the more general linear form as:

$$q(t) = \sum_{j=1}^{N} P_j C_j(t) \tag{1.2.2}$$

where $P_j, j = 1,\ldots, N$, are the unknown parameters and $C_j(t)$ are known basis functions. Therefore, the inverse problem of estimating the unknown function $q(t)$ is reduced to the problem of *estimating a finite number of parameters* P_j, where the number N of parameters is supposed to be chosen in advance. Another example of parameter estimation is the recovering of unknown constant thermophysical properties, such as the thermal conductivity k or the volumetric heat capacity ρc_p, appearing in equation (1.1.2a). If no prior information is available on the functional form of the unknown, the inverse problem can be regarded as a *function estimation approach in an infinite dimensional space of functions*. Techniques for the solution of inverse problems as a parameter estimation, as well as a function estimation approach, will be presented in the following chapters.

1.3   DIFFICULTIES IN THE SOLUTION OF INVERSE HEAT TRANSFER PROBLEMS

To illustrate inherent difficulties in the solution of IHTPs, we consider a semi-infinite solid ($0 < x < \infty$), initially at zero temperature. For times $t > 0$, the boundary surface at $x = 0$ is subjected to a periodically varying heat flux in the form:

$$q(t) = q_0 \cos \omega t \qquad (1.3.1)$$

where q_0 and ω are the amplitude and frequency of oscillations for the heat flux, respectively, and t is the time variable. Figure 1.1 illustrates this heat conduction problem.

After the initial transients have passed, the *quasi-stationary* temperature distribution in the solid is given by [56]:

$$T(x,t) = \frac{q_0}{k} \sqrt{\frac{\alpha}{\omega}} \, \exp\left(-x\sqrt{\frac{\omega}{2\alpha}}\right) \cos\left(\omega t - x\sqrt{\frac{\omega}{2\alpha}} - \frac{\pi}{4}\right) \qquad (1.3.2)$$

where α is the thermal diffusivity and k is the thermal conductivity of the solid.

Equation (1.3.2) shows that the temperature response is lagged and damped with respect to the heat flux excitation, due to heat conduction through the body. Lagging and damping are more pronounced for points located deep inside the body (large x) and for a large frequency ω. In order to illustrate these effects, consider an imposed heat flux of magnitude $q_0 = 10,000$ W/m^2, as shown in Figure 1.2 for frequencies $\omega = \pi$ and $\omega = 2\pi$ rad/s. The transient variations of temperatures at positions $x = 0.001$ m and $x = 0.005$ m for the heat fluxes presented in Figure 1.2, are shown by Figure 1.3, when heat conduction takes place in a carbon steel ($k = 43$ W/m°C, $\alpha = 1.2 \times 10^{-5}$ m^2/s). Figure 1.3 shows that the amplitude of the temperature decrease when the position x or the frequency ω increase. The effect of position on the temperature lag is also clear from Figure 1.3 by comparing the curves of same frequency at $x = 0.001$ m and $x = 0.005$ m. The effect of ω on the lag is noticed in Figures 1.2 and 1.3.

The results shown in Figure 1.3 correspond to a *direct problem*, since the heat flux is known to vary in accordance with equation (1.3.1) and all parameters in the analytical solution (1.3.2) are known. If the functional form of the imposed heat flux $q(t)$ is unknown, but transient temperature measurements are available from a sensor located at some position x_{meas} inside the medium, we have a *boundary IHTP of function estimation*.

The above solution of the direct problem, that is, the temperature response due to the periodic heat flux $q(t) = q_0 \cos \omega t$, reveals important aspects for the inverse analysis. The temperature lag with respect to the imposed heat flux indicates that meaningful measurements for the solution of the inverse problem must be taken after the moment that the heat flux is applied. It is also noticed that, in order to be able to estimate the boundary heat flux, a sensor must be located within a depth below the surface where the amplitude of the temperature oscillations is much larger than the measurement errors. Otherwise, it is impossible to distinguish if the measured temperature variation is due to changes in the boundary heat flux or due to measurement errors, thus resulting in the non-uniqueness of the inverse problem solution.

A much simpler inverse problem than the estimation of the function $q(t)$ can be considered, if the heat flux is known to vary in accordance with equation (1.3.1), but the amplitude q_0 is unknown.

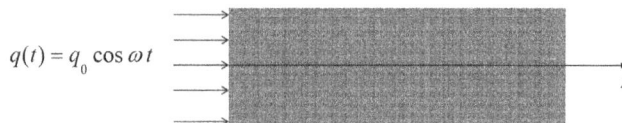

$q(t) = q_0 \cos \omega t$

FIGURE 1.1   Heat conduction in a semi-infinite medium.

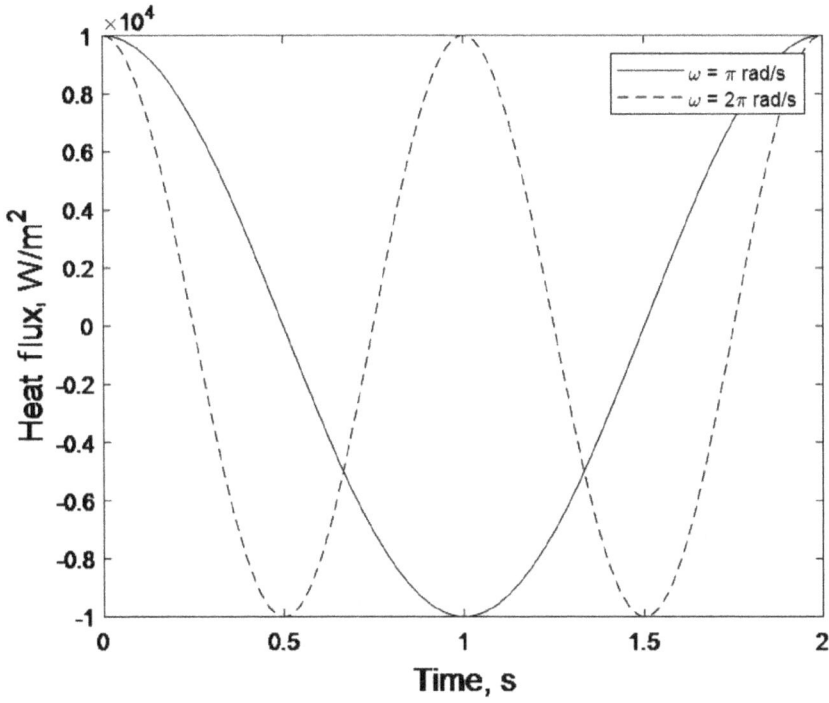

FIGURE 1.2 Periodic heat flux imposed at the surface of a semi-infinite medium.

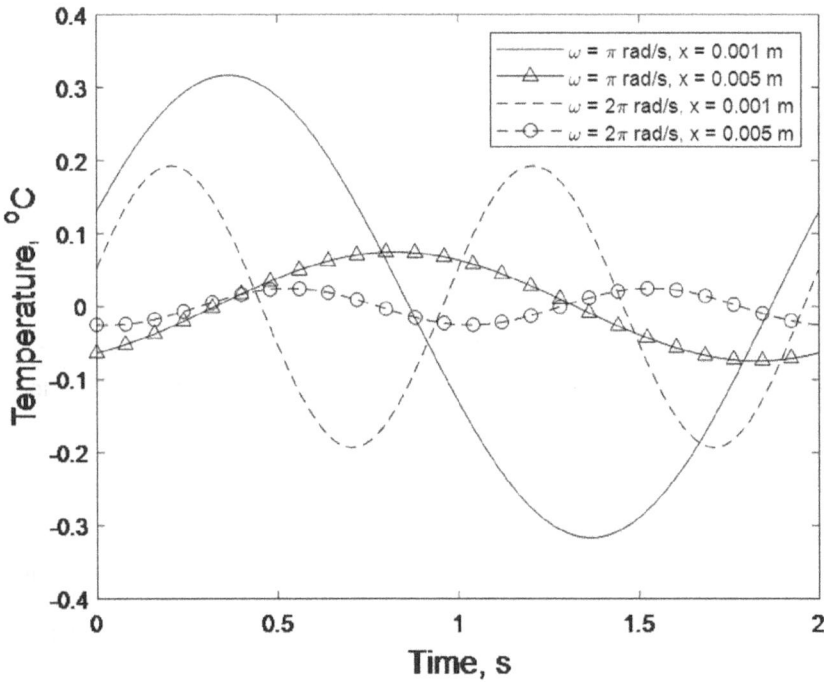

FIGURE 1.3 Transient variation of temperature for heat conduction in a carbon steel.

By examining equation (1.3.2), we note that the amplitude for the temperature oscillation at any location, $|\Delta T(x)|$, can be obtained by setting $\cos(\cdot) = 1$, that is,

$$|\Delta T(x)| = \frac{q_0}{k}\sqrt{\frac{\alpha}{\omega}}\ \exp\left(-x\sqrt{\frac{\omega}{2\alpha}}\right) \tag{1.3.3}$$

The reader can be tempted to estimate q_0 with measurements of $|\Delta T(x)|$ by rewriting equation (1.3.3) as:

$$q_0 = k|\Delta T(x)|\sqrt{\frac{\omega}{\alpha}}\exp\left(x\sqrt{\frac{\omega}{2\alpha}}\right) \tag{1.3.4}$$

However, note in equation (1.3.4) that any measurement error on $|\Delta T(x)|$ will be magnified exponentially with the position x and with the frequency ω, for the estimation of q_0. For example, if we calculate $|\Delta T(x)|$ for carbon steel with $q_0 = 10{,}000$ W/m², add to the calculated values a constant error of $+0.1°$C, and then use equation (1.3.4) to recover back q_0, we obtain the values presented in Table 1.1. As expected, the estimation errors increase with the frequency and with the measurement position. Moreover, Table 1.1 shows that the use of equation (1.3.4) with one single measurement of $|\Delta T(x)|$ did not result in acceptable estimates for q_0, since the minimum estimation error was about 31%.

One possibility to statistically reduce the estimation error is to repeat the experiment several times. Such as for the results presented in Table 1.1, we consider $q_0 = 10{,}000$ W/m² and heat conduction in a carbon steel. However, now the values calculated for $|\Delta T(x)|$ with equation (1.3.3) were modified by adding Gaussian random numbers with zero mean and constant standard deviation of $0.1°$C, in order to numerically *simulate experimental measurements*.

Simulated measurements are commonly obtained from the solution of the direct problem by prescribing values for the unknown parameters or functions. Measurement errors can be simulated, for example, by adding a Gaussian error term to the solution of the direct problem in the form:

$$Y = Y_{\text{exa}} + \gamma\sigma \tag{1.3.5}$$

where
  Y = simulated measurement containing random errors
  Y_{exa} = exact (errorless) simulated measurement, obtained from the solution of the direct problem
  σ = standard deviation of the measurement error
  γ = random variable with Gaussian distribution, zero mean and unitary standard deviation.

A successful solution of the inverse problem with simulated measurements basically consists in retrieving the same values prescribed for the parameters or functions, which were used in the direct problem to generate Y_{exa}.

The histogram of 1000 simulated measurements of $|\Delta T(x)|$ is presented by Figure 1.4, for $\omega = \pi$ rad/s and $x = 0.001$ m. Figure 1.5 presents the histogram of the heat flux amplitudes estimated

TABLE 1.1
Estimation of the Heat Flux with Equation (1.3.4)

Frequency (rad/s)	Measurement Position (m)	Estimated Heat Flux Amplitude (W/m²)
π	0.001	13,159
	0.005	23,431
2π	0.001	15,190
	0.005	50,182

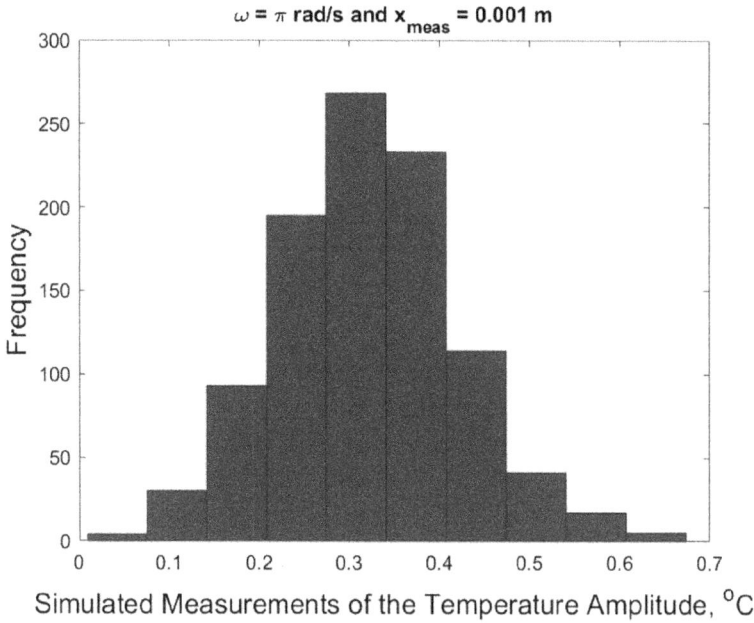

FIGURE 1.4   Histogram of the simulated measurements of $|\Delta T(x)|$.

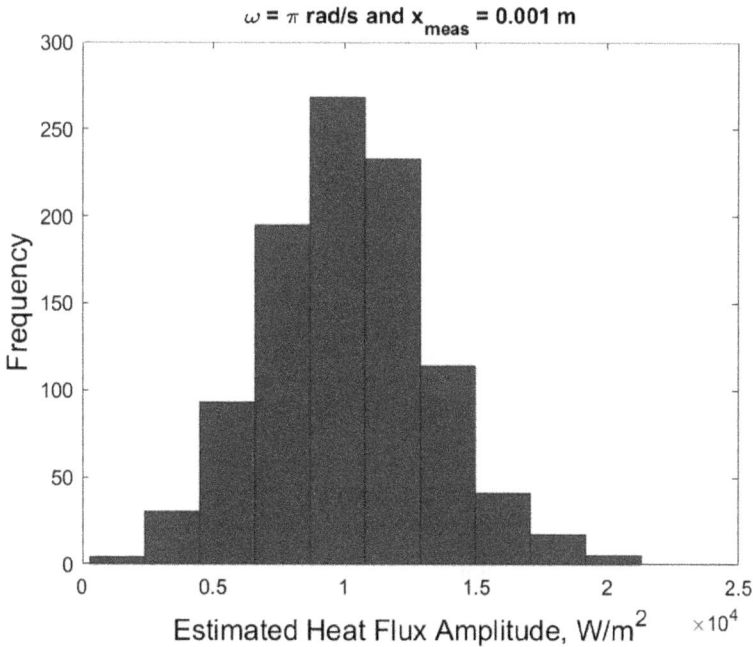

FIGURE 1.5   Histogram of the estimated q_0 with equation (1.3.4).

with these measurements by using equation (1.3.4). The histograms shown by these two figures have exactly the same format, since the distribution of $|\Delta T(x)|$ is linearly transformed to the distribution of q_0 with equation (1.3.4). Although the histogram of q_0 is centered around the exact value of 10,000 W/m², the estimation uncertainty (dispersion of the estimated values) is extremely large, as shown by Figure 1.5. Therefore, the straightforward use of equation (1.3.4), even if the experiment is repeated an enormous number of times like in this example, is not acceptable for the purpose of estimating the heat flux amplitude q_0 because of the large associated uncertainties.

The foregoing discussion reveals that, depending on the location of the sensor and the frequency of oscillations, the solution of the inverse problem may become very sensitive to measurement errors in the input data. Since the accuracy of the solution obtained by an inverse analysis is affected by the errors involved in the measurements, it is instructive to present the eight standard assumptions proposed by Beck [4,5,57], regarding the *statistical description* of such errors (basic statistical concepts are given in Note 1 at the end of this chapter). These assumptions are:

1. The errors are additive, that is,

$$Y_i = Y_{\text{exa},i} + \varepsilon_i$$

   where Y_i is the measured quantity, $Y_{\text{exa},i}$ is the exact measured quantity, ε_i is the random error and the subscript i is an index for the measurement. For example, for transient measurements obtained with a single sensor, i denotes a time t_i when the measurement is available.

2. The errors ε_i have a zero mean, that is,

$$E(\varepsilon_i) = 0$$

   where $E(\cdot)$ is the expected value operator. The errors are then said to be unbiased.

3. The errors have constant variance, that is,

$$\sigma_i^2 = E\left\{\left[Y_i - E(Y_i)\right]^2\right\} = \sigma^2 = \text{constant}$$

   which means that the variance of Y_i is independent of the measurement.

4. The errors associated with different measurements are uncorrelated. Two measurement errors, ε_i and ε_j, where $i \neq j$, are uncorrelated if the covariance of ε_i and ε_j is zero, that is,

$$\text{cov}(\varepsilon_i, \varepsilon_j) \equiv E\left\{\left[\varepsilon_i - E(\varepsilon_i)\right]\left[\varepsilon_j - E(\varepsilon_j)\right]\right\} = 0 \text{ for } i \neq j$$

   Such is the case if the errors ε_i and ε_j have no effect on or relationship to the other.

5. The measurement errors have a Gaussian distribution. By taking into consideration assumptions 2, 3 and 4 above, the probability distribution function of ε_i is given by:

$$f(\varepsilon_i) = \frac{1}{\sigma\sqrt{2\pi}} \exp\left(-\frac{\varepsilon_i^2}{2\sigma^2}\right)$$

6. The statistical parameters describing ε_i, such as σ, are known.

7. The only variables that contain random errors are the measurements. Measurement times, measurement positions and independent variables appearing in the formulation of the inverse problem are all *accurately* known.

8. There is no prior information regarding the quantities to be estimated, which can be either parameters or functions. If such information exists, it can be utilized to obtain improved estimates by working within the Bayesian framework of statistics.

All of the eight assumptions above rarely apply in actual experiments. For example, if the magnitudes of the measurements are quite unequal, the standard deviations σ_i are likely to be different. However, such assumptions are often convenient for the verification of the applicability of a method of solution to a specific inverse problem. In this context, the stability of the inverse problem solution with respect to measurement errors, number of sensors, sensor locations and experiment duration can be examined by using simulated measurements.

1.4 AN OVERVIEW OF SOLUTION TECHNIQUES FOR INVERSE HEAT TRANSFER PROBLEMS

A variety of techniques has been used to solve IHTPs. Therefore, it is useful to list some criteria proposed for the evaluation of the inverse problem solution techniques [4,5,57]:

1. The predicted quantity should be accurate if the measured data are of high accuracy.
2. The method should be stable with respect to measurement errors.
3. The method should have a statistical basis and permit various statistical assumptions for the measurement errors.
4. The method should not require the input data to be *a priori* smoothed.
5. The method should be stable for small time steps or intervals. This permits a better resolution of the time variation of the unknown quantity than is permitted by using large time steps.
6. Measurements of different quantities, from one or more sensors, should be permitted.
7. The method should not require continuous first derivatives of unknown functions. Furthermore, the method should be able to recover functions containing jump discontinuities.
8. Knowledge of the precise starting time of the application of an unknown excitation, such as a source term or surface heat flux, should not be required.
9. The method should not be restricted to any fixed number of measurements.
10. The method should be able to treat complex physical situations, including, among others, composite solids, moving boundaries, temperature-dependent properties, convective and radiative heat transfer, combined modes of heat transfer, multi-dimensional problems and irregular geometries.
11. The method should be easy for computer programming.
12. The computer cost should be moderate.
13. The user should not have to be highly skilled in mathematics in order to use the method.
14. The method should permit extension to more than one unknown.

Although the solution of inverse problems does not necessarily make use of optimization techniques, many popular methods are nowadays based on the minimization of an objective function. Despite their similarities, inverse and optimization problems are conceptually different. Inverse problems are concerned with the identification of unknown quantities appearing in the mathematical formulation of problems by using measurements of the system response. On the other hand, optimization problems generally deal with the minimization or maximization of a certain objective or cost function in order to find design variables that will result in desired state variables. Engineering applications of optimization techniques are very often concerned with the minimization or maximization of different quantities such as minimum weight (e.g., lighter airplanes), minimum fuel consumption (e.g., more economic cars) and maximum autonomy (e.g., longer range airplanes). In contrast to inverse problems, solution uniqueness may not be an important issue for optimization problems, as long as the solution obtained is physically feasible and can be practically implemented [58].

Consider the mathematical formulation of a heat transfer problem, which, for instance, can be linear or nonlinear, one or multi-dimensional, involve one single or coupled heat transfer modes. We denote the vector of parameters appearing in such formulation as:

$$\mathbf{P}^T \equiv \left[P_1, P_2, ..., P_N \right] \qquad (1.4.1)$$

where N is the number of parameters and the superscript T denotes the vector transpose. The parameters $[P_1, P_2, ..., P_N]$ can possibly be thermal conductivity components, heat transfer coefficients, heat sources and boundary heat fluxes. They can represent constant values of such quantities, or the parameters of the representation of a function in terms of known basis functions, such as given by equation (1.2.2).

Consider also that transient temperature measurements are available within the medium, or at its surface, where the heat transfer processes are being mathematically formulated. The vector containing the measured temperatures is written as:

$$\mathbf{Y}^T = \left[\vec{Y}_1, \vec{Y}_2, ..., \vec{Y}_I \right] \tag{1.4.2a}$$

where each vector $\vec{Y}_i$ contains the measurements of M sensors at time t_i, $i = 1, ..., I$, that is,

$$\vec{Y}_i = \left[Y_{i1}, Y_{i2}, ..., Y_{iM} \right] \quad \text{for} \quad i = 1, ..., I \tag{1.4.2b}$$

so that we have $D = MI$ measurements in total. Note that, in practice, the measured data are not limited to temperatures, but could also include heat fluxes and radiation intensities.

Throughout this book, the measurement errors are assumed to be additive, that is,

$$\mathbf{Y} = \mathbf{T}(\mathbf{P}) + \boldsymbol{\varepsilon} \tag{1.4.3}$$

where $\mathbf{T}(\mathbf{P})$ is the solution of the mathematical formulation of the heat transfer problem, obtained with the vector of parameters $\mathbf{P}$, that is,

$$\mathbf{T}^T(\mathbf{P}) = \left[\vec{T}_1(\mathbf{P}), \vec{T}_2(\mathbf{P}), ..., \vec{T}_I(\mathbf{P}) \right] \tag{1.4.4a}$$

where

$$\vec{T}_i(\mathbf{P}) = \left[T_{i1}(\mathbf{P}), T_{i2}(\mathbf{P}), ..., T_{iM}(\mathbf{P}) \right] \quad \text{for} \quad i = 1, ..., I \tag{1.4.4b}$$

The mathematical formulation is supposed to perfectly represent the phenomena under analysis. Similarly, the solution $\mathbf{T}(\mathbf{P})$ is supposed to be extremely accurate from the computational point of view. Anyhow, modeling errors can be appropriately taken into account within the Bayesian framework of statistics, as it will be apparent later in this book.

By further assuming that the measurement errors, $\boldsymbol{\varepsilon}$, are Gaussian random variables, with zero means, known covariance matrix $\mathbf{W}$ and independent of the parameters $\mathbf{P}$, their probability density function, $\pi(\boldsymbol{\varepsilon})$, is given by [4,17–19,22,48,49,59,60]:

$$\pi(\boldsymbol{\varepsilon}) = (2\pi)^{-D/2} |\mathbf{W}|^{-1/2} \exp\left\{ -\frac{1}{2} \boldsymbol{\varepsilon}^T \mathbf{W}^{-1} \boldsymbol{\varepsilon} \right\} \tag{1.4.5a}$$

where the number pi is represented in italics and the probability density function is given by the regular character π.

Due to the additive model for the measurement errors given by equation (1.4.3), we rewrite equation (1.4.5a) as:

$$\pi(\boldsymbol{\varepsilon}) = (2\pi)^{-D/2} |\mathbf{W}|^{-1/2} \exp\left\{ -\frac{1}{2} [\mathbf{Y} - \mathbf{T}(\mathbf{P})]^T \mathbf{W}^{-1} [\mathbf{Y} - \mathbf{T}(\mathbf{P})] \right\} \tag{1.4.5b}$$

We note that temperature measurement errors for thermocouples or infrared cameras can be fairly modeled by Gaussian distributions. For example, Figure 1.6a presents the histogram of the readings obtained with an infrared camera (see Figure 1.6b) of a plate maintained at a constant temperature about 22.3°C [61]. This histogram clearly approximates a Gaussian distribution.

(a)   (b)

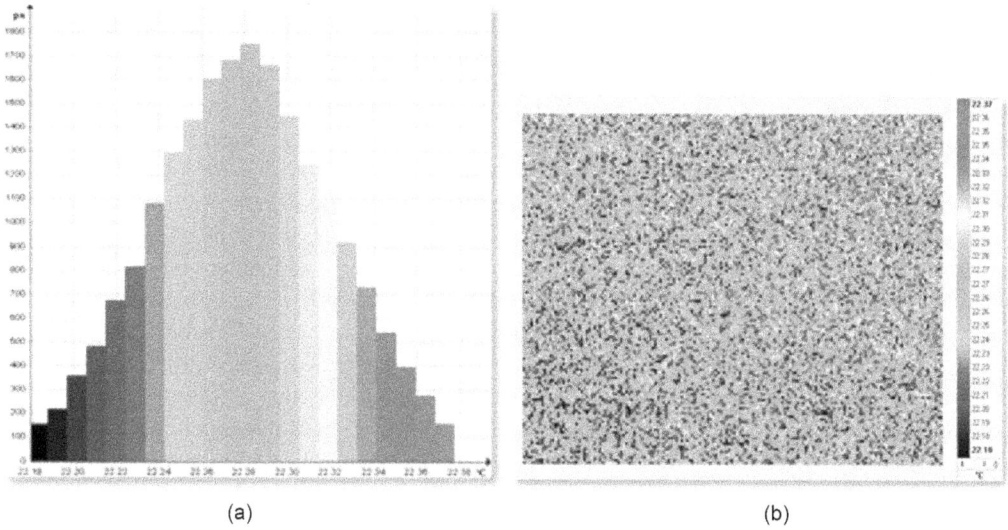

FIGURE 1.6   Histogram of the temperature measurements (a) corresponding to the thermal image with an infrared camera of an isothermal plate (b). (From Ref. [61].)

Note in equation (1.4.5b) that $\pi(\varepsilon)$ in fact represents the conditional probability density of different measurement outcomes $\mathbf{Y}$ with a fixed $\mathbf{P}$. This distribution, denoted as $\pi(\mathbf{Y}|\mathbf{P})$, is called the *likelihood function* [4,17–19,22,48,49,59,60].

A very common approach for the solution of inverse problems, dealing with the estimation of the parameters $\mathbf{P}$ with the measurements $\mathbf{Y}$, is to find a point estimate that maximizes the likelihood probability density. In equation (1.4.5b), this can be accomplished through the minimization of the following *maximum likelihood objective function*:

$$S_{ML}(\mathbf{P}) = \left[\mathbf{Y} - \mathbf{T}(\mathbf{P})\right]^{T} \mathbf{W}^{-1}\left[\mathbf{Y} - \mathbf{T}(\mathbf{P})\right] \qquad (1.4.6)$$

The *least squares norm* can be obtained as a particular case of equation (1.4.6), if the measurements are assumed uncorrelated and with constant variances, σ^2. In this case, the covariance matrix $\mathbf{W}$ is given by:

$$\mathbf{W} = \sigma^{2}\mathbf{I} \qquad (1.4.7)$$

where $\mathbf{I}$ is the identity matrix. Then, the minimization of equation (1.4.6) is equivalent to the minimization of:

$$S_{\text{OLS}}(\mathbf{P}) = \left[\mathbf{Y} - \mathbf{T}(\mathbf{P})\right]^{T}\left[\mathbf{Y} - \mathbf{T}(\mathbf{P})\right] \qquad (1.4.8a)$$

The least squares norm can also be written as:

$$S_{\text{OLS}}(\mathbf{P}) = \sum_{m=1}^{M} \sum_{i=1}^{I} \left[Y_{im} - T_{im}(\mathbf{P})\right]^{2} \qquad (1.4.8b)$$

Therefore, in order to make use of the minimization of the least squares norm for obtaining point estimates of the parameters $\mathbf{P}$ that have some statistical meaning (for example, that allows for estimates of the covariances of the estimated parameters), all the statistical hypotheses stated above need to be valid [4]. Namely, the measurement errors must be additive, Gaussian, uncorrelated, with zero mean and with constant variance. Such hypotheses are quite often overlooked in the literature, when an objective function is defined in terms of the least squares norm for the solution of an inverse problem via optimization techniques.

If the IHTP involves the estimation of only few unknown parameters, such as the estimation of a constant thermal conductivity value from the transient temperature measurements in a solid, the use of the maximum likelihood objective function, equation (1.4.6), can be stable. On the other hand, consider the minimization of such objective function to recover unknown transient heat flux components $q(t_i) \equiv P_i$ at times t_i, $i = 1,..., I = N$ (see Figure 1.7) by using measurements taken at the same times t_i. Note that this is a parameterization of the continuous heat flux function, $q(t)$, obtained through projection onto a domain with finite dimension, N. The estimation of this function is then reduced to the estimation of N heat flux components $q(t_i) \equiv P_i$, supposed constant in $t_i - \Delta t/2 < t < t_i + \Delta t/2$, where Δt is the time interval between two consecutive measurements. Therefore, the number of parameters to be estimated is equal to the number of measurements. For cases such as this one, the ill-posed nature of the inverse problem becomes apparent through excursions and oscillations of the solution. One classical approach to reduce such instabilities is to use *Tikhonov's regularization* [3,26–28], which modifies the maximum likelihood objective function by the addition of a penalty term.

In Tikhonov's regularization approach, the modified objective function is given by:

$$S(\mathbf{P}) = [\mathbf{Y} - \mathbf{T}(\mathbf{P})]^T \mathbf{W}^{-1} [\mathbf{Y} - \mathbf{T}(\mathbf{P})] + \alpha \|\mathbf{DP}\|^2 \tag{1.4.9a}$$

where for the *first-order* regularization, $\mathbf{D}$ is the following $(N-1) \times N$ difference matrix:

$$\mathbf{D} = \begin{bmatrix} -1 & 1 & 0 & \cdots & 0 \\ 0 & -1 & 1 & \cdots & 0 \\ \vdots & & \ddots & & \vdots \\ 0 & \cdots & 0 & -1 & 1 \end{bmatrix} \tag{1.4.9b}$$

and $\|\cdot\|$ is the L_2 vector norm.

In equation (1.4.9a), α (> 0) is the *regularization parameter* and $\alpha \|\mathbf{DP}\|_2$ is the *whole domain regularization term*. The values chosen for the regularization parameter α influence the stability

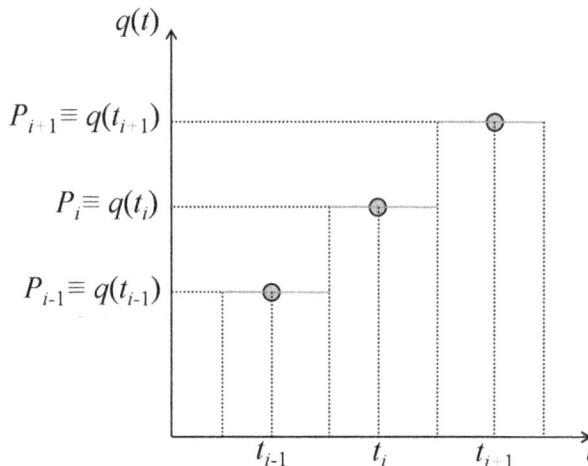

FIGURE 1.7   Parameters representing local values of a function that varies in time.

of the solution as the minimization of equation (1.4.9a) is performed. For $\alpha \to 0$, exact matching between estimated and measured temperatures is obtained with the minimization of $S(\mathbf{P})$ and the inverse problem solution becomes unstable because of its ill-posed nature. On the other hand, for large values of α and with $\mathbf{D}$ given by equation (1.4.9b), when the second term in equation (1.4.9a) is dominant, the heat flux components P_i tend to become constant for $i = 1, 2,..., I$, that is, the first derivative of the heat flux tends to zero. Note that $\mathbf{DP}$, with $\mathbf{D}$ given by equation (1.4.9b), is a finite difference approximation of the first derivative of the heat flux $q(t)$, represented by its discrete components $q(t_i) \equiv P_i$. Tikhonov has also proposed *zeroth-order* and *second-order* regularization terms [3,26–28].

Therefore, instabilities on the solution can be alleviated by proper selection of the value of α. Tikhonov [3,26–28] recommended that α should be selected so that the minimum value of the first term of the objective function would be equal to the sum of the squares of the errors expected for the measurements, which is known as *Morozov's discrepancy principle* [10,22]. Also, the use of the so-called L-curve [22,62] appears as a useful technique for the selection of the regularization parameter. The L-curve is the Pareto frontier for a multi-objective optimization problem involving two objective functions, given by the two terms of equation (1.4.9a).

An alternative approach for the regularization scheme described above is the use of *Alifanov's iterative regularization* [2,6,29–43]. In this approach, the number of iterations is so chosen that reasonably stable solutions are obtained by applying Morozov's discrepancy principle. Therefore, there is no need to modify the original objective function, as opposed to Tikhonov's approach. Another fundamental difference between Alifanov's and Tikhonov's regularization approaches is that, in the first, the unknown function is not required to be discretized *a priori*. In the example discussed above, the heat flux function, $q(t)$, was projected onto a domain with finite dimension, N, and then the inverse problem of estimating $q(t_i) \equiv P_i$ at times t_i, $i = 1,..., I=N$, would be solved with Tikhonov's regularization. On the other hand, with Alifanov's approach all the required mathematical derivations are made in a space of functions. The discretization of the function, resulting from the fact that measurements are taken at discrete times and positions, is then made during the implementation of the method [2,6,29–43]. Nevertheless, the iterative regularization approach is quite general and can also be applied to the estimation of parameterized functions. It is an extremely robust technique, which converges fast and is stable with respect to the measurement errors. Also, it can be systematically applied to different types of inverse problems by following the same basic steps, as will be apparent later in this book.

Classical techniques for the solution of inverse problems, such as Tikhonov's and Alifanov's regularization methods, are not based on the modeling of *prior* information about the unknown parameters or functions. On the other hand, in the *statistical inversion approach* [17] based on Bayesian statistics, the probability distributions for the measurements and for the unknowns are modeled separately and then applied in Bayes' theorem. The term *Bayesian* is commonly used to refer to techniques for the solution of inverse problems that fall within the framework of statistics developed by the Presbyterian minister Rev. Thomas Bayes (★1702-† 1761) [63]. Such framework was actually established after Bayes' death, when his friend, Richard Price, published Bayes' famous paper, which dealt with the following problem: *"Given the number of times in which an unknown event has happened and failed: Required the chance that the probability of its happening in a single trial lies somewhere between two degrees of probability that can be named."* [64]. On the other hand, it is attributed to Laplace the mathematical formulation that is known today as Bayes' theorem [65]. The term *Bayesian* was first used by R. A. Fisher, but in a pejorative context. Although born more than 120 years after the death of Bayes, Fisher was Bayes biggest intellectual rival [65]. The major issue of Fisher against Bayes and Laplace was that they used the concept of a *prior probability*, which represents the information about an unknown quantity before the measured data is available [65]. Fisher's theory relies solely on the measured data and on modeling of their associated uncertainty, aiming at unbiased inference and/or decision; therefore, it is usually referred to as the *frequentist* framework of statistics and takes into account only the *likelihood* function [63,65–67]. On the other hand, within the Bayesian framework, credit is also given to previous beliefs, in addition to that given to the measured data. Such previous information can even be qualitative, but needs to be represented in terms of a probability distribution function, and regretfully induces bias in the estimations [63,65,67]. Nevertheless, the use of prior information in the Bayesian framework does not mean that it completely overtakes the information provided by the measured data, unless the last one is too uncertain and/or useless to be really taken into account. In the Bayesian framework, regularization for the inverse problem solution is provided by the *prior model* for the unknown parameters [17–20,22]. One of the first applications of techniques within the Bayesian framework for the solution of IHTPs is the paper by Wang and Zabaras [66].

The use of simulated measurements to verify inverse problem solutions, like in the example shown in Section 1.3, may lead to analyses that do not represent actual situations. Consider, for example, that the same mathematical model, the same solution technique and the same solution accuracy are used to generate the simulated measurements $\mathbf{Y}$ with equation (1.3.5) and to compute $\mathbf{T}(\mathbf{P})$ that is required in equation (1.4.5b). Thus, modeling errors are not taken into account in such an idealized inverse analysis. Kaipio and Somersalo have popularized the term *inverse crime* to describe this undesirable condition for the inverse analysis, which makes the inverse problem less ill-posed than in actual situations [17]. Whenever using simulated measurements, such an inverse crime must be avoided by using different solution techniques/accuracies, or even different mathematical models, for the direct problem solutions used to generate the simulated measurements $\mathbf{Y}$ and to compute $\mathbf{T}(\mathbf{P})$ during the solution of the inverse problem [17]. A thorough analysis of the errors inherent to measurements in heat transfer and to the solution of inverse problems can be found in the work by Le Masson and Dal, presented in Chapter 16 of Reference [22].

1.5   BASIC STEPS FOR THE SOLUTION OF INVERSE HEAT TRANSFER PROBLEMS

The basic steps for the solution of inverse problems can be summarized as follows:

i. **Design of the experiment**: Before an attempt is made for the estimation of model parameters and/or functions, the experiment has to be designed in terms of which quantity should be measured, where the sensors should be located, how long the experiment should last and which should be the boundary conditions externally imposed, in order to obtain the most meaningful measurements for the inverse problem solution.

ii. **Estimation of the model parameters or functions**: Measurements and a technique to overcome ill-posedness are used for the solution of the inverse problem. The selection of the inverse problem solution technique to be used depends on several factors, including the number of parameters to be estimated, if they represent constant quantities in the formulation or distributed values of functions in the spatial or time domains and complexity of the model representing the phenomena.

iii. **Uncertainty quantification**: The solution of the inverse problem should not be limited to point estimates for the unknown parameters or functions, but also allow the quantification of uncertainties related to these estimates and to the mathematical model output. In this sense, Bayesian techniques naturally provide the framework for uncertainty quantification and sensitivity analysis, since the solution of the inverse problem is obtained as statistical inference.

The above steps are addressed in the following chapters of this book. After this introductory chapter, the remaining first part of the book is concerned with parameter estimation problems, solved through the minimization of objective functions as well as by stochastic simulation. The second part of the book is devoted to function estimation problems. Regarding the solution techniques examined in the first two parts of the book, we initially focus our attention on the application of the Levenberg-Marquardt method of parameter estimation [68,69], which is related to Tikhonov's regularization. Then, we consider Alifanov's method of iterative regularization and techniques within the Bayesian framework of statistics, for parameter and function estimations. These methods are quite stable, powerful, straightforward and can be applied to the solution of a large variety of IHTPs. They meet the majority of criteria enumerated above in this section regarding the evaluation of inverse problem solution procedures. The above methods for the solution of inverse problems are based on the whole time domain approach. The sequential solution of inverse problems is also addressed in the last part of the book, which deals with state estimation problems. Examples are provided throughout the book involving the solution of IHTPs.

PROBLEMS

1.1 Derive the analytical solution of the direct heat conduction problem given by equations (1.1.1a–d).

1.2 Use the analytical solution derived above in Problem 1.1 to plot the transient temperatures at different locations inside a steel slab [ρ = 7753 kg/m^3, c_p = 0.486 kJ/kg K and k = 36 W/m K] of thickness L = 5 cm, initially at the uniform temperature of 200°C. The boundary at x = 0 cm is kept insulated while the boundary at x = 5 cm is maintained at the constant temperature of 20°C.

1.3 Repeat Problem 1.2 for a slab made of brick [ρ = 1600 kg/m^3, c_p = 0.84 kJ/kg K and k = 0.69 W/m K] instead of steel. Compare the temperature variations in the steel and brick slabs at selected positions, say, x = 0, 2 and 4 cm.

1.4 Consider a physical problem involving one-dimensional heat conduction in a slab of thickness L, with initial temperature distribution $F(x)$. Assume constant thermophysical properties. A time-dependent heat flux $f(t)$ is supplied at the surface x = 0, while the surface at x = L is kept insulated. Energy is generated in the medium at a rate $g(x, t)$ per unit time and per unit volume. What is the mathematical formulation of this heat conduction problem?

1.5 Derive the analytical solution of the direct problem associated with the above heat conduction Problem 1.4.

1.6 Use the solution developed in Problem 1.5 to plot the transient temperatures at several locations inside an aluminum slab [ρ = 2707 kg/m^3, c_p = 0.896 kJ/kg K and k = 204 W/m K] of thickness L = 3 cm, initially at the uniform temperature of 20°C. No heat is generated inside the medium and a constant heat flux of 8000 W/m^2 is supplied at the surface x = 0 cm.

1.7 Consider a physical problem involving one-dimensional heat conduction in a slab of thickness $2L$, with initial temperature distribution $F(x)$. Assume constant thermophysical properties. Heat is lost by convection to an ambient at the temperature T_∞ with a heat transfer coefficient h, at the surfaces x = $-L$ and x = L. Energy is generated in the medium at a rate $g(x, t)$ per unit time and per unit volume. What is the mathematical formulation of this heat conduction problem?

1.8 Derive the analytical solution of the direct problem associated with the above heat conduction Problem 1.7.

1.9 Use the solution developed in Problem 1.8 to plot the transient temperatures at several locations inside an iron slab [ρ = 7850 kg/m^3, c_p = 0.460 kJ/kg K and k = 60 W/m K] of thickness $2L$ = 5 cm, initially at the uniform temperature of 250°C. No heat is generated inside the slab and the ambient temperature is 25°C. The heat transfer coefficient at both slab surfaces is 500 W/m^2 K.

1.10 Consider a physical problem involving one-dimensional heat conduction in a solid cylinder of radius b, with initial temperature distribution $F(r)$. Assume constant thermophysical properties. Heat is lost by convection to an ambient at the temperature T_∞ with a heat transfer coefficient h, at the surface r = b. Energy is generated in the medium at a rate

$g(r, t)$ per unit time and per unit volume. What is the mathematical formulation of this heat conduction problem?

1.11 Derive the analytical solution of the direct problem associated with the above heat conduction Problem 1.10.

1.12 Use the solution developed in Problem 1.11 to plot the transient temperatures at several locations inside an iron cylinder [$\rho = 7850$ kg/m^3, $c_p = 0.460$ kJ/kg K and $k = 60$ W/m K] of radius $b = 2.5$ cm, initially at the uniform temperature of 250°C. No heat is generated inside the cylinder and the ambient temperature is 25°C. The heat transfer coefficient at the cylinder surface is 500 W/m^2 K.

1.13 Repeat Problems 1.10, 1.11 and 1.12, for a solid sphere of radius $r = b$, instead of a solid cylinder.

1.14 Compare the transient temperature variations at $x = r = 0$ cm, in Problems 1.9, 1.12 and 1.13, for a slab, cylinder and sphere, respectively.

1.15 Consider a physical problem involving two-dimensional heat conduction in a plate of width a and height b, with initial temperature distribution $F(x, y)$. Assume constant thermophysical properties. Heat is lost by convection to an ambient at the temperature T_∞ with a heat transfer coefficient h, at all plate surfaces. Energy is generated in the medium at a rate $g(x, y, t)$ per unit time and per unit volume. What is the mathematical formulation of this heat conduction problem?

1.16 Derive the analytical solution of the direct problem associated with the above heat conduction Problem 1.15.

1.17 Use the solution developed in Problem 1.16 to plot the transient temperatures for the central point in a square iron plate [$\rho = 7850$ kg/m^3, $c_p = 0.460$ kJ/kg K and $k = 60$ W/m K] with sides $a = b = 5$ cm, initially at the uniform temperature of 250°C. No heat is generated inside the plate and the ambient temperature is 25°C. The heat transfer coefficient at the plate surfaces is 500 W/m^2 K.

1.18 Consider a physical problem involving a plate with width a and thickness b, which moves horizontally along the x direction with a constant velocity u. Assume constant physical properties. Also, suppose the plate to be infinitely long in the axial (x) direction. The plate loses heat by convection through its lateral surfaces, at $y = 0$ and $y = a$, to an ambient at temperature T_∞ with a heat transfer coefficient h. The bottom surface at $z = 0$ is supposed insulated, while a transient heat flux with distribution $q(x, y, t)$ is supplied at the top surface $z = b$, in the region $x > 0$. The initial temperature in the medium is $F(x, y, z)$ and the plate enters into the heated zone ($x > 0$) with a uniform temperature T_0 at $x = 0$. Heat is generated in the medium at a rate $g(x, y, z, t)$ per unit time and per unit volume. What is the mathematical formulation of this problem?

1.19 Simplify the formulation developed in Problem 1.18 for the steady-state heat transfer problem in a plate with negligible lateral heat losses and no heat generation. The heat flux at $z = b$ is a function of x only, say, $q(x)$.

1.20 Derive the analytical solution for the direct problem formulated in Problem 1.19.

1.21 By using the analytical solution derived in Problem 1.20, find the temperature field in a steel plate [$\rho = 7753$ kg/m^3, $c_p = 0.486$ kJ/kg K and $k = 36$ W/m K] of thickness $b = 2.5$ cm, moving with a velocity 0.15 m/s, for $T_0 = 20$°C, $q(x) = 50 \times 10^4$ W/m^2 in $1 < x < 2$ cm, and $q(x) = 0$ W/m^2 outside this region.

1.22 Review, in basic heat transfer books, the physics and formulation of heat transfer by radiation in non-participating and participating media.

1.23 For the heat transfer problems formulated above in Problems 1.4, 1.7, 1.10, 1.13, 1.15 and 1.18, devise inverse problems of:
   i. Boundary condition;
   ii. Initial condition;

    iii. Energy source-term;

    iv. Thermophysical properties.

    How would you address the solution of such inverse problems? In terms of parameter or of function estimation?

NOTE 1: STATISTICAL CONCEPTS

The purpose of this note is to present some basic statistical material, needed in the analysis and solution of IHTPs, that is generally not covered in regular courses in engineering. Readers should consult references [4,17,59,60,63,67,70–72] for a more in-depth discussion of such matters.

RANDOM VARIABLE

A *random variable* is a variable whose value is a numerical outcome of a random phenomenon. A *phenomenon is denoted random* if its individual outcomes are unpredictable, although a regular pattern of outcomes emerges in many repetitions. A collection of random variables uniquely associated with indexes is denoted as a *stochastic process* (the indexes can be, for example, integer numbers).

    Let the capital letter X denote a random variable. It is called a *discrete random variable* if it can only assume a set of discrete numbers x_n, $n = 1, 2, ..., N$. On the other hand, X is called a *continuous random variable* if it can assume all values in an interval of real numbers.

PROBABILITY DISTRIBUTION

The assignment of probabilities to the values of a random variable X gives the *probability distribution* of X. Depending on whether the random variable X is discrete or continuous, the probability distribution $\pi(x)$ is a non-negative number or function, respectively, satisfying:

$$\sum_{n=1}^{N} \pi(x_n) = 1 \text{ when } X \text{ is discrete} \tag{N1.1.1a}$$

$$\int_{-\infty}^{+\infty} \pi(x)\, dx = 1 \text{ when } X \text{ is continuous} \tag{N1.1.1b}$$

EXPECTED VALUE OF X

Let X be a random variable, discrete or continuous, with the corresponding probability distributions $\pi(x_n)$ or $\pi(x)$, respectively. The *expected value* of X, denoted by $E(X)$, is defined as:

$$E(X) = \begin{cases} \displaystyle\sum_{n=1}^{N} x_n \pi(x_n) & \text{when } X \text{ is discrete} \\[2em] \displaystyle\int_{-\infty}^{\infty} x\, \pi(x)\, dx & \text{when } X \text{ is continuous} \end{cases} \tag{N1.1.2}$$

The expected value of any random variable X is obtained by multiplying the random values by the corresponding probability distribution and then summing up the results if X is discrete, or integrating the results if X is continuous. Clearly, the expected value of X is a *weighted mean* of all possible values, with the weight factor $\pi(x)$. If the weights are equal for discrete X, that is, $\pi(x_n) = 1/N$, then the expected value becomes the *arithmetic average* of X. Usually, the expected value is simply referred to as the *mean* of the random variable X.

EXPECTED VALUE OF A FUNCTION $g(X)$

Consider a random variable X and the probability distribution $\pi(x)$ associated with it. The expected value of the function $g(X)$, denoted by $E[g(X)]$, is given by:

$$E[g(X)] = \begin{cases} \displaystyle\sum_{n=1}^{N} g(x_n)\pi(x_n) & \text{when } X \text{ is discrete} \\[2em] \displaystyle\int_{-\infty}^{\infty} g(x)\pi(x)\,dx & \text{when } X \text{ is continuous} \end{cases}$$

(N1.1.3)

VARIANCE OF A RANDOM VARIABLE X

The *variance* of a random variable X, denoted by σ^2, is a measure of the spread of X around its mean μ. It is defined by:

$$\sigma^2 \equiv E[(x-\mu)^2] \quad \text{where } \mu = E(x)$$

(N1.1.4a)

or an alternative form is obtained by expanding this expression, that is,

$$\sigma^2 = E(x^2) - \mu^2$$

(N1.1.4b)

$$\text{Since } E(\mu^2) = \mu^2$$

The positive square root, σ, of the variance is called the *standard deviation*.

COVARIANCE OF TWO RANDOM VARIABLES X AND Y

The *covariance* of two random variables X and Y is a measure of the linear dependence between them. It is defined as:

$$\text{cov}(X,Y) \equiv E\left[(x-\mu_x)(y-\mu_y)\right]$$

(N1.1.5)

where $\mu_x = E(x)$ and $\mu_y = E(y)$.
    The covariance cov(X, Y) is zero if X and Y are independent.

GAUSSIAN DISTRIBUTION

The most frequently used continuous probability distribution function is the *Gaussian or normal distribution*, which has a bell-shaped curve about its mean value. The Gaussian probability distribution function with mean μ and variance σ^2, denoted as $x \sim N(\mu,\sigma^2)$, is given by:

$$\pi(x) = \frac{1}{\sigma\sqrt{2\pi}} \exp\left[-\frac{1}{2}\left(\frac{x-\mu}{\sigma}\right)^2\right]$$

(N1.1.6)

The area below this function from $-\infty$ to x represents the probability $P(-\infty < X \le x)$ that a Gaussian variable X with mean μ and variance σ^2 assumes a value between $-\infty$ and x. Therefore, $P(-\infty < X \le x)$ is defined by:

$$P(-\infty < X \le x) = \frac{1}{\sigma\sqrt{2\pi}} \int_{-\infty}^{x} \exp\left[-\frac{1}{2}\left(\frac{X-\mu}{\sigma}\right)^2\right] dX \qquad (\text{N}1.1.7)$$

To alleviate the difficulty in the calculation of this integral for each given set of values of σ, μ and x, a new independent variable Z was defined as:

$$Z = \frac{X-\mu}{\sigma} \quad \text{or} \quad z = \frac{x-\mu}{\sigma} \qquad (\text{N}1.1.8)$$

Then, the integral in equation (N1.1.7) becomes:

$$P(-\infty < Z \le z) = \frac{1}{\sqrt{2\pi}} \int_{-\infty}^{z} e^{-Z^2/2} dZ \qquad (\text{N}1.1.9)$$

The results of this integration were tabulated, as given in Table 1.2.

Table 1.2 can be used as follows to determine the probability $P(x_1 \le X \le x_2)$, of a random variable X having a mean μ and variance σ^2, to assume a value between x_1 and x_2:

$$P(x_1 \le X \le x_2) = P\left(\frac{x_1-\mu}{\sigma} \le Z \le \frac{x_2-\mu}{\sigma}\right)$$

$$= P(z_1 \le Z \le z_2)$$

$$= P(-\infty < Z \le z_2) - P(-\infty < Z \le z_1) \qquad (\text{N}1.1.10)$$

where $P(-\infty < Z \le z_2)$ and $P(-\infty < Z \le z_1)$ are determined from Table 1.2.

UNIFORM DISTRIBUTION

A very simple distribution that allows lower and upper bounds for the random variable X is the *uniform distribution* $x \sim U(a,b)$ given by:

$$\pi(x) = \begin{cases} \dfrac{1}{(b-a)}, & a < x < b \\ 0, & \text{elsewhere} \end{cases} \qquad (\text{N}1.1.11)$$

Mean and variance for the uniform distribution are given by $\frac{1}{2}(a+b)$ and $\frac{1}{12}(b-a)^2$, respectively. In the uniform distribution, any value in $a < x < b$ is equally probable.

RAYLEIGH DISTRIBUTION

This distribution satisfies positivity constraints and depends on only one scale parameter (centerpoint), γ_0. The *Rayleigh distribution* $x \sim R(\gamma_0)$ is given by:

$$\pi(x) = \frac{x}{\gamma_0^2} \exp\left[-\frac{1}{2}\left(\frac{x}{\gamma_0}\right)^2\right] \quad \text{for } x > 0 \qquad (\text{N}1.1.12)$$

The mean and the variance of Rayleigh's distribution are given by $\gamma_0\sqrt{\dfrac{\pi}{2}}$ and $\dfrac{4-\pi}{2}\gamma_0^2$, respectively.

TABLE 1.2
Probability $P(-\infty < Z \leq z)$ Given by Equation (N1.1.9) for a Gaussian Distribution Function

Z	$P(-\infty < Z \leq z)$	Z	$P(-\infty < Z \leq z)$
−2.9	0.0019	0.0	0.5000
−2.8	0.0026	0.1	0.5398
−2.7	0.0035	0.2	0.5793
−2.6	0.0047	0.3	0.6179
−2.5	0.0062	0.4	0.6554
−2.4	0.0082	0.5	0.6915
−2.3	0.0107	0.6	0.7257
−2.2	0.0139	0.7	0.7580
−2.1	0.0179	0.8	0.7881
−2.0	0.0228	0.9	0.8159
−1.9	0.0287	1.0	0.8413
−1.8	0.0359	1.1	0.8643
−1.7	0.0446	1.2	0.8849
−1.6	0.0548	1.3	0.9032
−1.5	0.0668	1.4	0.9192
−1.4	0.0808	1.5	0.9332
−1.3	0.0968	1.6	0.9452
−1.2	0.1151	1.7	0.9554
−1.1	0.1357	1.8	0.9641
−1.0	0.1587	1.9	0.9713
−0.9	0.1841	2.0	0.9772
−0.8	0.2119	2.1	0.9821
−0.7	0.2420	2.2	0.9861
−0.6	0.2743	2.3	0.9893
−0.5	0.3085	2.4	0.9918
−0.4	0.3446	2.5	0.9938
−0.3	0.3821	2.6	0.9953
−0.2	0.4207	2.7	0.9965
−0.1	0.4602	2.8	0.9974
		2.9	0.9981

GAMMA DISTRIBUTION

The *Gamma distribution* with parameters α and β, denoted as $x \sim G(\alpha, \beta)$, has the following density:

$$\pi(x) = \frac{1}{\beta^\alpha \Gamma(\alpha)} x^{\alpha-1} \exp\left(-\frac{x}{\beta}\right) \quad \text{for } x > 0 \qquad (N1.1.13)$$

with mean $\alpha\beta$ and variance $\alpha\beta^2$, where $\Gamma(\cdot)$ is the gamma function defined as:

$$\Gamma(n) = \int_{x=0}^{\infty} e^{-x} x^{n-1} \, dx, \quad n > 0 \qquad (N1.1.14a)$$

and, for n integer, we have:

$$\Gamma(n+1) = n! \tag{N1.1.14b}$$

For $\beta = 1$, the so-called one-parameter gamma distribution is obtained. The density that results by making $\alpha = 1$ is called exponential distribution.

BETA DISTRIBUTION

The *Beta distribution* $x \sim \mathrm{Be}(\alpha,\beta)$ has support in $0 < x < 1$. The density of this distribution is given by:

$$\pi(x) = \frac{\Gamma(\alpha+\beta)}{\Gamma(\alpha)\Gamma(\beta)} x^{\alpha-1}(1-x)^{\beta-1} \quad \text{in } 0 < x < 1 \tag{N1.1.15}$$

with mean $\dfrac{\alpha}{\alpha+\beta}$ and variance $\dfrac{\alpha\beta}{(\alpha+\beta)^2(\alpha+\beta+1)}$.

Figure 1.8 shows the probability distributions U(0,1), N(0.5,0.5²), R(0.5), G(1.5,1.5) and Be(1.5,1.5). These distributions were normalized by their maximum values to allow the comparison among them.

CHI-SQUARE DISTRIBUTION

Let $Z_1, Z_2, \ldots, Z_N$ be independent Gaussian random variables with mean zero and unitary standard deviation. In this case, the summation:

$$\chi_N^2 = \sum_{n=1}^{N} Z_n^2 \tag{N1.1.16}$$

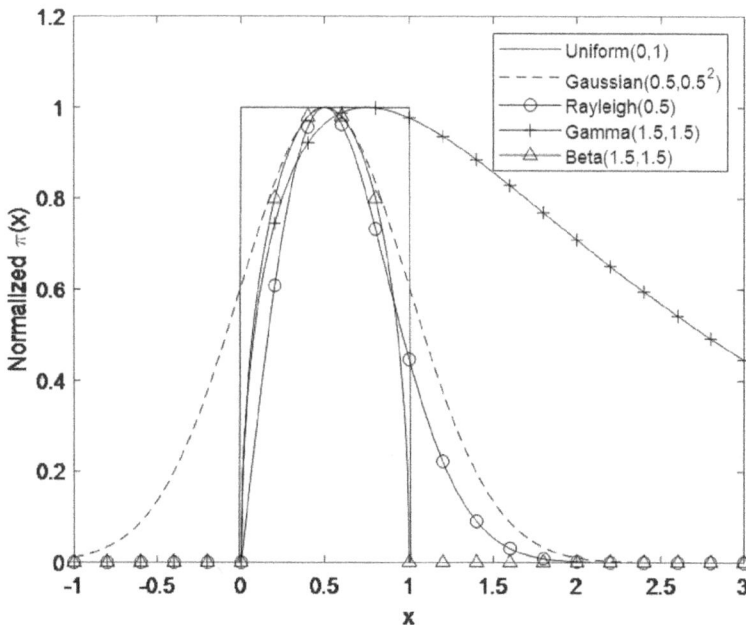

FIGURE 1.8   Probability distributions.

has a *chi-square probability distribution with N degrees of freedom*, given by:

$$\pi\left(\chi_N^2\right) = \frac{\left(\chi_N^2\right)^{\frac{1}{2}(N-2)}}{2^{N/2}\Gamma\left(\dfrac{N}{2}\right)} e^{-\chi_N^2/2} \quad \text{for } 0 < \chi_N^2 < \infty \tag{N1.1.17}$$

The mean and the variance of the chi-square distribution are N and $2N$, respectively. Such distribution is skewed to the right, but it tends to the normal distribution as $N \to \infty$. This behavior is shown in Figure 1.9.

The probability of having a value x smaller than χ_N^2 is obtained by integrating equation (N1.1.17) from zero to χ_N^2, that is,

$$P\left(0 \le x \le \chi_N^2\right) = \int_0^{\chi_N^2} \frac{(x)^{\frac{1}{2}(N-2)}}{2^{N/2}\Gamma(N/2)} e^{-x/2} dx \tag{N1.1.18}$$

Table 1.3 shows the values of χ_N^2 for various probabilities $\pi(\chi_N^2)$, as a function of the number of degrees of freedom N. The values of χ_N^2 shown in Table 1.3 are useful in obtaining confidence regions for estimated parameters. A discussion on confidence regions and other quantities of importance to assess the accuracy of the estimated parameters will be presented in the following chapters.

COVARIANCE MATRIX

Consider now that the random variable $\mathbf{X}$ is the following vector:

$$\mathbf{X}^T \equiv [X_1, X_2, \ldots, X_J] \tag{N1.1.19}$$

The elements of the *covariance matrix* of $\mathbf{X}$ are the covariances between each pair of the components of $\mathbf{X}$, that is,

$$[\text{cov}(\mathbf{X})]_{i,j} \equiv \text{cov}(x_i, x_j) \quad \text{for } i, j = 1, \ldots, J \tag{N1.1.20}$$

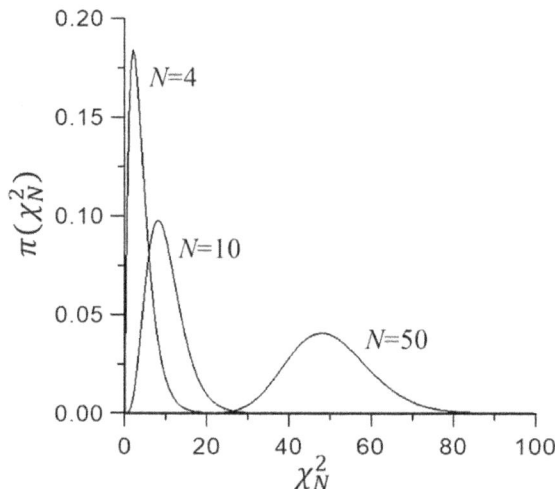

FIGURE 1.9 Chi-square probability distribution function given by equation (N1.1.17).

TABLE 1.3

Values of χ_N^2 for Various Degrees of Freedom N and Probabilities $\pi\left(\chi_N^2\right) \equiv P\left(0 \le x \le \chi_N^2\right)$

N	$\pi\left(\chi_N^2\right)$ 0.900	0.950	0.975	0.990	0.995
1	2.71	3.84	5.02	6.63	7.88
2	4.61	5.99	7.38	9.21	10.6
3	6.25	7.81	9.35	11.3	12.8
4	7.78	9.49	11.1	13.3	14.9
5	9.24	11.1	12.8	15.1	16.7
6	10.6	12.6	14.4	16.8	18.5
7	12.0	14.1	16.0	18.5	20.3
8	13.4	15.5	17.5	20.1	22.0
9	14.7	16.9	19.0	21.7	23.6
10	16.0	18.3	20.5	23.2	25.2
11	17.3	19.7	21.9	24.7	26.8
12	18.5	21.0	23.3	26.2	28.3
13	19.8	22.4	24.7	27.7	29.8
14	21.1	23.7	26.1	29.1	31.3
15	22.3	25.0	27.5	30.6	32.8
16	23.5	26.3	28.8	32.0	34.3
17	24.8	27.6	30.2	33.4	35.7
18	26.0	28.9	31.5	34.8	37.2
19	27.2	30.1	32.9	36.2	38.6
20	28.4	31.4	34.2	37.6	40.0

or

$$\text{cov}(\mathbf{X}) \equiv \begin{bmatrix} \text{cov}(x_1,x_1) & \text{cov}(x_1,x_2) & \cdots & \text{cov}(x_1,x_{J-1}) & \text{cov}(x_1,x_J) \\ \text{cov}(x_2,x_1) & \text{cov}(x_2,x_2) & \cdots & \text{cov}(x_2,x_{J-1}) & \text{cov}(x_2,x_J) \\ \vdots & \vdots & \ddots & \vdots & \vdots \\ \text{cov}(x_{J-1},x_1) & \text{cov}(x_{J-1},x_2) & \cdots & \text{cov}(x_{J-1},x_{J-1}) & \text{cov}(x_{J-1},x_J) \\ \text{cov}(x_J,x_1) & \text{cov}(x_J,x_2) & \cdots & \text{cov}(x_J,x_{J-1}) & \text{cov}(x_J,x_J) \end{bmatrix} \qquad \text{(N1.1.21)}$$

Therefore, the covariance matrix is symmetric and its diagonal elements are the variances:

$$\sigma_i^2 \equiv \text{cov}(x_i,x_i), \text{ for } i = 1,\ldots,J.$$

The covariance matrix can also be defined as:

$$\text{cov}(\mathbf{X}) \equiv E\left\{[\mathbf{x} - E(\mathbf{x})][\mathbf{x} - E(\mathbf{x})]^T\right\} \qquad \text{(N1.1.22)}$$

MULTIVARIATE GAUSSIAN DISTRIBUTION

The Gaussian distribution for the random vector $\mathbf{X}$, with mean $\boldsymbol{\mu} = E(\mathbf{x})$ and covariance matrix $\mathbf{W}$, denoted as $\mathbf{x} \sim N(\boldsymbol{\mu}, \mathbf{W})$, is given by:

$$\pi(\mathbf{x}) = (2\pi)^{-J/2} |\mathbf{W}|^{-1/2} \exp\left\{-\frac{1}{2}[\mathbf{x} - \boldsymbol{\mu}]^T \mathbf{W}^{-1}[\mathbf{x} - \boldsymbol{\mu}]\right\} \qquad \text{(N1.1.23)}$$

where J is the dimension of $\mathbf{X}$, and $|\mathbf{W}|$ is the determinant of $\mathbf{W}$.

2 Parameter Estimation

Minimization of an Objective Function without Prior Information about the Unknown Parameters

In the previous chapter we discussed general principles related to the formulation and solution of inverse heat transfer problems. The main objective of this chapter is to provide the necessary mathematical background needed in the use of three powerful techniques for solving inverse parameter estimation problems.

The following techniques are considered here:

Technique I: Levenberg-Marquardt Method
Technique II: Conjugate Gradient Method
Technique III: Conjugate Gradient Method with Adjoint Problem for Parameter Estimation

Technique I is an iterative method for solving nonlinear parameter estimation problems. The technique was first derived by Levenberg [68] in 1944 by modifying the ordinary least squares norm. Later, in 1963, Marquardt [69] derived basically the same technique by using a different approach. Marquardt's intention was to obtain a method that would tend to the Gauss method in the neighborhood of the minimum of the ordinary least squares norm and would tend to the steepest descent method in the neighborhood of the initial guess used for the iterative procedure. The so called *Levenberg-Marquardt method* [4,22,68,69,71,73,74] has been applied to the solution of a variety of inverse problems involving the estimation of unknown parameters.

Technique II is also an iterative method, which utilizes the *conjugate gradient method of minimization* to solve parameter estimation problems [6,7,22,58,73,75–79]. This method originates from the work of Hestenes and Stiefel [78] for the solution of linear systems of equations [75].

Technique III utilizes the *conjugate gradient method of minimization with adjoint problem* [2,6,7,22,29–43]. **Technique III** is especially suitable for problems involving the estimation of the coefficients of basis functions used to approximate an unknown function. The use of the adjoint problem in Technique III results in an expression for the gradient direction involving a Lagrange multiplier, thus alleviating the need for the computation of the sensitivity matrix.

The solution of inverse parameter estimation problems by Techniques I and II requires the computation of the *sensitivity matrix*, $\mathbf{J}$, the elements of which are the *sensitivity coefficients*, J_{ij}, defined as:

$$J_{ij} = \frac{\partial T_i(\mathbf{P})}{\partial P_j}$$

where, for the case of transient measurements of one single sensor (see also equations 1.4.2 and 1.4.4):

$$i = 1,2,\ldots,I$$

$$j = 1,2,\ldots,N$$

I = number of transient measurements
N = number of unknown parameters

$T_i(\mathbf{P})$ is the ith estimated temperature

P_j is the jth unknown parameter in the vector $\mathbf{P} \equiv \left[P_1, P_2, ..., P_N \right]^T$

It is assumed in this chapter that no prior information is available regarding the values of the unknown parameters before the solution of the inverse problem. If this information is available, either from previous experience or from literature data, it can be taken into account for the solution of the parameter estimation problem, as presented in Chapter 3.

In this chapter, we describe the basic steps and present the solution algorithms for the Levenberg-Marquardt method and for the conjugate gradient method by using a whole time domain approach. We discuss the physical and mathematical significance of sensitivity coefficients and describe three different methods for their computation. The design of optimum experiments and the estimation of uncertainties related to the estimated parameters are also addressed here.

2.1   OBJECTIVE FUNCTION

The objective function used in this chapter is derived from the likelihood function, which is the statistical model of the measurement errors. The measurement errors are assumed additive, as given by equation (1.4.3), and modeled as Gaussian random variables, which is a pertinent hypothesis for temperature measurements commonly used in heat transfer, as illustrated by Figures 1.6a and b. The calibrations of sensors and of the data acquisition system are assumed to be performed before the experiment, in order to eliminate any bias from the measurement errors, which can then be assumed with zero mean. The covariance matrix $\mathbf{W}$ of the measurement errors is also supposed to be known from the calibration process. The measurement errors are assumed independent of the parameters $\mathbf{P}$. By referring to Section 1.3, these hypotheses correspond to the statistical assumptions 1, 2, 5, 6, 7 and 8.

We seek here parameter values that maximize the associated likelihood function (equation 1.4.5b) through the minimization of the objective function given by (see also equation 1.4.6):

$$S_{ML}(\mathbf{P}) = \left[\mathbf{Y} - \mathbf{T}(\mathbf{P}) \right]^T \mathbf{W}^{-1} \left[\mathbf{Y} - \mathbf{T}(\mathbf{P}) \right] \tag{2.1.1}$$

For simplicity in the analysis, derivations are first performed in this chapter by considering transient measurements Y_i, $i = 1, ..., I$, of one single sensor, that is (see equations 1.4.2a and b):

$$\mathbf{Y}^T = \left[Y_1, Y_2, ... , Y_I \right] \tag{2.1.2}$$

By analogy, the solution of the mathematical formulation of the physical problem with the vector of parameters $\mathbf{P}$, $\mathbf{T}(\mathbf{P})$, is given for one single sensor by (see equations 1.4.4a and b):

$$\mathbf{T}^T(\mathbf{P}) = [T_1(\mathbf{P}), T_2(\mathbf{P}), ..., T_I(\mathbf{P})] \tag{2.1.3}$$

The equations derived for one single sensor are extended later for multiple sensors in this chapter.

Note that equation (2.1.1) can be simplified if the measurements are uncorrelated, that is, the off-diagonal elements of $\mathbf{W}$ (the covariance of the measurements) are null. In this case, $\mathbf{W}$ becomes diagonal and, for one single sensor, it is given by:

$$\mathbf{W} = \begin{bmatrix} \sigma_1^2 & 0 & \cdots & 0 & 0 \\ 0 & \sigma_2^2 & \cdots & 0 & 0 \\ \vdots & \vdots & \ddots & \vdots & \vdots \\ 0 & 0 & \cdots & \sigma_{I-1}^2 & 0 \\ 0 & 0 & \cdots & 0 & \sigma_I^2 \end{bmatrix} \tag{2.1.4}$$

where σ_i^2 is the variance of measurement Y_i, $i = 1, \ldots, I$. Thus, for the case of uncorrelated measurements of one single sensor, equation (2.1.1) can be written as:

$$S_{ML}(\mathbf{P}) = \sum_{i=1}^{I} \frac{[Y_i - T_i(\mathbf{P})]^2}{\sigma_i^2} \qquad (2.1.5)$$

If, in addition, the measurements can be considered with constant variance, that is, $\sigma_i^2 = \sigma^2 = $ constant and $\mathbf{W} = \sigma^2\mathbf{I}$, where $\mathbf{I}$ is the identity matrix, the minimization of equation (2.1.5) is equivalent to the minimization of:

$$S_{\text{OLS}}(\mathbf{P}) = \sum_{i=1}^{I} [Y_i - T_i(\mathbf{P})]^2 \qquad (2.1.6a)$$

or, in matrix form,

$$S_{\text{OLS}}(\mathbf{P}) = [\mathbf{Y} - \mathbf{T}(\mathbf{P})]^T [\mathbf{Y} - \mathbf{T}(\mathbf{P})] \qquad (2.1.6b)$$

which is the ordinary least squares norm.

The methods examined in this chapter rely on the gradient of the objective function to iteratively obtain directions of descent along which the minimization is performed. For this reason, these methods are called *gradient* methods. Other methods could also be considered for the minimization procedure, such as those referred to as *stochastic* and *hybrid* [58].

2.2   TECHNIQUE I: THE LEVENBERG-MARQUARDT METHOD

The Levenberg-Marquardt method, originally devised for application to nonlinear parameter estimation problems, has also been successfully applied to the solution of linear problems that are ill-conditioned.

The solution of inverse heat transfer problems with the Levenberg-Marquardt method can be suitably arranged in the following basic steps:

- Direct problem
- Inverse problem
- Analysis of the sensitivity coefficients and design of optimum experiments
- Iterative procedure
- Stopping criteria
- Computational algorithm
- Statistical analysis

Other sections in this chapter will be dedicated to the analysis of the sensitivity coefficients, design of optimum experiments and statistical analysis. We present below the details of each of the other steps of Technique I, as applied to the solution of an inverse heat conduction test problem, involving the following physical situation:

Consider transient linear heat conduction in a plate of unitary dimensionless thickness. The plate is initially at zero temperature and both boundaries at $x = 0$ and $x = 1$ are kept insulated. For times $t > 0$, a plane heat source of strength $g_p(t)$ per unit area, placed in the mid-plane $x = 0.5$, releases its energy as shown by Figure 2.1.

The mathematical formulation of this heat conduction problem is given in dimensionless form by:

$$\frac{\partial^2 T(x,t)}{\partial x^2} + g_p(t)\delta(x - 0.5) = \frac{\partial T(x,t)}{\partial t} \text{ in } 0 < x < 1, \text{ for } t > 0 \qquad (2.2.1a)$$

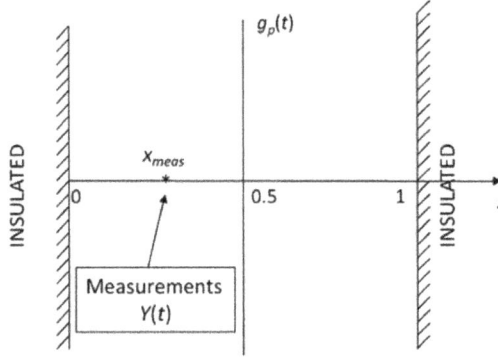

FIGURE 2.1   Geometry and coordinates for a plane heat source $g_p(t)$.

$$\frac{\partial T(0, t)}{\partial x} = 0 \text{ at } x = 0, \text{ for } t > 0 \tag{2.2.1b}$$

$$\frac{\partial T(1, t)}{\partial x} = 0 \text{ at } x = 1, \text{ for } t > 0 \tag{2.2.1c}$$

$$T(x,0) = 0 \text{ for } t = 0, \text{ in } 0 < x < 1 \tag{2.2.1d}$$

where $\delta(\cdot)$ is the Dirac delta function.

THE DIRECT PROBLEM

In the *direct problem* associated with the heat transfer problem described above, the time-varying strength $g_p(t)$ of the plane heat source is known. The objective of the direct problem is then to determine the transient temperature field $T(x, t)$ in the plate.

THE INVERSE PROBLEM

For the *inverse problem* considered of interest here, the time-varying strength $g_p(t)$ of the plane heat source is regarded as unknown. The additional information obtained from transient temperature measurements taken at a location $x = x_{meas}$, at times t_i, $i = 1, 2, ..., I$, is then used for the estimation of $g_p(t)$.

   For the solution of the present inverse problem in this chapter, we consider the unknown energy generation function $g_p(t)$ to be parameterized in the following general linear form:

$$g_p(t) = \sum_{j=1}^{N} P_j C_j(t) \tag{2.2.2}$$

where P_j are unknown parameters and $C_j(t)$ are known basis functions (e.g., polynomials, B-Splines, etc.). In addition, the total number of parameters, N, is specified. It is important to notice that the number of parameters to be estimated, N, must be small in comparison to the number of measurements, I, for stability of the solution of the inverse problem by the minimization of the objective functions shown in the previous section. Therefore, the techniques examined in this chapter do not apply for functions parameterized by piecewise constant functions, such as illustrated in Figure 1.7, which will be the subject of Part II of this book.

The problem given by equations (2.2.1a-d) with $g_p(t)$ unknown, but parameterized as given by equation (2.2.2), is an inverse heat conduction problem in which the coefficients P_j are to be estimated. The solution of this inverse heat conduction problem for the estimation of the N unknown parameters $P_j, j = 1,..., N$, is treated in this chapter as the minimization of the *maximum likelihood objective function* given by equation (2.1.1).

THE ITERATIVE PROCEDURE FOR TECHNIQUE I

To minimize the objective function (2.1.1), we need to equate to zero the derivatives of $S_{ML}(\mathbf{P})$ with respect to each of the unknown parameters $[P_1, P_2, ..., P_N]$, that is,

$$\frac{\partial S_{ML}(\mathbf{P})}{\partial P_1} = \frac{\partial S_{ML}(\mathbf{P})}{\partial P_2} = \cdots = \frac{\partial S_{ML}(\mathbf{P})}{\partial P_N} = 0 \qquad (2.2.3)$$

Such a necessary condition for the minimization of $S_{ML}(\mathbf{P})$ can be represented in matrix notation by equating the *gradient of $S_{ML}(\mathbf{P})$* with respect to the vector of parameters $\mathbf{P}$ to zero, that is,

$$\nabla S_{ML}(\mathbf{P}) = -2\,\mathbf{J}^T \mathbf{W}^{-1}[\mathbf{Y} - \mathbf{T}(\mathbf{P})] = 0 \qquad (2.2.4)$$

where $\mathbf{J}$ is the *sensitivity matrix* defined by [4,22]:

$$\mathbf{J} = \left[\frac{\partial \mathbf{T}^T(\mathbf{P})}{\partial \mathbf{P}}\right]^T \qquad (2.2.5)$$

and

$$\frac{\partial \mathbf{T}^T(\mathbf{P})}{\partial \mathbf{P}} = \begin{bmatrix} \dfrac{\partial}{\partial P_1} \\[6pt] \dfrac{\partial}{\partial P_2} \\[4pt] \vdots \\[4pt] \dfrac{\partial}{\partial P_N} \end{bmatrix} \left[T_1(\mathbf{P}), T_2(\mathbf{P}),..., T_I(\mathbf{P})\right] \qquad (2.2.6)$$

In explicit form, the sensitivity matrix is written as:

$$\mathbf{J}(\mathbf{P}) = \left[\frac{\partial \mathbf{T}^T(\mathbf{P})}{\partial \mathbf{P}}\right]^T = \begin{bmatrix} \dfrac{\partial T_1(\mathbf{P})}{\partial P_1} & \dfrac{\partial T_1(\mathbf{P})}{\partial P_2} & \dfrac{\partial T_1(\mathbf{P})}{\partial P_3} & \cdots & \dfrac{\partial T_1(\mathbf{P})}{\partial P_N} \\[10pt] \dfrac{\partial T_2(\mathbf{P})}{\partial P_1} & \dfrac{\partial T_2(\mathbf{P})}{\partial P_2} & \dfrac{\partial T_2(\mathbf{P})}{\partial P_3} & \cdots & \dfrac{\partial T_2(\mathbf{P})}{\partial P_N} \\[10pt] \vdots & \vdots & \vdots & & \vdots \\[10pt] \dfrac{\partial T_I(\mathbf{P})}{\partial P_1} & \dfrac{\partial T_I(\mathbf{P})}{\partial P_2} & \dfrac{\partial T_I(\mathbf{P})}{\partial P_3} & \cdots & \dfrac{\partial T_I(\mathbf{P})}{\partial P_N} \end{bmatrix} \qquad (2.2.7)$$

where N = total number of unknown parameters
I = total number of measurements

The elements of the sensitivity matrix are called the *sensitivity coefficients*. The sensitivity coefficient J_{ij} is thus defined as the first derivative of the estimated temperature at time t_i with respect to the unknown parameter P_j, that is,

$$J_{ij} = \frac{\partial T_i}{\partial P_j} \tag{2.2.8}$$

For *linear inverse problems*, the sensitivity matrix is not a function of the unknown parameters and we can write:

$$\mathbf{T(P) = JP} \tag{2.2.9}$$

In such case, equation (2.2.4) can be solved in explicit form for the vector of unknown parameters **P** as [4,22]:

$$\mathbf{P} = \left[\mathbf{J}^T \mathbf{W}^{-1} \mathbf{J}\right]^{-1} \left[\mathbf{J}^T \mathbf{W}^{-1} \mathbf{Y}\right] \tag{2.2.10}$$

In the case of a *nonlinear inverse problem*, the sensitivity matrix has some functional dependence on the vector of unknown parameters **P** and relation (2.2.9) is not valid. The solution of equation (2.2.4) for nonlinear estimation problems then requires an iterative procedure, which is obtained by linearizing the vector of estimated temperatures, **T(P)**, with a Taylor series expansion around the current solution $\mathbf{P}^k$ at iteration k. Such a linearization is given by:

$$\mathbf{T(P)} \approx \mathbf{T}\left(\mathbf{P}^k\right) + \mathbf{J}^k \left(\mathbf{P} - \mathbf{P}^k\right) \tag{2.2.11}$$

where $\mathbf{T}(\mathbf{P}^k)$ and $\mathbf{J}^k$ are the estimated temperatures and the sensitivity matrix evaluated at iteration k, respectively. Equation (2.2.11) is substituted into equation (2.2.4) and the resulting expression is rearranged to yield the following iterative procedure to obtain the vector of unknown parameters **P** [4,22]:

$$\mathbf{P}^{k+1} = \mathbf{P}^k + \left[\mathbf{J}^T \mathbf{W}^{-1} \mathbf{J}\right]^{-1} \mathbf{J}^T \mathbf{W}^{-1} \left[\mathbf{Y} - \mathbf{T}(\mathbf{P}^k)\right] \tag{2.2.12}$$

The iterative procedure given by equation (2.2.12) is called the Gauss method. Such method is actually an approximation for the Newton (or Newton-Raphson) method [4,22,71].

We note that equation (2.2.10), as well as the implementation of the iterative procedure given by equation (2.2.12), require the matrix $[\mathbf{J}^T \mathbf{W}^{-1} \mathbf{J}]$ to be non-singular. Such a condition can be accomplished by requiring [4]:

$$\left|\mathbf{J}^T \mathbf{J}\right| \neq 0 \tag{2.2.13}$$

where | . | is the determinant.

Equation (2.2.13) is the *identifiability condition,* that is, if the determinant of $\mathbf{J}^T \mathbf{J}$ is zero, or even very small, the parameters P_j, for $j = 1,..., N$, cannot be estimated by using the iterative procedure of equation (2.2.12) for the nonlinear case or equation (2.2.10) for the linear case. Problems satisfying $\left|\mathbf{J}^T \mathbf{J}\right| \approx 0$ are *ill-conditioned*. The ill-conditioning of matrix $\mathbf{J}^T \mathbf{J}$ may be caused by small magnitudes of the sensitivity coefficients or linearly dependent columns of the sensitivity matrix [4,22].

Inverse heat transfer problems are, in most situations, nonlinear and ill-conditioned (especially near the initial guess used for the unknown parameters), thus creating difficulties in the application of equation (2.2.12). The *Levenberg-Marquardt method* [4,22,68,69,71,73,74] alleviates such difficulties by utilizing an iterative procedure in the form:

$$\mathbf{P}^{k+1} = \mathbf{P}^k + \left[\mathbf{J}^T \mathbf{W}^{-1} \mathbf{J} + \rho^k \, \mathbf{\Omega}^k\right]^{-1} \mathbf{J}^T \mathbf{W}^{-1} \left[\mathbf{Y} - \mathbf{T}(\mathbf{P}^k)\right] \tag{2.2.14}$$

where ρ^k is a positive scalar named *damping parameter* and $\mathbf{\Omega}^k$ is a *diagonal matrix*.

The purpose of the matrix term $\rho^k \Omega^k$ is to damp oscillations and instabilities due to the ill-conditioned character of the problem by making its components large as compared to those of $[\mathbf{J}^T \mathbf{W}^{-1} \mathbf{J}]$ if necessary. The damping parameter is made large in the beginning of the iterations, since the problem is generally ill-conditioned in the region around the initial guess used for the iterative procedure, which can be quite far from the exact parameters. With such an approach, the matrix $[\mathbf{J}^T \mathbf{W}^{-1} \mathbf{J}]$ is not required to be non-singular in the beginning of iterations and the Levenberg-Marquardt method tends to the *steepest descent method*, that is, a very small step is taken in the negative gradient direction. The parameter ρ^k is gradually reduced as the iteration procedure advances to the solution of the parameter estimation problem and the problem becomes better conditioned. Then, the Levenberg-Marquardt method tends to the *Gauss method* given by equation (2.2.12) [4,22].

The penalization term used by Tikhonov (see equation 1.4.9a) results in an iterative procedure similar to equation (2.2.14), but with a constant damping parameter (given by Tikhonov's regularization parameter, α). For this reason, Levenberg-Marquardt's method is also related to the so-called *iterated Tikhonov's methods* [80,81].

THE STOPPING CRITERIA FOR TECHNIQUE I

The following criteria were suggested by Dennis and Schnabel [73] to stop the iterative procedure of the Levenberg-Marquardt method given by equation (2.2.14):

$$S_{ML}\left(\mathbf{P}^{k+1}\right) < \varepsilon_1 \tag{2.2.15a}$$

$$\left\|\left(\mathbf{J}^k\right)^T \mathbf{W}^{-1}[\mathbf{Y} - \mathbf{T}(\mathbf{P}^k)]\right\| < \varepsilon_2 \tag{2.2.15b}$$

$$\left\|\mathbf{P}^{k+1} - \mathbf{P}^k\right\| < \varepsilon_3 \tag{2.2.15c}$$

where ε_1, ε_2 and ε_3 are user prescribed tolerances and $\| \cdot \|$ is the vector Euclidean norm, i.e., $\left\|\mathbf{x}\right\| = (\mathbf{x}^T \mathbf{x})^{1/2}$, where the superscript T denotes transpose.

The criterion given by equation (2.2.15a) tests if the maximum likelihood objective function is sufficiently small, which is expected to be in the neighborhood of the solution for the inverse problem. Similarly, equation (2.2.15b) checks if the norm of the gradient of $S_{ML}(\mathbf{P})$ is small, since it is expected to vanish at the point where $S_{ML}(\mathbf{P})$ is minimum. The last criterion, given by equation (2.2.15c), results from the fact that changes in the vector of parameters are very small when the method has converged. The use of a stopping criterion based on small changes of $S_{ML}(\mathbf{P})$ could also be used, but with extreme caution. It may happen that the method stalls for a few iterations and then starts advancing to the point of minimum afterward [4,71,73].

The tolerance for the stopping criterion given by equation (2.2.15a) can be conveniently selected by using Morozov's discrepancy principle [2–22]. This principle relies on the fact that the expected minimum value of the objective function (2.1.1) is obtained when the differences between measured and estimated temperatures are of the same order of magnitude of the measurement errors. It is thus recognized that, with the minimization of the objective function beyond this point, that is, when equation (2.1.1) tends to zero, the estimated temperatures tend to the measurements that contain experimental errors. Such experimental errors can be amplified for the solution of the inverse problem because of its ill-posed character, if the final value of the objective function becomes too small [2–22].

The tolerance ε_1 based on Morozov's discrepancy principle is thus obtained by assuming:

$$\left| Y_i - T_i(\mathbf{P}) \right| \approx \sigma_i \tag{2.2.16}$$

where σ_i is the standard deviation of the measurement error at time t_i.

For uncorrelated measurements (see equation 2.1.5), we obtain:

$$\varepsilon_1 = I \tag{2.2.17}$$

If the measurements are uncorrelated and with a constant standard deviation, i.e., $\sigma_i = \sigma = $ constant, and the ordinary least squares function can be used, ε_1 is obtained by substituting equation (2.2.16) into equation (2.1.6a), that is,

$$\varepsilon_1 = \sum_{i=1}^{I} \sigma^2 = I\sigma^2 \tag{2.2.18}$$

On the other hand, the use of the discrepancy principle may not be required to provide the Levenberg-Marquardt method with the regularization necessary to obtain stable solutions for the inverse problem. Computational experiments [82] revealed that the Levenberg-Marquardt method, through its automatic control of the damping parameter ρ^k, drastically reduces the increment in the vector of estimated parameters at the iteration where the measurement errors start to cause instabilities on the inverse problem solution. The iterative procedure of the Levenberg-Marquardt method is then naturally stopped by the criterion given by equation (2.2.15c).

THE COMPUTATIONAL ALGORITHM FOR TECHNIQUE I

Different versions of the Levenberg-Marquardt method can be found in the literature depending on the choice of the diagonal matrix Ω^k and on the form chosen for the variation of the damping parameter ρ^k [4,22,68,69,71,73,74,80,81,83]. We illustrate here a procedure with the matrix Ω^k taken as:

$$\Omega^k = \text{diag}\left[(\mathbf{J}^k)^T \mathbf{W}^{-1} \mathbf{J}^k \right] \tag{2.2.19}$$

Suppose that temperature measurements $\mathbf{Y} = \left[Y_1, Y_2, \dots, Y_I \right]^T$ are given at times t_i, $i = 1,\dots, I$. Also, suppose an initial guess $\mathbf{P}^0$ is available for the vector of unknown parameters $\mathbf{P}$. Choose a value for ρ^0, say, $\rho^0 = 0.001$, and set $k = 0$. Then [83]:

Step 1: Solve the direct heat transfer problem given by equations (2.2.1a-d) with the available estimate $\mathbf{P}^k$ in order to obtain the temperature vector $\mathbf{T}(\mathbf{P}^k) = [T_1(\mathbf{P}^k), T_2(\mathbf{P}^k),\dots,T_I(\mathbf{P}^k)]^T$.
Step 2: Compute $S_{ML}(\mathbf{P}^k)$ from equation (2.1.1).
Step 3: Compute the sensitivity matrix $\mathbf{J}^k$ defined by equation (2.2.7) and then the matrix Ω^k given by equation (2.2.19) by using the current values of $\mathbf{P}^k$.
Step 4: Solve the following linear system of algebraic equations obtained from the iterative procedure of the Levenberg-Marquardt Method, equation (2.2.14):

$$\left[(\mathbf{J}^k)^T \mathbf{W}^{-1} \mathbf{J}^k + \rho^k \Omega^k \right] \Delta \mathbf{P}^k = (\mathbf{J}^k)^T \mathbf{W}^{-1} \left[\mathbf{Y} - \mathbf{T}(\mathbf{P}^k) \right] \tag{2.2.20}$$

in order to compute $\Delta \mathbf{P}^k = \mathbf{P}^{k+1} - \mathbf{P}^k$.
Step 5: Compute the new estimate $\mathbf{P}^{k+1}$ as:

$$\mathbf{P}^{k+1} = \mathbf{P}^k + \Delta \mathbf{P}^k \tag{2.2.21}$$

Step 6: Solve the direct problem (2.2.1) with the new estimate $\mathbf{P}^{k+1}$ in order to find $\mathbf{T}(\mathbf{P}^{k+1})$. Then compute $S_{ML}(\mathbf{P}^{k+1})$, as defined by equation (2.1.1).
Step 7: If $S_{ML}(\mathbf{P}^{k+1}) \geq S_{ML}(\mathbf{P}^k)$, replace ρ^k by $10\rho^k$ and return to step 4.

Step 8: If $S_{ML}(\mathbf{P}^{k+1}) < S_{ML}(\mathbf{P}^k)$, accept the new estimate $\mathbf{P}^{k+1}$ and replace ρ^k by $0.1\rho^k$.

Step 9: Check the stopping criteria given by equation (2.2.15). Stop the iterative procedure if any of them is satisfied. Otherwise, replace k by $k + 1$ and return to step 3.

This simple algorithm of the Levenberg-Marquardt method works quite well for the estimation of parameters that have the same order of magnitude. The reader is referred to [71,73,74,80,81] for more robust and computationally more efficient versions of the Levenberg-Marquardt method, especially the one by Moré [74] that is available in most computational packages.

2.3 TECHNIQUE II: THE CONJUGATE GRADIENT METHOD FOR PARAMETER ESTIMATION

We present in this section an alternative technique for the estimation of unknown parameters by the minimization of the objective function (2.1.1). **Technique II**, the *conjugate gradient method*, is a straightforward and powerful iterative technique for solving linear and nonlinear inverse problems of parameter estimation. In the iterative procedure of the conjugate gradient method, at each iteration a suitable step size is taken along a direction of descent in order to minimize the objective function. The direction of descent is obtained as a linear combination of the negative gradient direction at the current iteration with directions of descent of previous iterations. The linear combination is such that the resulting angle between the direction of descent and the negative gradient direction is less than 90° and the minimization of the objective function is assured. Theorems regarding the convergence of the conjugate gradient method can be found in references [6,7,73,75–79,84]. The conjugate gradient method with an appropriate stopping criterion belongs to the class of iterative regularization techniques, in which the number of iterations is chosen so that stable solutions are obtained for the inverse problem [2,6,7,22,29–43].

Similarly to Technique I, the application of Technique II to inverse heat transfer problems of parameter estimation can be conveniently organized in the following steps:

- Direct problem
- Inverse problem
- Analysis of the sensitivity coefficients and design of optimum experiments
- Iterative procedure
- Stopping criterion
- Computational algorithm
- Statistical analysis

Except for the analysis of the sensitivity coefficients, design of optimum experiments and statistical analysis, the steps of Technique II are presented in this section, as applied to the heat conduction test problem described in Section 2.2, involving the estimation of the unknown source term function $g_p(t)$.

THE DIRECT PROBLEM

In the *direct problem* mathematically formulated by equations (2.2.1a-d), the time-varying strength $g_p(t)$ of the plane heat source is known. The objective of the direct problem is then to determine the transient temperature field $T(x, t)$ in the region.

THE INVERSE PROBLEM

In the *inverse problem* considered here, the time-varying strength $g_p(t)$ of the plane heat source is regarded as unknown and transient temperature measurements taken at a location $x = x_{meas}$, at times t_i, $i = 1, 2, ..., I$, are considered available for the analysis.

For the solution of such inverse problem, we consider the unknown energy generation function $g_p(t)$ to be parameterized in the general linear form given by equation (2.2.2). The estimation of the unknown function $g_p(t)$ then reduces to the estimation of the N unknown parameters P_j, $j = 1, ..., N$. Such parameter estimation problem is solved by the minimization of the maximum likelihood objective function given by equation (2.1.1).

THE ITERATIVE PROCEDURE FOR TECHNIQUE II

The iterative procedure of the conjugate gradient method, as well as of the Levenberg-Marquardt method, can be written in the following general form [2,6,7,15,22,29–43,73,84]:

$$\mathbf{P}^{k+1} = \mathbf{P}^k + \beta^k \mathbf{d}^k \tag{2.3.1}$$

where β^k is the step size and $\mathbf{d}^k$ is the direction of descent. By comparing equations (2.2.14) and (2.3.1), we note that the effect of the matrix term $\rho^k \mathbf{\Omega}^k$ in the Levenberg-Marquardt method is to change both the step size and the direction of descent at each iteration.

An iteration step is *acceptable* if $S_{ML}(\mathbf{P}^{k+1}) < S_{ML}(\mathbf{P}^k)$. The direction of descent $\mathbf{d}^k$ will generate acceptable steps if and only if there exists a positive definite matrix $\mathbf{R}$ such that $\mathbf{d}^k = -\mathbf{R}\nabla S_{ML}(\mathbf{P}^k)$ [71].

In the conjugate gradient method, the step size is selected as the one that minimizes the objective function at each iteration along the direction $\mathbf{d}^k$. From equation (2.1.1), we can write:

$$\min_{\beta^k} S_{ML}(\mathbf{P}^{k+1}) = \min_{\beta^k} \left\{ [\mathbf{Y} - \mathbf{T}(\mathbf{P}^{k+1})]^T \mathbf{W}^{-1} [\mathbf{Y} - \mathbf{T}(\mathbf{P}^{k+1})] \right\} \tag{2.3.2}$$

By substituting equation (2.3.1) into equation (2.3.2), we obtain:

$$\min_{\beta^k} S_{ML}(\mathbf{P}^{k+1}) = \min_{\beta^k} \left\{ [\mathbf{Y} - \mathbf{T}(\mathbf{P}^k + \beta^k \mathbf{d}^k)]^T \mathbf{W}^{-1} [\mathbf{Y} - \mathbf{T}(\mathbf{P}^k + \beta^k \mathbf{d}^k)] \right\} \tag{2.3.3}$$

Then, by performing the minimization the following expression results:

$$\beta^k = \frac{[\mathbf{J}^k \mathbf{d}^k]^T \mathbf{W}^{-1} [\mathbf{Y} - \mathbf{T}(\mathbf{P}^k)]}{[\mathbf{J}^k \mathbf{d}^k]^T \mathbf{W}^{-1} [\mathbf{J}^k \mathbf{d}^k]} \tag{2.3.4}$$

The derivation of equation (2.3.4) is presented in Note 1 at the end of this chapter.

The *direction of descent* $\mathbf{d}^k$ of the conjugate gradient method is given in the following general form [2,6,7,15,22,29–43,73,75–79,84]:

$$\mathbf{d}^k = -\nabla S_{ML}(\mathbf{P}^k) + \gamma^k \mathbf{d}^{k-1} + \psi^k \mathbf{d}^q \tag{2.3.5}$$

where γ^k and ψ^k are conjugation coefficients and $\nabla S_{ML}(\mathbf{P})$ is the gradient vector given by (see equation 2.2.4):

$$\nabla S_{ML}(\mathbf{P}) = -2 \mathbf{J}^T \mathbf{W}^{-1} [\mathbf{Y} - \mathbf{T}(\mathbf{P})] \tag{2.3.6}$$

The superscript q in equation (2.3.5) denotes the iteration number where a *restarting strategy* is applied to the iterative procedure of the conjugate gradient method. Restarting strategies were suggested for the conjugate gradient method in order to improve its convergence rate [79]. Different versions of the conjugate gradient method can be found in the literature depending on the form used for the computation of the direction of descent given by equation (2.3.5) [2,6,7,15,22,29–43,73, 75–79,84]. These versions are equivalent for linear problems, but not for nonlinear ones.

In the *Fletcher-Reeves* version [76], the conjugation coefficients γ^k and ψ^k are obtained from the following expressions:

$$\gamma^k = \frac{[\nabla S_{ML}(\mathbf{P}^k)]^T[\nabla S_{ML}(\mathbf{P}^k)]}{[\nabla S_{ML}(\mathbf{P}^{k-1})]^T[\nabla S_{ML}(\mathbf{P}^{k-1})]} \quad \text{with } \gamma^0 = 0 \text{ for } k = 0 \tag{2.3.7a}$$

$$\psi^k = 0 \quad \text{for } k = 0,1,2,\ldots \tag{2.3.7b}$$

In the *Polak-Ribiere* version [77] of the conjugate gradient method, the conjugation coefficients are given by:

$$\gamma^k = \frac{[\nabla S_{ML}(\mathbf{P}^k)]^T[\nabla S_{ML}(\mathbf{P}^k) - \nabla S_{ML}(\mathbf{P}^{k-1})]}{[\nabla S_{ML}(\mathbf{P}^{k-1})]^T[\nabla S_{ML}(\mathbf{P}^{k-1})]} \quad \text{with } \gamma^0 = 0 \text{ for } k = 0 \tag{2.3.8a}$$

$$\psi^k = 0 \quad \text{for } k = 0,1,2,\ldots \tag{2.3.8b}$$

For the *Hestenes-Stiefel* version [78] of the conjugate gradient method we have:

$$\gamma^k = \frac{[\nabla S_{ML}(\mathbf{P}^k)]^T[\nabla S_{ML}(\mathbf{P}^k) - \nabla S_{ML}(\mathbf{P}^{k-1})]}{[\mathbf{d}^{k-1}]^T[\nabla S_{ML}(\mathbf{P}^k) - \nabla S_{ML}(\mathbf{P}^{k-1})]} \quad \text{with } \gamma^0 = 0 \text{ for } k = 0 \tag{2.3.9a}$$

$$\psi^k = 0 \quad \text{for } k = 0,1,2,\ldots \tag{2.3.9b}$$

Powell [79] suggested the following expressions for the conjugation coefficients, which gives the so-called *Powell-Beale's* version of the conjugate gradient method:

$$\gamma^k = \frac{[\nabla S_{ML}(\mathbf{P}^k)]^T[\nabla S_{ML}(\mathbf{P}^k) - \nabla S_{ML}(\mathbf{P}^{k-1})]}{[\mathbf{d}^{k-1}]^T[\nabla S_{ML}(\mathbf{P}^k) - \nabla S_{ML}(\mathbf{P}^{k-1})]} \quad \text{with } \gamma^0 = 0 \text{ for } k = 0 \tag{2.3.10a}$$

$$\psi^k = \frac{[\nabla S_{ML}(\mathbf{P}^k)]^T[\nabla S_{ML}(\mathbf{P}^{q+1}) - \nabla S_{ML}(\mathbf{P}^q)]}{[\mathbf{d}^q]^T[\nabla S_{ML}(\mathbf{P}^{q+1}) - \nabla S_{ML}(\mathbf{P}^q)]} \quad \text{with } \psi^0 = 0 \text{ for } k = 0 \tag{2.3.10b}$$

In accordance with Powell [79], the application of the conjugate gradient method with the conjugation coefficients given by equations (2.3.10a, b) requires restarting when gradients at successive iterations tend to be non-orthogonal (which is a measure of the local non-linearity of the problem) or when the direction of descent is not sufficiently downhill. Restarting is performed by making $\psi^k = 0$ in equation (2.3.5).

The non-orthogonality of gradients at successive iterations is tested by using:

$$\text{ABS}\big([\nabla S_{ML}(\mathbf{P}^{k-1})]^T[\nabla S_{ML}(\mathbf{P}^k)]\big) \geq 0.2[\nabla S_{ML}(\mathbf{P}^k)]^T[\nabla S_{ML}(\mathbf{P}^k)] \tag{2.3.11}$$

where ABS(.) denotes the absolute value.

A direction of descent that is not sufficiently downhill (i.e., the angle between the direction of descent and the negative gradient direction is too large) is identified if either of the following inequalities are satisfied:

$$[\mathbf{d}^k]^T[\nabla S_{ML}(\mathbf{P}^k)] \leq -1.2[\nabla S_{ML}(\mathbf{P}^k)]^T[\nabla S_{ML}(\mathbf{P}^k)] \tag{2.3.12a}$$

$$[\mathbf{d}^k]^T[\nabla S_{ML}(\mathbf{P}^k)] \geq -0.8[\nabla S_{ML}(\mathbf{P}^k)]^T[\nabla S_{ML}(\mathbf{P}^k)] \tag{2.3.12b}$$

In Powell-Beale's version of the conjugate gradient method, the direction of descent given by equation (2.3.5) is computed in accordance with the following algorithm, for $k \geq 1$ [79]:

  i. Test the inequality (2.3.11). If it is true set $q = k - 1$.
  ii. Compute γ^k with equation (2.3.10a).
  iii. If $k = q + 1$, set $\psi^k = 0$. Otherwise, compute ψ^k with equation (2.3.10b).
  iv. Compute the search direction $\mathbf{d}^k$ with equation (2.3.5).
  v. If $k \neq q + 1$, test the inequalities (2.3.12a and b). If either one of them is satisfied set $q = k - 1$ and $\psi^k = 0$; then recompute the search direction with equation (2.3.5).

The computation of the direction of descent with the other versions of the conjugate gradient method is straightforward by utilizing the conjugation coefficients given by equations (2.3.7a and b), (2.3.8a and b) or (2.3.9a and b).

THE STOPPING CRITERION FOR TECHNIQUE II

The iterative procedure given by equations (2.3.1) and (2.3.4–2.3.10) does not necessarily result in stable solutions for the minimization of the objective function (2.1.1). As the estimated temperatures approach the measured temperatures containing errors, during the minimization of the function (2.1.1), large oscillations may appear in the inverse problem solution. However, stable solutions can be obtained with the conjugate gradient method if Morozov's discrepancy principle [2–22] is used to select the tolerance to stop the iterative procedure. In the discrepancy principle, the iterative procedure is stopped when the following criterion is satisfied:

$$S_{ML}(\mathbf{P}^k) < \varepsilon_1 \qquad (2.3.13)$$

where the value of the tolerance ε_1 is chosen so that sufficiently stable solutions are obtained. In this case, we stop the iterative procedure when the residuals between measured and estimated temperatures are of the same order of magnitude of the measurement errors, that is,

$$\left| Y_i - T_i(\mathbf{P}) \right| \approx \sigma_i \qquad (2.3.14)$$

where σ_i is the standard deviation of the measurement Y_i, $i = 1,\dots, I$. Equation (2.2.17) or (2.2.18) is then used for ε_1 if the measurements are uncorrelated, or uncorrelated and with constant standard deviation, respectively.

THE COMPUTATIONAL ALGORITHM FOR TECHNIQUE II

Suppose that temperature measurements $\mathbf{Y} = \left[Y_1, Y_2,\dots,Y_I \right]^T$ are given at times t_i, $i = 1,\dots, I$. Also, suppose an initial guess $\mathbf{P}^0$ is available for the vector of unknown parameters $\mathbf{P}$. Set $k = 0$ and then [2,6,7,15,22,29–43,73,75–79,84]:

  Step 1: Solve the direct heat transfer problem (2.2.1) by using the available estimate $\mathbf{P}^k$ and obtain the vector of estimated temperatures $\mathbf{T}(\mathbf{P}^k) = [T_1(\mathbf{P}^k), T_2(\mathbf{P}^k),\dots, T_I(\mathbf{P}^k)]^T$.
  Step 2: Check the stopping criterion given by equation (2.3.13). Continue if not satisfied.
  Step 3: Compute the sensitivity matrix $\mathbf{J}^k$ defined by equation (2.2.7).
  Step 4: Compute the gradient direction $\nabla S_{ML}(\mathbf{P}^k)$ from equation (2.3.6) and then the conjugation coefficients with one of the pair of equations (2.3.7a, b), (2.3.8a, b), (2.3.9a, b) or (2.3.10a, b). Note that Powell-Beale's version requires restarting if any of the inequalities (2.3.11) or (2.3.12a and b) are satisfied.

Step 5: Compute the direction of descent $\mathbf{d}^k$ by using equation (2.3.5).

Step 6: Compute the search step size β^k from equation (2.3.4).

Step 7: Compute the new estimate $\mathbf{P}^{k+1}$ with equation (2.3.1). Replace k by $k + 1$ and return to step 1.

2.4 SENSITIVITY COEFFICIENTS

The sensitivity matrix (2.2.7) plays an important role in parameter estimation problems. Therefore, we present in this section a discussion of the physical and mathematical significance of the sensitivity coefficients and the methods for their computation.

The sensitivity coefficient J_{ij}, as defined in equation (2.2.8), is a measure of the sensitivity of the estimated temperature T_i with respect to changes in the parameter P_j. A small value of the magnitude of J_{ij} indicates that large changes in P_j yield small changes in T_i. It can be easily noticed that the estimation of the parameter P_j is extremely difficult in such case, because basically the same value for temperature would be obtained for a wide range of values of P_j. In fact, when the sensitivity coefficients are small, we have $|\mathbf{J}^T\mathbf{J}| \approx 0$ and the inverse problem is ill-conditioned. It can also be shown that $|\mathbf{J}^T\mathbf{J}|$ is null if any column of $\mathbf{J}$ can be expressed as a linear combination of the other columns [4]. Therefore, *it is desirable to have linearly independent sensitivity coefficients J_{ij} with large magnitudes*, so that the inverse problem is not very sensitive to measurement errors and accurate estimates of the parameters can be obtained. The maximization of $|\mathbf{J}^T\mathbf{J}|$ is generally aimed in order to *design optimum experiments* for the estimation of the unknown parameters, because the confidence region of the estimates is then minimized. Details on such an approach are presented in Section 2.5.

Generally, *the variations in time of the sensitivity coefficients and of $|\mathbf{J}^T\mathbf{J}|$ must be examined before a solution for the inverse problem is attempted.* Such examinations give an indication of the best sensor location and measurement times to be used in the inverse analysis, which correspond to linearly independent sensitivity coefficients with large absolute values and large magnitudes of $|\mathbf{J}^T\mathbf{J}|$ [85]. Moreover, they reveal which group of parameters can be estimated via inverse analysis.

METHODS OF DETERMINING THE SENSITIVITY COEFFICIENTS

There are several different approaches for the computation of the sensitivity coefficients. We present below, with illustrative examples, three such approaches, including: (i) The direct analytic solution, (ii) the boundary value problem and (iii) the finite difference approximation.

Direct Analytic Solution for Determining Sensitivity Coefficients

If the direct heat transfer problem is linear and an analytic solution is available for the temperature field, the sensitivity coefficient with respect to an unknown parameter P_j is determined by differentiating the solution with respect to P_j. This approach is illustrated in the following two examples.

Example 2.1

Consider the dimensionless test problem given by equations (2.2.1a-d). The analytic solution of this problem at the measurement position is obtained as [56]:

$$T(x_{\text{meas}}, t) = \int_{t'=0}^{t} g_p(t') \, dt' + 2\sum_{m=1}^{\infty} e^{-\beta_m^2 t} \cos(\beta_m x_{\text{meas}}) \cos(0.5\beta_m) \int_{t'=0}^{t} e^{\beta_m^2 t'} g_p(t') \, dt' \quad (2.4.1a)$$

where $\beta_m = m\pi$ are the eigenvalues.

The first integral term on the right-hand side of equation (2.4.1a) is due to the fact that both boundary conditions for the problem are homogeneous of the second kind. Suppose $g_p(t)$ is parameterized in the general linear form as:

$$g_p(t) = \sum_{j=1}^{N} P_j C_j(t) \qquad (2.4.1b)$$

Find an analytic expression for the sensitivity coefficient $J_j \equiv \dfrac{\partial T}{\partial P_j}$ with respect to the parameter P_j.

Solution: By substituting the strength of the source term $g_p(t)$ given by equation (2.4.1b) into equation (2.4.1a) and differentiating the resulting expression with respect to P_j, we find the expression for the sensitivity coefficient for the parameter P_j as:

$$J_j \equiv \frac{\partial T}{\partial P_j} = \int_{t'=0}^{t} C_j(t')\, dt' + 2\sum_{m=1}^{\infty} e^{-\beta_m^2 t} \cos(\beta_m x_{\text{meas}}) \cos(0.5\beta_m) \int_{t'=0}^{t} e^{\beta_m^2 t'} C_j(t')\, dt' \qquad (2.4.1c)$$

The above inverse problem is linear because the sensitivity coefficients do not depend on P_j.

Figure 2.2 presents the timewise variation of the sensitivity coefficients given by equation (2.4.1c), for a sensor located at $x_{\text{meas}} = 1$ and for a case involving $N = 5$ unknown parameters, where the basis functions were taken in the form of polynomials as:

$$C_j(t) = t^{(j-1)} \qquad (2.4.1d)$$

Figure 2.2 shows that the sensitivity coefficients J_2, J_3, J_4 and J_5, with respect to the parameters P_2, P_3, P_4 and P_5, respectively, are proportional (thus, linearly dependent) in the time interval $0 < t < 1$. Therefore, the estimation of the five coefficients of the polynomial used to approximate the unknown source function is not possible in such a case. This figure also shows that the sensitivity coefficient J_1 with respect to the parameter P_1 is not linearly dependent with the others in this time interval. Hence, the estimation of any pair of parameters, which necessarily includes P_1 as one of them, appears to be feasible with a sensor located at $x_{\text{meas}} = 1$ and with measurements taken

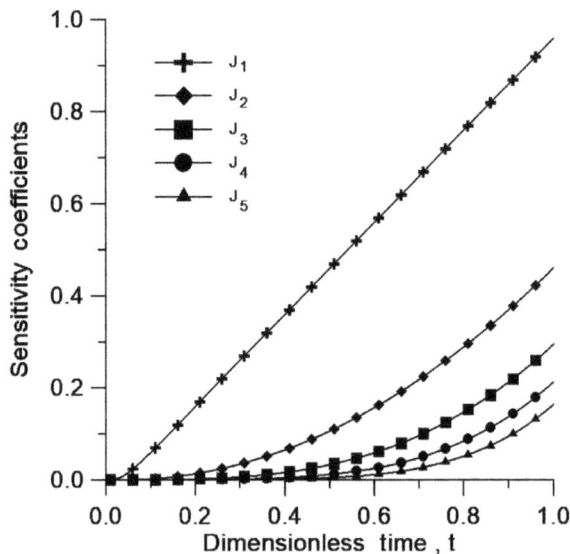

FIGURE 2.2 Sensitivity coefficients for polynomial trial functions given by equation (2.4.1d).

in the time interval $0<t<1$. In fact, for 100 transient measurements taken in this time interval, the determinant of $\mathbf{J}^T\mathbf{J}$ assumes the values 7.7 and 3.2×10^{-11} for 2 (P_1 and P_2) and 5 (P_1 through P_5) unknown parameters, respectively, indicating that linearly dependent sensitivity coefficients yield small values of $\left|\mathbf{J}^T\mathbf{J}\right|$.

Figure 2.3 shows the sensitivity coefficients for a sensor located at $x_{meas}=1$ and for basis functions in the form:

$$C_j(t) = \cos\left[(j-1)\frac{\pi}{2}t\right] \text{ for } j = 1,3 \text{ and } 5 \qquad (2.4.1e)$$

$$C_j(t) = \sin\left[j\frac{\pi}{2}t\right] \text{ for } j = 2 \text{ and } 4 \qquad (2.4.1f)$$

where the source function was approximated by a Fourier series.

We notice in Figure 2.3 that the sensitivity coefficients are not linearly dependent in the time interval $0.3<t<2$. Linear dependence is noticed among the sensitivity coefficients J_1, J_3 and J_5 for $t<0.3$. Similarly, J_2 and J_4 are linearly dependent for $t<0.3$. Therefore, the conditions for the estimation of the five unknown parameters are not adequate if measurements taken only in the interval $0<t<0.3$ are used in the analysis; but it appears that the parameters can be estimated if the measurements are taken up to $t_f=2$.

Figure 2.4 illustrates the time variation of the determinant of $\mathbf{J}^T\mathbf{J}$ up to a final experimental time $t_f=5$ by considering $I=100$, 250 and 500 measurements available from a sensor located at $x_{meas}=1$. The basis functions are given by equations (2.4.1e and f). For the three numbers of measurements considered, we notice a large increase in the magnitude of $\left|\mathbf{J}^T\mathbf{J}\right|$ up to about $t_f=2$. The magnitude of $\left|\mathbf{J}^T\mathbf{J}\right|$ continues to grow for larger times, but at a much smaller rate. As expected, $\left|\mathbf{J}^T\mathbf{J}\right|$ increases with the number of measurements, since more information is available for the estimation of the unknown parameters. However, such increase is not as significant as increasing the experimental time from $t_f=1$ to $t_f=2$. In the example, $t_f=2$ is a suitable duration for the experiment, since the value of $\left|\mathbf{J}^T\mathbf{J}\right|$ has already approached a reasonable large magnitude and the experiment duration is not too long.

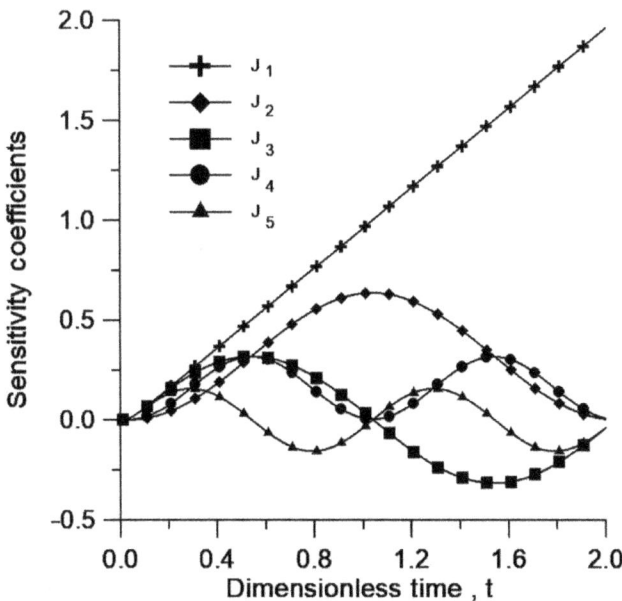

FIGURE 2.3 Sensitivity coefficients for the trial functions given by equations (2.4.1e and f).

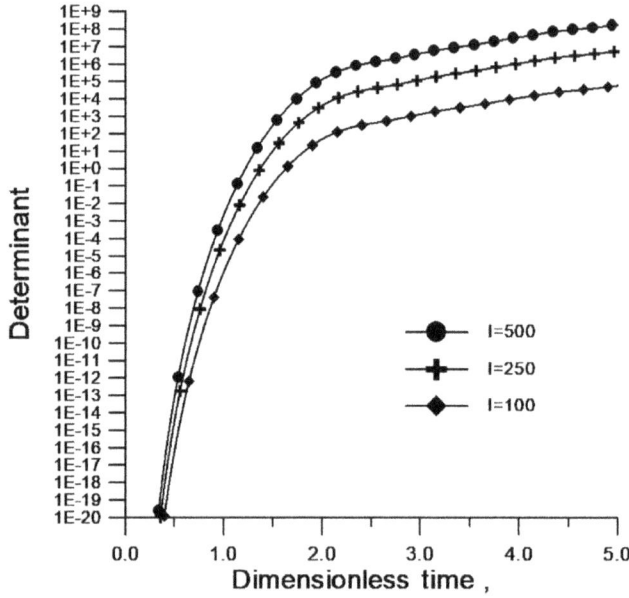

FIGURE 2.4   Determinant of $\mathbf{J}^T\mathbf{J}$ for the trial functions given by equations (2.4.1e and f).

Example 2.2

Consider a semi-infinite medium initially at zero temperature. For times $t>0$, the boundary surface at $x = 0$ is subjected to a constant heat flux q_0 W/m^2. Develop analytic expressions for the sensitivity coefficient $Jq_0(0,t)$ with respect to the applied heat flux q_0 and for the sensitivity coefficient $J_\alpha(0,t)$ with respect to the thermal diffusivity α, based on the temperature at the boundary surface $x = 0$. Examine the behavior of these sensitivity coefficients for carbon steel ($k = 43$ W/m °C, $\alpha = 1.2 \times 10^{-5}$ m^2/s) with $q_0 = 10{,}000$ W/m^2.

Solution: The temperature of the boundary surface at $x = 0$ is given by [56]:

$$T(0,t) = \frac{2q_0}{k}\left(\frac{\alpha t}{\pi}\right)^{1/2} \tag{2.4.2a}$$

Then, the sensitivity coefficient with respect to q_0 is determined from its definition as:

$$Jq_0(0,t) \equiv \frac{\partial T(0,t)}{\partial q_0} = \frac{2}{k}\left(\frac{\alpha t}{\pi}\right)^{1/2} \tag{2.4.2b}$$

which is independent of the applied heat flux q_0. Then, the inverse problem of estimating q_0 is linear.

The sensitivity coefficient with respect to α is determined as:

$$J_\alpha(0,t) \equiv \frac{\partial T(0,t)}{\partial \alpha} = \frac{q_0}{k}\left(\frac{t}{\pi\alpha}\right)^{1/2} \tag{2.4.2c}$$

which depends on α, and hence, the inverse problem of estimating α is nonlinear. Therefore, the inverse problem for the simultaneous estimation of $\mathbf{P} = \left[q_0, \alpha\right]^T$ is nonlinear.

The analysis of the sensitivity coefficients for nonlinear problems depends on reference values for the unknown parameters. In this case, we use $q_0 = 10{,}000$ W/m^2, $\alpha = 1.2 \times 10^{-5}$ m^2/s and $k = 43$ W/m °C. Figures 2.5a-c present the variations of temperature and sensitivity coefficients with respect to q_0 and α, respectively, up to time 120 s. These figures show that the curves of temperature and sensitivity coefficients differ by several orders of magnitude, but they basically have the same shape. The units of these quantities are different, that is, °C for temperature, °C/(W/m²) for the heat flux sensitivity coefficient and °C/(m²/s) for the thermal diffusivity sensitivity coefficient. Thus, different from Example 2.1, where the heat conduction problem was dimensionless, a comparison of the sensitivity coefficients cannot be directly performed in this example, because they do not have the same units. On the other hand, the curves shown by Figures 2.5b and c indicate that the sensitivity coefficients with respect to q_0 and α are linearly dependent, that is, these parameters are correlated.

In problems involving parameters with different orders of magnitude, the sensitivity coefficients with respect to the various parameters may also differ by several orders of magnitude, such as in this example, creating difficulties in their comparison and identification of linear dependence. These difficulties can be alleviated through the analysis of either dimensionless sensitivity coefficients or the *reduced sensitivity coefficients* defined as:

$$X_j \equiv P_j \frac{\partial T}{\partial P_j} \tag{2.4.3}$$

Therefore, the reduced sensitivity coefficients are obtained by multiplying the original sensitivity coefficients, J_j, by reference values of the corresponding parameter P_j, $j = 1, \ldots, N$. Note that the

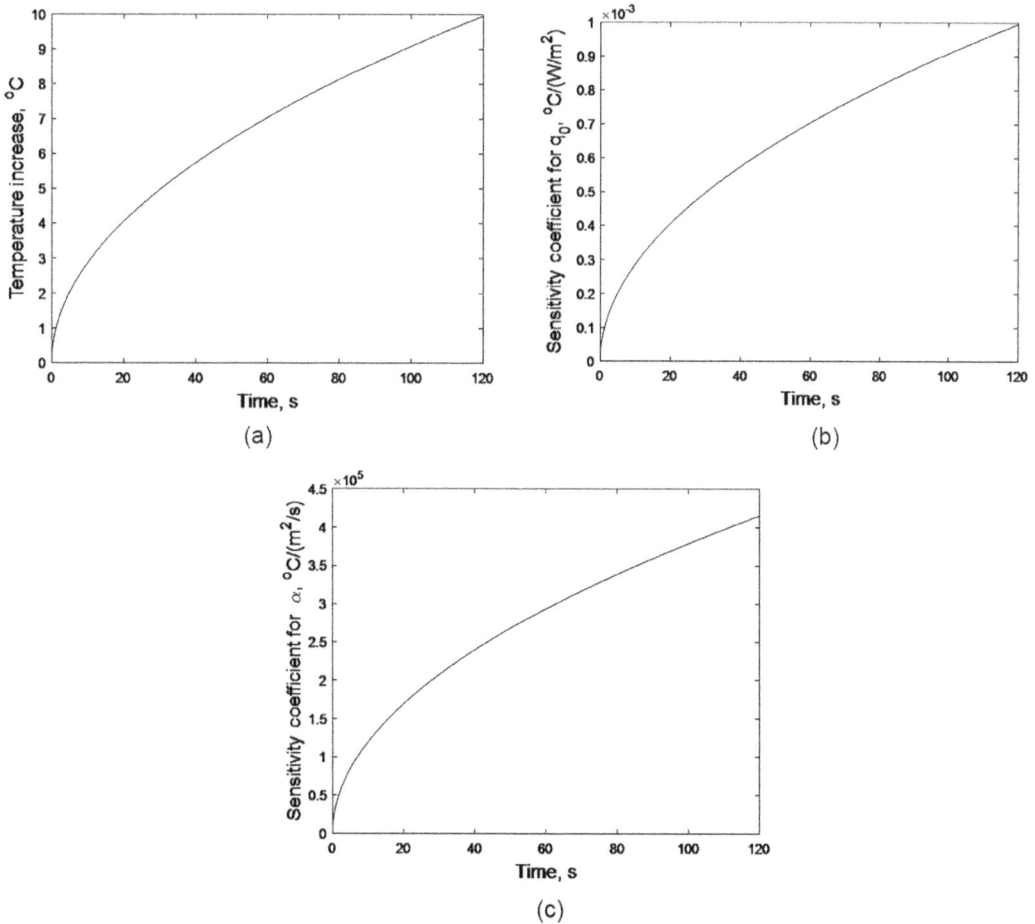

(a)

(b)

(c)

Figure 2.5 (a) Temperature variation; (b) sensitivity coefficient with respect to q_0; (c) sensitivity coefficient with respect to α.

reduced sensitivity coefficients have the units of temperature; hence, they can be compared as having the magnitude of the temperature as a basis.

In Example 2.2, the reduced sensitivity coefficients are given by:

$$X_{q_0}(0,t) \equiv q_0 \frac{\partial T(0,t)}{\partial q_0} = \frac{2q_0}{k}\left(\frac{\alpha t}{\pi}\right)^{1/2} \tag{2.4.4a}$$

$$X_{\alpha}(0,t) \equiv \alpha \frac{\partial T(0, t)}{\partial \alpha} = \frac{q_0 \alpha}{k}\left(\frac{t}{\pi\alpha}\right)^{1/2} \tag{2.4.4b}$$

The reduced sensitivity coefficients with respect to q_0 and α are presented in Figure 2.6, together with the temperature variation. The reduced sensitivity coefficient with respect to q_0 coincides with the temperature variation, since it is obtained from the multiplication of equation (2.4.2b) by q_0 (see equations 2.4.2a and 2.4.4a). Although smaller, the sensitivity coefficient with respect to α is of the same order of magnitude of the temperature variation. Therefore, the temperature is similarly affected by perturbations on both parameters. Figure 2.6 reveals that the curves for the reduced sensitivity coefficients are proportional. In fact, from equations (2.4.4a and b), we note that $X_{q_0}(0,t)/X_{\alpha}(0,t) = 2$.

It is also interesting to consider in this example the definition of the thermal diffusivity, that is,

$$\alpha = \frac{k}{\rho c} \tag{2.4.5}$$

and reduce equation (2.4.2a) to:

$$T(0,t) = \frac{2q_0}{e}\left(\frac{t}{\pi}\right)^{1/2} \tag{2.4.6}$$

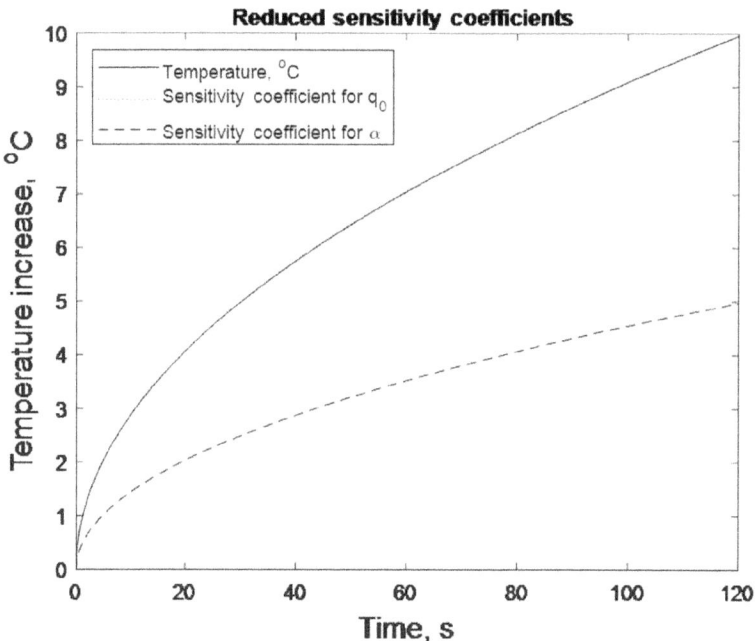

Figure 2.6 Temperature variation and reduced sensitivity coefficients.

where $e = \sqrt{k\rho c}$ = thermal effusivity, ρ is the density and c is the specific heat. By rewriting equation (2.4.2a) in terms of the effusivity, it becomes clear the linear dependence of the heat flux q_0 and the thermophysical properties of the heat conducting material. Note in equation (2.4.6) that q_0 and e do not appear independently in the solution, but in terms of one single parameter, that is, q_0/e. The determinant $|\mathbf{J}^T \mathbf{J}|$ computed with the sensitivity coefficients for q_0 and e is zero.

We now illustrate the behavior of the objective function for the case of two correlated parameters, such as in this example. Figure 2.7a shows the objective function (2.1.5) by considering a constant standard deviation of 0.1°C for the measurement errors. This figure was obtained with k = 43 W/m °C, α = 1.2×10^{-5} m²/s and q_0 = 10,000 W/m². Thus, the effusivity is 12413.03 J/(°C m² s^{1/2}). Figure 2.7a clearly shows that the objective function does not present a well-defined point of minimum, because of the linear dependence between q_0 and e. Instead, the objective function has a valley that leads to infinite minima over a straight line, where the ratio q_0/e is constant. The valley where the infinite minima are located is more clearly visualized with the contour plot of the objective function that is presented in Figure 2.7b. Therefore, the gradient $\nabla S_{ML}(\mathbf{P}) = -2\mathbf{J}^T \mathbf{W}^{-1}[\mathbf{Y} - \mathbf{T}(\mathbf{P})]$ is zero and the matrix $[\mathbf{J}^T \mathbf{W}^{-1}\mathbf{J}]$ is singular at the bottom of the valley and the minimization problem does not have an unique solution.

We note that a similar valley would occur if the sensitivity coefficient of one of the parameters would be very small or zero. In such case, the line where the minima are located is parallel to the axis of the parameter with small sensitivity coefficient.

The Boundary Value Problem Approach for Determining the Sensitivity Coefficients

A boundary value problem can be developed for the determination of the sensitivity coefficients by differentiating the original direct problem with respect to the unknown coefficients. If the direct heat transfer problem is linear, the construction of the corresponding sensitivity problem is a relatively simple and straightforward matter. To illustrate this approach, we use the following examples.

Example 2.3

For the test problem given by equations (2.2.1a-d) with $g_p(t)$ parameterized by equation (2.2.2), find the boundary value problem for the sensitivity coefficient with respect to the parameter P_j.

Solution: By using equation (2.2.2), differentiating equations (2.2.1a-d) with respect to the parameter P_j and noting that $J_j = \dfrac{\partial T}{\partial P_j}$, we obtain the sensitivity problem governing the sensitivity coefficients $J_j(x,t)$ as:

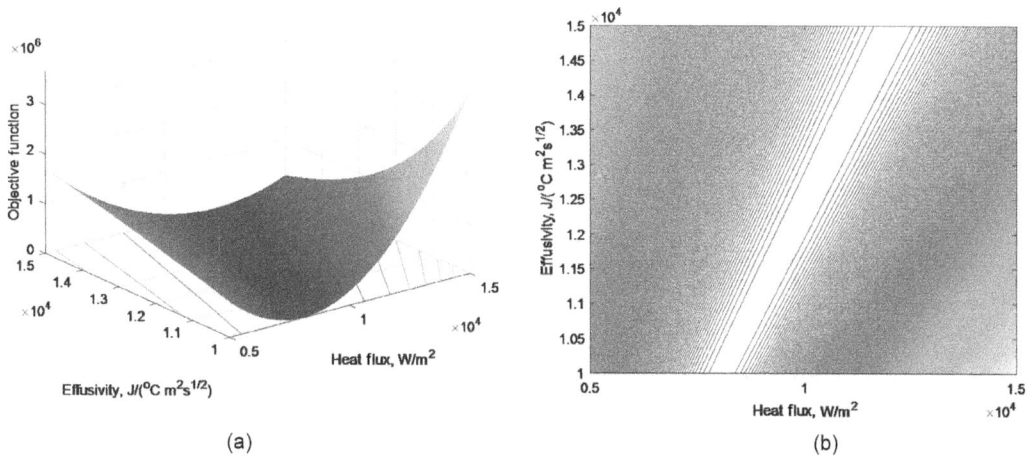

(a)

(b)

Figure 2.7 Objective function (2.1.5) for Example 2.2: (a) 3D surface; (b) contour plot.

$$\frac{\partial^2 J_j(x,t)}{\partial x^2} + C_j(t)\,\delta(x-0.5) = \frac{\partial J_j(x,t)}{\partial t} \quad \text{in } 0 < x < 1, \text{ for } t > 0 \tag{2.4.7a}$$

$$\frac{\partial J_j}{\partial x} = 0 \text{ at } x = 0, \text{ for } t > 0 \tag{2.4.7b}$$

$$\frac{\partial J_j}{\partial x} = 0 \text{ at } x = 1, \text{ for } t > 0 \tag{2.4.7c}$$

$$J_j(x,0) = 0 \text{ for } t = 0, \text{ in } 0 < x < 1 \tag{2.4.7d}$$

Note that problem (2.4.7) is similar to problem (2.2.1). Problem (2.4.7) needs to be solved N times in order to compute the sensitivity coefficients with respect to each parameter P_j, $j = 1,..., N$. For this particular case, the analytical solution of problem (2.4.7) can be easily obtained with equation (2.4.1c). For more involved cases, the solution of the boundary value problem for determining the sensitivity coefficients may require numerical techniques, such as finite differences. Thus, the computation of the sensitivity coefficients may become very time consuming.

Example 2.4

Consider the following heat conduction problem:

$$k\frac{\partial^2 T}{\partial x^2} = C\frac{\partial T}{\partial t} \quad \text{in } 0 < x < L, \text{ for } t > 0 \tag{2.4.8a}$$

$$-k\frac{\partial T}{\partial x} = q_0 \text{ at } x = 0, \text{ for } t > 0 \tag{2.4.8b}$$

$$\frac{\partial T}{\partial x} = 0 \text{ at } x = L, \text{ for } t > 0 \tag{2.4.8c}$$

$$T = T_i \text{ for } t = 0, \text{ in } 0 < x < L \tag{2.4.8d}$$

where $C \equiv \rho c$, heat capacity, $q_0 =$ applied heat flux and $k =$ thermal conductivity.
    Construct the *sensitivity problem* for determining the sensitivity coefficient with respect to thermal conductivity, i.e.,

$$J_k \equiv \frac{\partial T}{\partial k} \tag{2.4.9}$$

Solution: By differentiating problem (2.4.8) with respect to k and utilizing the definition of J_k, we obtain the following boundary value problem for determination of the sensitivity coefficient:

$$k\frac{\partial^2 J_k}{\partial x^2} + \frac{\partial^2 T}{\partial x^2} = C\frac{\partial J_k}{\partial t} \quad \text{in } 0 < x < L, \text{ for } t > 0 \tag{2.4.10a}$$

$$k\frac{\partial J_k}{\partial x} + \frac{\partial T}{\partial x} = 0 \text{ at } x = 0, \text{ for } t > 0 \tag{2.4.10b}$$

$$\frac{\partial J_k}{\partial x} = 0 \text{ at } x = L, \text{ for } t > 0 \tag{2.4.10c}$$

$$J_k = 0 \text{ for } t = 0, \text{ in } 0 < x < L \tag{2.4.10d}$$

We note that problem (2.4.10) contains the non-homogeneous terms $\partial^2 T/\partial x^2$ and $\partial T/\partial x$ in equations (2.4.10a and b), respectively. Also, the unknown parameter k appears in these two equations; thus, the inverse problem of estimating k is nonlinear. The solution of problem (2.4.10) yields the sensitivity coefficients J_k with respect to thermal conductivity k. By following a similar procedure, the sensitivity problem for determining the sensitivity coefficient J_C, with respect to heat capacity C, can be developed.

Finite Difference Approximation for Determining Sensitivity Coefficients

The first derivative appearing in the definition of the sensitivity coefficient, equation (2.2.8), can be computed by finite differences. If a *forward difference* is used, the sensitivity coefficient with respect to the parameter P_j is approximated by [86]:

$$J_{ij} \cong \frac{T_i(P_1,P_2,\dots,P_j + \varepsilon P_j,\dots,P_N) - T_i(P_1,P_2,\dots,P_j,\dots,P_N)}{\varepsilon P_j} \qquad (2.4.11)$$

If the forward difference approximation given by equation (2.4.11) is not sufficiently accurate, the sensitivity coefficients can be approximated by using *central differences* in the form [86]:

$$J_{ij} \cong \frac{T_i(P_1,P_2,\dots,P_j + \varepsilon P_j,\dots,P_N) - T_i(P_1,P_2,\dots,P_j - \varepsilon P_j,\dots,P_N)}{2\varepsilon P_j} \qquad (2.4.12)$$

We note that the approximation of the sensitivity coefficients given by equation (2.4.11) requires the computation of N additional solutions of the direct problem, while equation (2.4.12) requires $2N$ additional solutions of this problem. Therefore, the computation of the sensitivity coefficients by using finite differences can be very time consuming. On the other hand, the solution of the direct problem should have already been duly verified before the solution of the inverse problem. Thus, no additional code or solution verifications are necessary if the sensitivity coefficients are computed by finite differences, which is not the case when a sensitivity problem is used for the calculation of the sensitivity coefficients.

As expected, the accuracy of the finite difference approximation for the sensitivity coefficients is highly dependent on the increment ε. The orders of the approximation errors are (εP_j) and $(\varepsilon P_j)^2$ for equations (2.4.11) and (2.4.12), respectively [86]. Thus, the approximation error is large if ε is large. On the other hand, numerical errors may also be large if the value selected for ε is too small, in which case the numerators and the denominators of equations (2.4.11) and (2.4.12) would be very small numbers. It is recommended that different values of ε be tested and the corresponding sensitivity coefficients compared in order to avoid both approximation and numerical errors. Practical experience shows that usually $\varepsilon \approx 10^{-5}$ or 10^{-6}.

2.5   DESIGN OF OPTIMUM EXPERIMENTS

Optimum experiments are usually designed by minimizing the hypervolume of the confidence region of the estimated parameters in order to ensure minimum variance for the estimates. The minimization of the confidence region can be obtained by maximizing the determinant of $\mathbf{J}^T\mathbf{J}$, in the so-called *D-optimum design* [4,6,8,22,40,85,87]. Therefore, experimental variables such as the duration of the experiment, location and number of sensors are chosen based on the criterion:

$$\max|\mathbf{J}^T\mathbf{J}| \qquad (2.5.1)$$

By using the definition of the sensitivity matrix for the case involving a single sensor, equation (2.2.7), each element $\mathbf{F}_{p,n}$, $p, n = 1, \dots, N$, of the matrix $\mathbf{F} \equiv \mathbf{J}^T\mathbf{J}$ is given by:

$$\mathbf{F}_{p,n} \equiv \left[\mathbf{J}^T\mathbf{J}\right]_{p,n} = \sum_{i=1}^{I} \left(\frac{\partial T_i}{\partial P_p}\right)\left(\frac{\partial T_i}{\partial P_n}\right) \qquad (2.5.2)$$

where I is the number of measurements and N is the number of unknown parameters.
  Two particular cases of the general criterion (2.5.1) are now examined [85].

Case 1: A large but fixed number of equally spaced measurements is available. Then, each
element $\mathbf{F}_{p,n}$ can be written as:

$$\mathbf{F}_{p,n} = \frac{1}{\Delta t} \sum_{i=1}^{I} \left(\frac{\partial T_i}{\partial P_p} \right) \left(\frac{\partial T_i}{\partial P_n} \right) \Delta t \approx \frac{I}{t_f} \int_{t=0}^{t_f} \left(\frac{\partial T}{\partial P_p} \right) \left(\frac{\partial T}{\partial P_n} \right) dt \quad \text{for } p,n=1,\dots,N \qquad (2.5.3a)$$

where t_f is the duration of the experiment and Δt is the constant time interval between two
consecutive measurements. Since the number of measurements, I, is fixed, we can choose
to maximize the determinant of $\mathbf{F}_I$ instead of maximizing the determinant of $\mathbf{F}$, where the
elements of $\mathbf{F}_I$ are given by:

$$[\mathbf{F}_I]_{p,n} = \frac{1}{t_f} \int_{t=0}^{t_f} \left(\frac{\partial T}{\partial P_p} \right) \left(\frac{\partial T}{\partial P_n} \right) dt \quad \text{for } p,n=1,\dots,N \qquad (2.5.3b)$$

Case 2: In addition to a large and fixed number of equally spaced measurements, the maximum value for the temperature in the region, $T_{\max}$, is known. Thus, equation (2.5.3b) can
be written as:

$$[\mathbf{F}_I]_{p,n} = \frac{T_{\max}^2}{t_f P_p P_n} \int_{t=0}^{t_f} \left(\frac{P_p}{T_{\max}} \frac{\partial T}{\partial P_p} \right) \left(\frac{P_n}{T_{\max}} \frac{\partial T}{\partial P_n} \right) dt \quad \text{for } p,n=1,\dots,N \qquad (2.5.4a)$$

Note that the quantities inside parentheses in equation (2.5.4a) are dimensionless. However, it is
possible that T^*, and not $T_{\max}$, is the variable suitable for the non-dimensionalization of the temperature T. In such a case, equation (2.5.4a) can be written as:

$$[\mathbf{F}_I]_{p,n} = \frac{T_{\max}^2}{t_f P_p P_n} \int_{t=0}^{t_f} \left(\frac{P_p}{T^*} \frac{\partial T}{\partial P_p} \right) \left(\frac{P_n}{T^*} \frac{\partial T}{\partial P_n} \right) \left(\frac{T^*}{T_{\max}} \right)^2 dt \quad \text{for } p,n=1,\dots,N \qquad (2.5.4b)$$

and the design of optimum experiments is then based on the maximization of the determinant of the
dimensionless form of $\mathbf{F}_I$, $\mathbf{F}_I^*$, the elements of which are given by:

$$[\mathbf{F}_I^*]_{p,n} = \frac{1}{t_f} \int_{t=0}^{t_f} \left(\frac{P_p}{T^*} \frac{\partial T}{\partial P_p} \right) \left(\frac{P_n}{T^*} \frac{\partial T}{\partial P_n} \right) \left(\frac{T^*}{T_{max}} \right)^2 dt \quad \text{for } p,n=1,\dots,N \qquad (2.5.4c)$$

We note that for nonlinear parameter estimation problems, the sensitivity matrix, and thus $\left|\mathbf{J}^T\mathbf{J}\right|$,
depends on the unknown parameters $P_j, j = 1, \dots, N$. In such cases, only a local optimum experimental design is possible by using some *a priori* information regarding the expected values for the
unknown parameters.

2.6   THE USE OF MULTIPLE SENSORS

For simplicity in the analysis and to facilitate the reader's understanding, all equations presented
above considered transient measurements of one single sensor. However, the use of multiple sensors
can be easily accommodated without major modifications for the implementations of Techniques
I and II.

In cases where the measurements of M sensors are available, the vector containing the differences between measured and estimated temperatures is written as (see equations 1.4.2 and 1.4.4):

$$[\mathbf{Y} - \mathbf{T}(\mathbf{P})]^T = [\vec{Y}_1 - \vec{T}_1(\mathbf{P}) , \ \vec{Y}_2 - \vec{T}_2(\mathbf{P}) ,..., \ \vec{Y}_I - \vec{T}_I(\mathbf{P})] \tag{2.6.1a}$$

where $[\vec{Y}_i - \vec{T}_i(\mathbf{P})]$ is a row vector that contains the differences between measured and estimated temperatures for each of the M sensors at time t_i, $i = 1, ..., I$. It is given in the form:

$$[\vec{Y}_i - \vec{T}_i(\mathbf{P})] = [Y_{i1} - T_{i1}(\mathbf{P}) , \ Y_{i2} - T_{i2}(\mathbf{P}) ,..., \ Y_{iM} - T_{iM}(\mathbf{P})] \ \text{for} \ i = 1,...,I \tag{2.6.1b}$$

In the vector element $[Y_{im} - T_{im}(\mathbf{P})]$, the subscript i refers to time t_i, while the subscript m refers to the sensor number, where $i = 1, ..., I$ and $m = 1, ..., M$.

The sensitivity matrix defined by equation (2.2.7) needs to be modified in order to accommodate the measurements of M sensors. The transpose of the sensitivity matrix, defined by equation (2.2.6), is then written as:

$$\frac{\partial \mathbf{T}^T(\mathbf{P})}{\partial \mathbf{P}} = \begin{bmatrix} \dfrac{\partial}{\partial P_1} \\[6pt] \dfrac{\partial}{\partial P_2} \\[6pt] \vdots \\ \vdots \\[6pt] \dfrac{\partial}{\partial P_N} \end{bmatrix} [\vec{T}_1(\mathbf{P}) , \ \vec{T}_2(\mathbf{P}) ,..., \ \vec{T}_I(\mathbf{P})] \tag{2.6.2a}$$

where

$$\vec{T}_i(\mathbf{P}) = \left[T_{i1}(\mathbf{P}) , \ T_{i2}(\mathbf{P}) ,... , \ T_{iM}(\mathbf{P}) \right] \ \text{for} \ i = 1,...,I \tag{2.6.2b}$$

Therefore, we can write the sensitivity matrix in the form:

$$\mathbf{J}(\mathbf{P}) \equiv \left[\frac{\partial \mathbf{T}^T(\mathbf{P})}{\partial \mathbf{P}} \right]^T = \begin{bmatrix} \dfrac{\partial \vec{T}_1^T(\mathbf{P})}{\partial P_1} & \dfrac{\partial \vec{T}_1^T(\mathbf{P})}{\partial P_2} & \dfrac{\partial \vec{T}_1^T(\mathbf{P})}{\partial P_3} & \cdots & \dfrac{\partial \vec{T}_1^T(\mathbf{P})}{\partial P_N} \\[10pt] \dfrac{\partial \vec{T}_2^T(\mathbf{P})}{\partial P_1} & \dfrac{\partial \vec{T}_2^T(\mathbf{P})}{\partial P_2} & \dfrac{\partial \vec{T}_2^T(\mathbf{P})}{\partial P_3} & \cdots & \dfrac{\partial \vec{T}_2^T(\mathbf{P})}{\partial P_N} \\[10pt] \vdots & \vdots & \vdots & & \vdots \\[10pt] \dfrac{\partial \vec{T}_I^T(\mathbf{P})}{\partial P_1} & \dfrac{\partial \vec{T}_I^T(\mathbf{P})}{\partial P_2} & \dfrac{\partial \vec{T}_I^T(\mathbf{P})}{\partial P_3} & \cdots & \dfrac{\partial \vec{T}_I^T(\mathbf{P})}{\partial P_N} \end{bmatrix} \tag{2.6.3a}$$

where

$$\frac{\partial \vec{T}_i^T(\mathbf{P})}{\partial P_j} = \begin{bmatrix} \dfrac{\partial T_{i1}(\mathbf{P})}{\partial P_j} \\[2ex] \dfrac{\partial T_{i2}(\mathbf{P})}{\partial P_j} \\[1ex] \vdots \\[1ex] \dfrac{\partial T_{iM}(\mathbf{P})}{\partial P_j} \end{bmatrix} \quad \text{for } i = 1,...,I \text{ and } j = 1,...,N \tag{2.6.3b}$$

I = number of transient measurements per sensor
M = number of sensors
N = number of unknown parameters

Equations (2.6.1a and b) show that the vector $[\mathbf{Y} - \mathbf{T}(\mathbf{P})]^T$ can be conveniently rewritten in terms of one single index, k, which defines its column number as a function of the time and sensor indexes, that is,

$$[\mathbf{Y} - \mathbf{T}(\mathbf{P})]^T = \left[Y_1 - T_1(\mathbf{P}) ,..., Y_k - T_k(\mathbf{P}) ,..., Y_D - T_D(\mathbf{P}) \right] \tag{2.6.4}$$

where $D = M\,I$ is the total number of measurements. The index k is calculated as:
    For $i = 1$ to I
        For $m = 1$ to M
            $k = (i-1)M + m$
        End loop for m
    End loop for i

The elements of the sensitivity matrix can also be written in terms of the index k, that is,

$$J_{kj} = \frac{\partial T_k}{\partial P_j} \tag{2.6.5}$$

By also writing the covariance matrix of the measurements, $\mathbf{W}$, in terms of the index k, the computational algorithms for Techniques I and II can be readily applied to cases involving the measurements of multiple sensors.

For the design of experiments with the transient measurements of M sensors, the elements $[\mathbf{F}_I^*]_{p,n}$ of the matrix given by equation (2.5.4c) is written as:

$$[\mathbf{F}_I^*]_{p,n} = \frac{1}{M t_f} \sum_{m=1}^{M} \int_{t=0}^{t_f} \left(\frac{P_p}{T^*} \frac{\partial T_m}{\partial P_p} \right) \left(\frac{P_n}{T^*} \frac{\partial T_m}{\partial P_n} \right) \left(\frac{T^*}{T_{\max}} \right)^2 dt \quad \text{for } p,n = 1,...,N \tag{2.6.6}$$

2.7   STATISTICAL ANALYSIS

By performing a statistical analysis it is possible to assess the accuracy of $\hat{P}_j$, which are the values estimated for the unknown parameters $P_j, j = 1, ..., N$.

Assuming that the measurement errors are additive, Gaussian, with zero mean and known covariance matrix $\mathbf{W}$, the *covariance matrix*, Φ, of the estimated parameters $\hat{P}_j, j = 1, ..., N$, which is obtained with the minimization of the maximum likelihood objective function (equation 2.1.1), is given by [4]:

$$\Phi = \mathrm{cov}(\hat{\mathbf{P}}) \equiv \begin{bmatrix} \mathrm{cov}(\hat{P}_1,\hat{P}_1) & \mathrm{cov}(\hat{P}_1,\hat{P}_2) & \cdots & \mathrm{cov}(\hat{P}_1,\hat{P}_N) \\ \mathrm{cov}(\hat{P}_2,\hat{P}_1) & \mathrm{cov}(\hat{P}_2,\hat{P}_2) & \cdots & \mathrm{cov}(\hat{P}_2,\hat{P}_N) \\ \vdots & \vdots & \vdots & \vdots \\ \vdots & \vdots & \vdots & \vdots \\ \mathrm{cov}(\hat{P}_N,\hat{P}_1) & \mathrm{cov}(\hat{P}_N,\hat{P}_2) & \cdots & \mathrm{cov}(\hat{P}_N,\hat{P}_N) \end{bmatrix} = (\mathbf{J}^T \mathbf{W}^{-1} \mathbf{J})^{-1} \quad (2.7.1a)$$

Equation (2.7.1a) reduces to:

$$\Phi = \mathrm{cov}(\hat{\mathbf{P}}) = (\mathbf{J}^T \mathbf{J})^{-1} \sigma^2 \tag{2.7.1b}$$

if, in addition to the previous hypotheses, the measurement errors are also uncorrelated and with constant variances, that is, $\mathbf{W} = \sigma^2 \mathbf{I}$, and the maximum likelihood objective function reduces to the ordinary least squares norm given by equation (2.1.6b). Equations (2.7.1a and b) are exact for linear estimation problems, but can be used as approximations for nonlinear problems.

We note that equation (2.7.1b) gives the covariance matrix of parameters estimated with the minimization of the least squares norm, only if all the above statistical hypotheses for the measurement errors are valid. However, if the estimation problem is linear, the measurement errors are additive, with zero mean, and with a covariance matrix that is positive definite and known to within a multiplicative constant σ^2, that is,

$$\mathbf{W} = \hat{\mathbf{W}} \sigma^2 \tag{2.7.2}$$

the Gauss-Markov theorem [4,22] states that minimum variance estimates can be obtained with the minimization of:

$$S_{GM}(\mathbf{P}) = \left[\mathbf{Y} - \mathbf{T}(\mathbf{P}) \right]^T \hat{\mathbf{W}}^{-1} \left[\mathbf{Y} - \mathbf{T}(\mathbf{P}) \right] \tag{2.7.3}$$

even if the measurement errors are not Gaussian. In such a case, if $\hat{\mathbf{W}} = \mathbf{I}$, the minimization of the ordinary least squares norm provides minimum variance estimates. On the other hand, the covariance matrix of the values estimated for the parameters $\mathbf{P}$ cannot be computed with equation (2.7.1b), since σ^2 is not known.

The *standard deviations* for the estimated parameters can be obtained from the diagonal elements of Φ as:

$$\sigma_{\hat{P}_j} \equiv \sqrt{\mathrm{cov}(\hat{P}_j,\hat{P}_j)} = \sqrt{\Phi_{jj}} \quad \text{for } j = 1,...,N \tag{2.7.4}$$

where Φ_{jj} is the jth element in the diagonal of Φ.

Confidence intervals at the 99% confidence level for the estimated parameters are obtained as:

$$\hat{P}_j - 2.576\sigma_{\hat{P}_j} \leq P_j \leq \hat{P}_j + 2.576\sigma_{\hat{P}_j} \quad \text{for } j = 1,...,N \tag{2.7.5}$$

The factor 2.576 appearing in the expression above comes from Table 1.2 in Chapter 1 for the normal distribution, so that the probability of the actual parameter P_j be in the interval $\hat{P}_j \pm 2.576\sigma_{\hat{P}_j}$

is 99%. For other confidence levels, this factor is changed accordingly. For example, for the 95% confidence level, 2.576 should be replaced by 1.96.

Confidence intervals do not provide a good approximation for a *joint confidence region* for the estimated parameters. In fact, the confidence interval is obtained for each parameter regardless of the linear dependence of the parameters, which is measured by the off-diagonal elements of Φ. Confidence regions built from confidence intervals may include areas outside the actual confidence region and not include areas that belong to the actual confidence region [4,22].

The *joint confidence region* for the estimated parameters is given by [4,22]:

$$(\hat{\mathbf{P}} - \mathbf{P})^T \Phi^{-1}(\hat{\mathbf{P}} - \mathbf{P}) \le \chi_N^2 \tag{2.7.6}$$

where Φ = covariance matrix of the estimated parameters
$\hat{\mathbf{P}} = [\hat{P}_1, \hat{P}_2, \dots, \hat{P}_N]$ is the vector with the values estimated for the parameters
$\mathbf{P} = [P_1, P_2, \dots, P_N]$ is the vector of unknown parameters
N = the number of parameters
χ_N^2 = value of the chi-square distribution with N degrees of freedom for a given probability, obtained from Table 1.3.

The confidence region given by equation (2.7.6) is thus the interior of a hyperellipsoid centered at the origin and with coordinates $(\hat{P}_1 - P_1)$, $(\hat{P}_2 - P_2), \dots, (\hat{P}_N - P_N)$. The surface of the hyperellipsoid is a constant probability density surface obtained from the chi-square distribution for a chosen confidence level. For a case involving the estimation of two parameters, the values of χ_2^2 obtained from Table 1.3 in Chapter 1 for the 95% and 99% confidence levels are 5.99 and 9.21, respectively.

If the statistical hypotheses required to obtain the maximum likelihood function given by equation (2.1.1) are valid, the parameters estimated with its minimization are *unbiased*, that is, the expected value of $\hat{\mathbf{P}}$ are the actual parameters $\mathbf{P}$ [4].

2.8 ESTIMATION OF THERMAL CONDUCTIVITY COMPONENTS OF AN ORTHOTROPIC HEAT CONDUCTING MEDIUM

The objective of this section is to present the implementation of the solution of an inverse problem of parameter estimation, including all steps described above, by using Techniques I and II. These techniques are also compared in this section, in terms of solution accuracy and convergence rate [82,88].

The physical problems considered here involve the three-dimensional linear heat conduction in an orthotropic solid, with thermal conductivity components k_1^*, k_2^* and k_3^* in the x^*, y^* and z^* directions, respectively. The solid is a parallelepiped with sides a^*, b^* and c^*, initially at temperature T_0^*. For times $t^* > 0$, uniform heat fluxes $q_1^*(t)$, $q_2^*(t)$ and $q_3^*(t)$ are supplied at the surfaces $x^* = a^*$, $y^* = b^*$ and $z^* = c^*$, respectively.

The following experimental arrangements are considered here, depending on the boundary conditions used for the other three remaining surfaces at $x^* = 0$, $y^* = 0$ and $z^* = 0$: (i) insulated boundaries and (ii) boundaries maintained at constant temperature (equal to the initial temperature). Such arrangements are hereafter designated as *Experiments I and II*, respectively. We note that constant temperature boundary conditions can be more easily implemented for low thermal conductivity materials. This can be accomplished by placing a block of a high thermal conductivity material in contact with the surface. On the other hand, insulated boundaries can be more easily implemented for high thermal conductivity materials by placing a low thermal conductivity material in contact with the surface.

The mathematical formulation of such physical problems are given in dimensionless form by:

$$k_1 \frac{\partial^2 T}{\partial x^2} + k_2 \frac{\partial^2 T}{\partial y^2} + k_3 \frac{\partial^2 T}{\partial z^2} = \frac{\partial T}{\partial t} \quad \text{in } 0 < x < a, 0 < y < b, 0 < z < c \,; t > 0 \qquad (2.8.1a)$$

$$\alpha \frac{\partial T}{\partial x} + \beta T = 0 \text{ at } x = 0; \quad k_1 \frac{\partial T}{\partial x} = q_1(t) \quad \text{at } x = a, \text{ for } t > 0 \qquad (2.8.1b,c)$$

$$\alpha \frac{\partial T}{\partial y} + \beta T = 0 \text{ at } y = 0; \quad k_2 \frac{\partial T}{\partial y} = q_2(t) \quad \text{at } y = b, \text{ for } t > 0 \qquad (2.8.1.d,e)$$

$$\alpha \frac{\partial T}{\partial z} + \beta T = 0 \text{ at } z = 0; \quad k_3 \frac{\partial T}{\partial z} = q_3(t) \quad \text{at } z = c, \text{ for } t > 0 \qquad (2.8.1f,g)$$

$$T = 0 \text{ for } t = 0; \text{ in } 0 < x < a, 0 < y < b, 0 < z < c \qquad (2.8.1h)$$

where for *Experiment I* $\alpha = 1$ and $\beta = 0$, while for *Experiment II* $\alpha = 0$ and $\beta = 1$. The following dimensionless groups were introduced:

$$t = \frac{k_{ref}^* \, t^*}{\rho^* C^* L^{*2}} \quad T = \frac{T^* - T_0^*}{\dfrac{q_{ref}^* \, L^*}{k_{ref}^*}} \quad x = \frac{x^*}{L^*} \quad y = \frac{y^*}{L^*} \quad z = \frac{z^*}{L^*} \qquad (2.8.2a\text{-}e)$$

$$a = \frac{a^*}{L^*} \quad b = \frac{b^*}{L^*} \quad c = \frac{c^*}{L^*} \quad k_1 = \frac{k_1^*}{k_{ref}^*} \quad k_2 = \frac{k_2^*}{k_{ref}^*} \qquad (2.8.2f\text{-}j)$$

$$k_3 = \frac{k_3^*}{k_{ref}^*} \quad q_1 = \frac{q_1^*}{q_{ref}^*} \quad q_2 = \frac{q_2^*}{q_{ref}^*} \quad q_3 = \frac{q_3^*}{q_{ref}^*} \qquad (2.8.2k\text{-}n)$$

where the superscript * above denotes dimensional variables, while L^* is a characteristic length, and q_{ref}^* and k_{ref}^* are reference values for heat flux and thermal conductivity, respectively.

THE DIRECT PROBLEM

In the *direct problem* associated with the physical problems described above, the three thermal conductivity components k_1, k_2 and k_3, as well as the solid geometry, initial and boundary conditions, are known. The objective of the direct problem is then to determine the transient temperature distribution $T(x, y, z, t)$ in the body.

The boundary heat fluxes are assumed to be pulses of finite duration t_h, that is [82,85,88,89]:

$$q_j(t) = \begin{cases} \bar{q}_j, & \text{for} \quad 0 < t \le t_h \\ 0, & \text{for} \quad t > t_h \end{cases} \quad \text{for} \quad j = 1,2,3 \qquad (2.8.3)$$

where $\bar{q}_j$ is the dimensionless magnitude of the applied heat flux.

The solution of the direct problem (2.8.1) for *Experiment I*, with boundary heat fluxes given by equation (2.8.3), can be obtained analytically as a superposition of three one-dimensional solutions, by using the split-up procedure [90]. We obtain for $0 < t \le t_h$:

$$T(x,y,z,t) = \frac{\bar{q}_1\, a}{k_1}\, \theta_1\!\left(\frac{x}{a}, \frac{k_1\, t}{a^2}\right) + \frac{\bar{q}_2\, b}{k_2}\, \theta_1\!\left(\frac{y}{b}, \frac{k_2\, t}{b^2}\right) + \frac{\bar{q}_3\, c}{k_3}\, \theta_1\!\left(\frac{z}{c}, \frac{k_3\, t}{c^2}\right) \qquad (2.8.4a)$$

where

$$\theta_1(\xi,\tau) = -\frac{1}{6} + \frac{\xi^2}{2} + \tau + \sum_{i=1}^{\infty} (-1)^{(i+1)}\, \frac{2}{(i\pi)^2}\, \cos(i\,\pi\,\xi)\, \exp[-\tau\,(i\pi)^2] \qquad (2.8.4b)$$

After the heating period, problem (2.8.1) for *Experiment I* becomes homogeneous with initial temperature distribution given by equation (2.8.4a) at $t = t_h$. Hence, it can be easily solved by separation of variables [56]. Since the initial condition at $t = t_h$ obtained from equation (2.8.4a) is a superposition of three one-dimensional solutions, such is also the case of the solution for $t > t_h$. The temperature distribution for $t > t_h$ is given by:

$$T(x,y,z,t) = \frac{\bar{q}_1\, a}{k_1}\, \theta_2\!\left(\frac{x}{a}, \frac{k_1\, t}{a^2}, \frac{k_1\, t_h}{a^2}\right) + \frac{\bar{q}_2\, b}{k_2}\, \theta_2\!\left(\frac{y}{b}, \frac{k_2\, t}{b^2}, \frac{k_2\, t_h}{b^2}\right) + \frac{\bar{q}_3\, c}{k_3}\, \theta_2\!\left(\frac{z}{c}, \frac{k_3\, t}{c^2}, \frac{k_3\, t_h}{c^2}\right) \qquad (2.8.5a)$$

where

$$\theta_2(\xi,\tau,\tau_h) = \tau_h + 2 \sum_{i=1}^{\infty} \frac{(-1)^i \cos(i\,\pi\,\xi)\exp[-(\tau-\tau_h)(i\pi)^2]\{1-\exp[-(i\pi)^2\tau_h]\}}{(i\pi)^2} \qquad (2.8.5b)$$

On the other hand, the analytical solution of problem (2.8.1) for *Experiment II* cannot be obtained as a superposition of three one-dimensional solutions, because of the first-kind boundary conditions at $x = 0$, $y = 0$ and $z = 0$. The solution of problem (2.8.1) for *Experiment II* for $0 < t \le t_h$ can be obtained with the classical integral transform technique [56,90] as:

$$T(x,y,z,t) = \frac{8}{a\,b\,c} \sum_{o=1}^{\infty}\sum_{n=1}^{\infty}\sum_{m=1}^{\infty} \left\{ \Theta(\lambda_m,x)\, \Omega(\beta_n,y)\, \Psi(\gamma_o,z)\, \hat{\bar{\hat{T}}}(\lambda_m,\beta_n,\gamma_o) \right.$$

$$\left. \left[1-\exp(-t(k_1\lambda_m^2 + k_2\beta_n^2 + k_3\gamma_o^2))\right]\right\} \qquad (2.8.6a)$$

where

$$\hat{\bar{\hat{T}}}(\lambda_m,\beta_n,\gamma_o) = \frac{\dfrac{(-1)^{m+1}\bar{q}_1}{\beta_n\,\gamma_o} + \dfrac{(-1)^{n+1}\bar{q}_2}{\lambda_m\,\gamma_o} + \dfrac{(-1)^{o+1}\bar{q}_3}{\lambda_m\,\beta_n}}{k_1\,\lambda_m^2 + k_2\,\beta_n^2 + k_3\,\gamma_o^2} \qquad (2.8.6b)$$

and the eigenfunctions are given by:

$$\Theta(\lambda_m,x) = \sin(\lambda_m x)$$

$$\Omega(\beta_n,y) = \sin(\beta_n y) \qquad (2.8.6c\text{-}e)$$

$$\Psi(\gamma_o,z) = \sin(\gamma_o z)$$

with eigenvalues

$$\lambda_m = \frac{(2m-1)}{2a}\pi; \quad \beta_n = \frac{(2n-1)}{2b}\pi; \quad \gamma_o = \frac{(2o-1)}{2c}\pi \qquad (2.8.6\text{f-h})$$

After the heating is stopped, that is, for $t > t_h$, the analytical solution of problem (2.8.1) for *Experiment II* can also be obtained with the classical integral transform technique as [56,90]:

$$T(x,y,z,t) = \frac{8}{abc}\sum_{o=1}^{\infty}\sum_{n=1}^{\infty}\sum_{m=1}^{\infty}\left\{\Theta(\lambda_m,x)\,\Omega(\beta_n,y)\,\Psi(\gamma_o,z)\left[1-\exp(-t_h(k_1\lambda_m^2 + k_2\beta_n^2 + k_3\gamma_o^2))\right]\right.$$

$$\left. \times \hat{\bar{T}}(\lambda_m\,\beta_n\,\gamma_o)\exp(-(t-t_h)(k_1\lambda_m^2 + k_2\beta_n^2 + k_3\gamma_o^2))\right\} \qquad (2.8.7)$$

with $\hat{\bar{T}}$, eigenfunctions and eigenvalues given by equation (2.8.6b-h), respectively.

For the results presented below, we assume the solid with unknown thermal conductivities to be available in the form of a cube, so that $a = b = c = 1$. Also, the heat fluxes applied on the boundaries $x = a = 1$, $y = b = 1$ and $z = c = 1$ are assumed to be of equal magnitudes during the heating period $0 < t \le t_h$, so that $\bar{q}_j = 1$ in equation (2.8.3), for $j = 1, 2, 3$.

THE INVERSE PROBLEM

In the *inverse problem* of interest in this section, the three thermal conductivity components k_1, k_2 and k_3 are unknown, while the solid geometry, initial and boundary conditions are known. The measurements of three non-intrusive sensors ($M = 3$) are assumed available for the inverse analysis. The objective of the inverse problem is then to determine the vector of thermal conductivity components $\mathbf{P}^T = [k_1, k_2, k_3]$ by using the transient temperature measurements of the sensors.

In *Experiment I*, with insulated boundaries at $x = 0$, $y = 0$ and $z = 0$, the sensors can be located at any of the six cube surfaces, since all of them involve second-kind boundary conditions. On the other hand, the sensors need necessarily to be located at the heated surfaces for *Experiment II*, because the sensitivity coefficients at the constant temperature surfaces ($x = 0$, $y = 0$ and $z = 0$) are null. Hence, we planned the study of three test cases involving *Experiments I* and *II*, as well as different sensor locations. Such test cases are summarized in Table 2.1. Test cases 1 and 2 deal with *Experiment I*, with each of the sensors located at each of the insulated surfaces at $x = 0$, $y = 0$ and $z = 0$ (Test case 1) and at each of the heated surfaces at $x = 1$, $y = 1$ and $z = 1$ (Test case 2), respectively. Test case 3 involves *Experiment II*, with each of the sensors located at each of the heated surfaces at $x = 1$, $y = 1$ and $z = 1$. For these three test cases, we used the following dimensionless values for the thermal conductivity components in order to generate the simulated measurements: $k_1 = 1$, $k_2 = 15$ and $k_3 = 15$, which are related to carbon-carbon composites [91].

TABLE 2.1
Test Cases

Test Case	Experiment	Sensors at
1	I	$x = 0$, $y = 0$ and $z = 0$
2	I	$x = 1$, $y = 1$ and $z = 1$
3	II	$x = 1$, $y = 1$ and $z = 1$

Source: From Refs. [82,88].

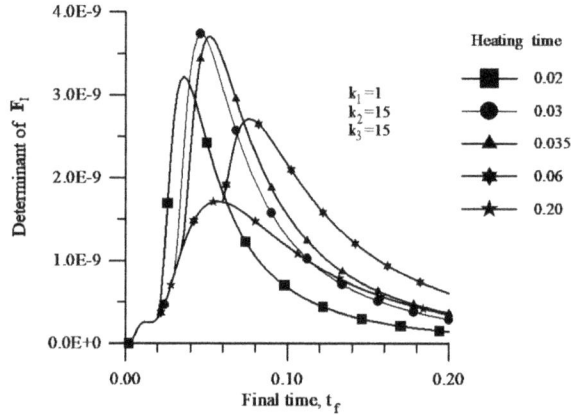

FIGURE 2.8 Determinant of $\mathbf{F}_I^*$ for Test case 1. (From Refs. [82,88].)

ANALYSIS OF THE SENSITIVITY COEFFICIENTS AND DESIGN OF OPTIMUM EXPERIMENTS

For the maximization of the determinant of $(\mathbf{J}^T\mathbf{J})$, we take into account the restrictions of a large but fixed number of measurements for each sensor and the maximum temperature in the region, so that the matrix $\mathbf{F}_I^*$ with elements given by equation (2.6.6) is used in the analysis.

Figure 2.8 presents the timewise variation of the determinant of $\mathbf{F}_I^*$ for different heating times, for sensors located at (0,0.9,0.9), (0.9,0,0.9) and (0.9,0.9,0) in Test case 1. Such locations for the sensors are the ones that resulted on the largest values for the determinant of $\mathbf{F}_I^*$ for this test case. We note in Figure 2.8 that, as observed by Taktak et al. [89] for a one-dimensional property estimation problem, the use of heating times smaller than the final experimental time yielded larger values for the determinant of $\mathbf{F}_I^*$ for Test case 1. Such is the case because any tendency of the sensitivity coefficients to become linearly dependent is reversed after the heating is stopped, that is, when a new transient

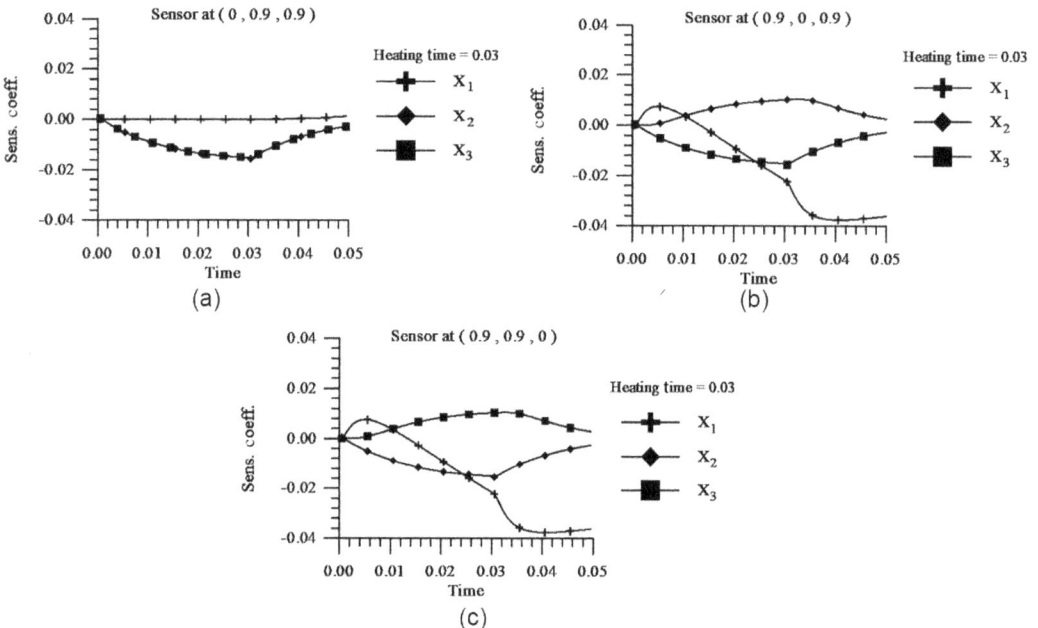

FIGURE 2.9 Reduced sensitivity coefficients for a sensor at: (0,0.9,0.9) for Test case 1 (a), (0.9,0,0.9) for Test case 1 (b), (0.9,0.9,0) for Test case 1 (c). (From Refs. [82,88].)

period is created. This can be clearly noticed in Figures 2.9a-c for the reduced sensitivity coefficients at each of the measurement locations, respectively, for Test case 1, with $t_h = 0.03$ and $t_f = 0.05$, which are the heating and final times, respectively, resulting on the maximum determinant of $\mathbf{F}_I^*$.

A behavior similar to that observed in Test case 1 for the determinant of $\mathbf{F}_I^*$ is also noticed in Test case 2 when the heating is stopped, as shown by Figure 2.10, where the sensors are located at (1,0.5,0.5), (0.5,1,0.5) and (0.5,0.5,1). These positions for the sensors are the ones resulting in the largest values for $\det(\mathbf{F}_I^*)$ for Test case 2. We can also notice in Figure 2.10 that the maximum value of the determinant of $\mathbf{F}_I^*$ for Test case 2 takes place at much smaller dimensionless times than for Test case 1. Such maximum value for Test case 2 is about one order of magnitude larger than that for Test case 1. Hence, we can expect more accurate estimates for the parameters with the experimental conditions of Test case 2 than those of Test case 1. Also, the use of sensors located at the heated surfaces (Test case 2), instead of at the insulated surfaces (Test case 1), results in much faster experiments for the estimation of the thermal conductivity components of the orthotropic cube. The optimum values of heating and final times for Test case 2 are $t_h = 0.010$ and $t_f = 0.012$, respectively.

Let us consider now the analysis of the determinant of $\mathbf{F}_I^*$ for Test case 3. Figure 2.11 presents the variation of $\det(\mathbf{F}_I^*)$ for two different heating times and for sensors located at (1, 0.5, 0.5), (0.5, 1, 0.5) and (0.5, 0.5, 1). We can note in Figure 2.11 that, differently from Test cases 1 and 2, the determinant does not undergo a sudden increase at the moment that the heating is stopped. In fact, $\det(\mathbf{F}_I^*)$ decreases very fast after the heating is stopped for this case. Also, $\det(\mathbf{F}_I^*)$ tends to a constant value if the body is heated continuously. Such behaviors are due to the fact that the sensitivity coefficients tend abruptly to zero after the heating is stopped and they tend to constant values as the steady state is approached, as illustrated is Figures 2.12a-c for the reduced sensitivity coefficients at the positions (1,0.5,0.5), (0.5,1,0.5) and (0.5,0.5,1), respectively. Also, as the steady state is approached, T_{max} in equation (2.6.6) tends to a constant value, so that $\det(\mathbf{F}_I^*)$ does not decrease. It is interesting to note that there exists no steady state in Test cases 1 and 2 if the body is heated continuously, as a result of the insulated boundaries at $x = y = z = 0$. The foregoing analysis of Figure 2.11 reveals that, for Test case 3, the heating and final times should be equal for the maximization of $\det(\mathbf{F}_I^*)$. A comparison of Figures 2.8, 2.10 and 2.11 reveals that the largest

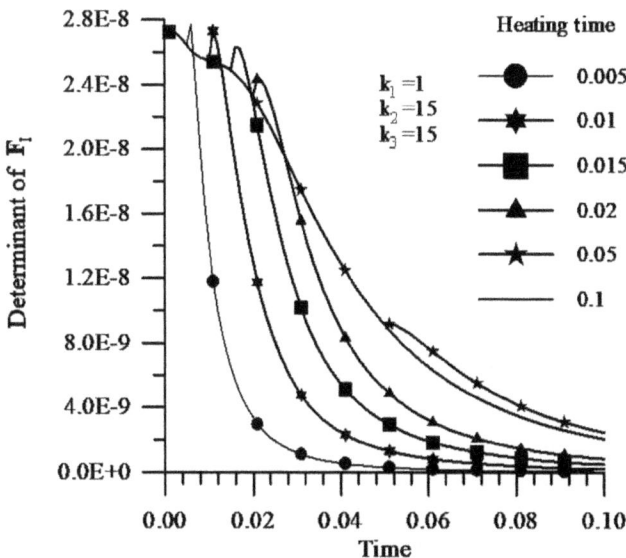

FIGURE 2.10 Determinant of $\mathbf{F}_I^*$ for Test case 2. (From Refs. [82,88].)

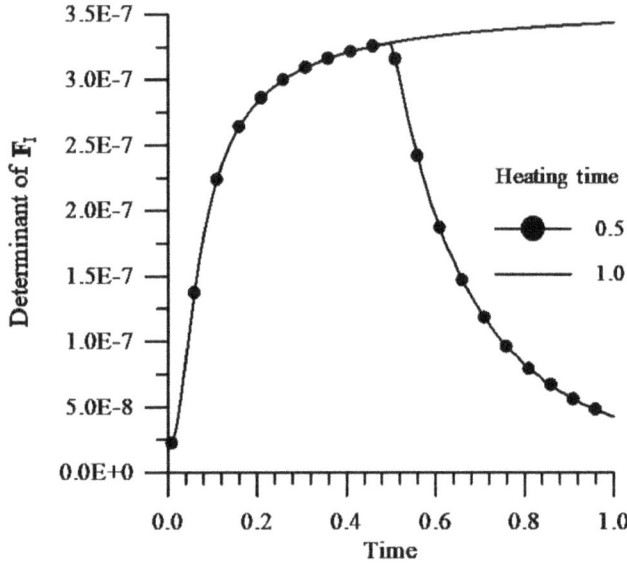

FIGURE 2.11 Determinant of $\mathbf{F}_I^*$ for Test case 3. (From Refs. [82,88].)

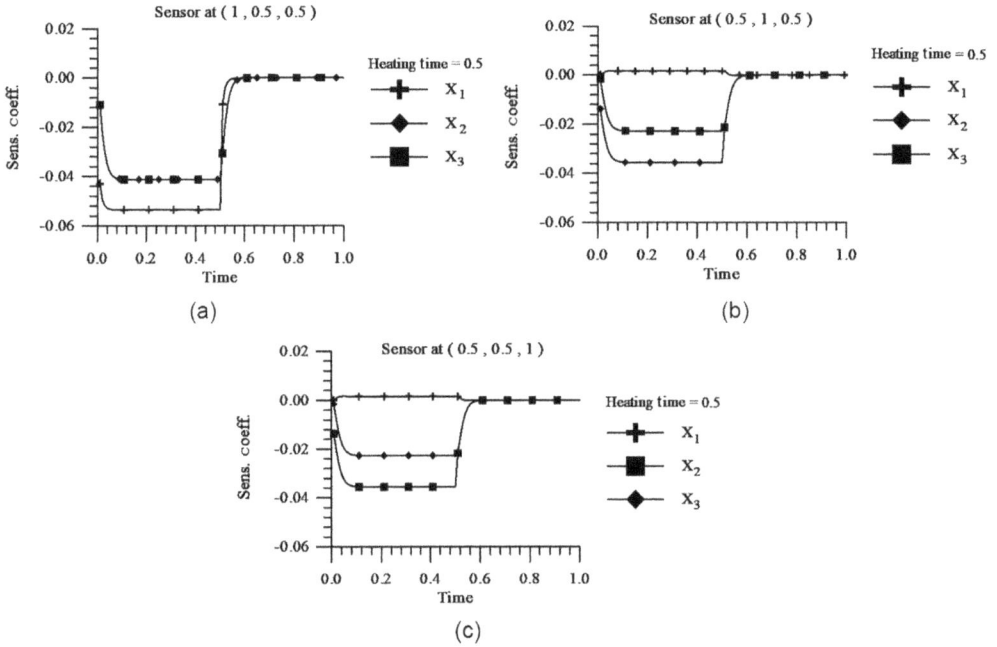

FIGURE 2.12 Reduced sensitivity coefficients for a sensor at: (1,0.5,0.5) for Test case 3 (a), (0.5,1,0.5) for Test case 3 (b), (0.5,0.5,1) for Test case 3 (c). (From Refs. [82,88].)

values of $\det(\mathbf{F}_I^*)$ are obtained with Test case 3. Hence, more accurate estimates for the parameters are expected with such a test case.

The optimal experimental variables estimated with the foregoing analysis of $\det(\mathbf{F}_I^*)$ are summarized in Table 2.2. Dimensional values for the heating and final times are also included in Table 2.2, based on a carbon-carbon cubic specimen of side 0.076 m, with volumetric heat capacity $1.52 \times 10^6 \, \text{J/m}^3$ and reference thermal conductivity of 3.89 W/m °C [91].

TABLE 2.2
Optimal Experimental Variables

Test Case	Sensors' Positions (x,y,z)	Heating Time t_h	$t_h^*(s)$	Final Time t_f	$t_f^*(s)$
1	(0,0.9,0.9) (0.9,0,0.9) (0.9,0.9,0)	0.03	68.1	0.05	113.4
2	(1,0.5,0.5) (0.5,1,0.5) (0.5,0.5,1)	0.01	22.7	0.012	27.2
3	(1,0.5,0.5) (0.5,1,0.5) (0.5,0.5,1)	1	2268.8	1	2268.8

Source: From Refs. [82,88].

PARAMETER ESTIMATION AND STATISTICAL ANALYSIS

We now illustrate the use of the Levenberg-Marquardt method for the estimation of the three thermal conductivity components with simulated temperature measurements for Test case 3. The optimal experimental variables estimated with the above analysis of $\det(\mathbf{F}_I^*)$ were used to generate the simulated measurements. One hundred simulated measurements per sensor, containing additive, uncorrelated and Gaussian errors, with zero mean and constant standard deviation, were assumed available for the estimation procedure. With such hypotheses, the ordinary least squares norm (equation 2.1.6) was used as the objective function. The version of the Levenberg-Marquardt method proposed by Moré [74] was used as the minimization procedure. The standard deviation of the simulated measurements was $\sigma = 0.01Y_{max}$, where Y_{max} is the maximum measured temperature. The results presented below were obtained with an initial guess $\mathbf{P}^0 = [0.5,0.5,0.5]^T$ and were averaged over 50 different runs of the Levenberg-Marquardt method, in order to reduce the effects of the random number generator utilized. The values estimated for the unknown parameters, with their 99% confidence intervals (see equation 2.7.5), are presented in Table 2.3. For the sake of comparison, the parameters estimated with the experimental conditions of Test case 1 are also included in this table. Table 2.3 clearly shows that more accurate estimates were obtained with Test case 3 as compared to Test case 1, as a result of the larger values of $\det(\mathbf{F}_I^*)$.

TABLE 2.3
Parameters Estimated with 99% Confidence Intervals

Test Case	Parameter	Estimated Value	99% Confidence Interval
1	k_1	1.00	±0.01
	k_2	15.0	±0.4
	k_3	15.0	±0.4
3	k_1	1.000	±0.001
	k_2	15.0	±0.3
	k_3	15.0	±0.3

Source: From Refs. [82,88].

We note that the behavior observed above for the case involving $k_1 = 1$, $k_2 = k_3 = 15$, regarding the choice of optimum boundary conditions, was also observed with other values for the thermal conductivity components, such as $k_1 = 1$, $k_2 = 1.5$, $k_3 = 2$ and $k_1 = 1$, $k_2 = 2$, $k_3 = 3$. The confidence region for this last set of parameters (see equation 2.7.6) is given by:

$$46936 + 24465\, k_1^2 + 4034.82\, k_2^2 - k_2\,(14685.3 + 399.28\, k_3)$$

$$- (48383.4 + 128.615\, k_2 + 245.66\, k_3)k_1 - 5240.9\, k_3 + 1036.41\, k_3^2 \le 0 \qquad (2.8.8)$$

The exact values $k_1 = 1$, $k_2 = 2$, $k_3 = 3$ fall inside the confidence region given by equation (2.8.8), that is, the inequality is satisfied for these parameter values.

A comparison of different versions of Technique I (Levenberg-Marquardt method) and Technique II (conjugate gradient method) is now presented. The experimental conditions used for this comparison are those of Test case I, with the sensors' locations shown in Table 2.2. The cases examined here are summarized in Table 2.4, together with their respective optimal values of heating and final times that were used in the analysis. Table 2.5 summarizes the versions of Techniques I and II used in the present comparison.

Tables 2.6–2.8 present the results obtained for the average rate of reduction of the objective function with respect to the number of iterations (r) and RMS error (e_{RMS}), obtained with each of the techniques summarized in Table 2.5, for cases (i) to (iii), respectively. The results presented in Tables 2.6 (case i) and 2.7 (case ii) were obtained with an initial guess $\mathbf{P}^0 = [0.1,0.1,0.1]^T$ for the unknown parameters, while an initial guess of $\mathbf{P}^0 = [0.5,0.5,0.5]^T$ was used for case (iii) (Table 2.8). Convergence was not obtained with any of the methods for $\mathbf{P}^0 = [0.1,0.1,0.1]^T$ in case (iii), even when errorless measurements ($\sigma = 0$) were used in the analysis. For those cases involving simulated measurements with random errors ($\sigma = 0.01Y_{max}$), the results shown in Tables 2.6–2.8 were averaged over 500 different runs in order to reduce the effects of the random number generator utilized.

The average rates of reduction of the objective function (r) were obtained from the following approximation for the variation of the ordinary least squares norm, $S_{OLS}(\mathbf{P})$, with the number of iterations (K):

$$S_{OLS}(\mathbf{P}) = CK^{-r} \qquad (2.8.9)$$

where C is a constant depending on the data.

The RMS errors were computed as:

$$e_{RMS} = \sqrt{\frac{1}{3}\sum_{j=1}^{3}(k_{ex,j} - k_{est,j})^2} \qquad (2.8.10)$$

TABLE 2.4

Cases with Corresponding Dimensionless Heating and Final Times

Case	Exact Parameters			t_h	t_f
	k_1	k_2	k_3		
(i)	1	1.5	2	0.2	0.28
(ii)	1	2	3	0.15	0.2
(iii)	1	15	15	0.03	0.05

Source: From Refs. [82,88].

TABLE 2.5

Techniques Used for the Estimation of the Unknown Parameters

Technique	Method	Version
I-a	Levenberg-Marquardt	Section 2.2
I-b	Levenberg-Marquardt	Reference [74]
II-a	Conjugate gradient method	Fletcher-Reeves, Section 2.3
II-b	Conjugate gradient method	Polak-Ribiere, Section 2.3
II-c	Conjugate gradient method	Hestenes-Stiefel, Section 2.3
II-d	Conjugate gradient method	Powell-Beale, Section 2.3

Source: From Refs. [82,88].

TABLE 2.6

Results for Case (*i*)

Technique	σ	r	e_{RMS}
I-a	0	18	0.00
	$0.01Y_{max}$	3	0.01
I-b	0	33	0.00
	$0.01Y_{max}$	4	0.01
II-a	0	14	0.00
	$0.01Y_{max}$	3	0.01
II-b	0	14	0.00
	$0.01Y_{max}$	3	0.01
II-c	0	13	0.00
	$0.01Y_{max}$	3	0.01
II-d	0	13	0.00
	$0.01Y_{max}$	3	0.01

Source: From Refs. [82,88].

TABLE 2.7

Results for Case (*ii*)

Technique	σ	r	e_{RMS}
I-a	0	23	0.00
	$0.01Y_{max}$	4	0.01
I-b	0	30	0.00
	$0.01Y_{max}$	4	0.01
II-a	0	12	0.00
	$0.01Y_{max}$	3	0.02
II-b	0	13	0.00
	$0.01Y_{max}$	3	0.02
II-c	0	14	0.00
	$0.01Y_{max}$	2	0.54
II-d	0	12	0.00
	$0.01Y_{max}$	2	0.35

Source: From Refs. [82,88].

TABLE 2.8

Results for Case (*iii*)

Technique	σ	r	e_{RMS}
I-a	0	23	0.00
	$0.01Y_{max}$	3	0.13
I-b	0	30	0.00
	$0.01Y_{max}$	4	0.12
II-a	0	12	0.00
	$0.01Y_{max}$	3	0.18
II-b	0	10	0.00
	$0.01Y_{max}$	2	0.15
II-c	0	9	0.00
	$0.01Y_{max}$	3	1.01
II-d	0	9	0.00
	$0.01Y_{max}$	3	0.41

Source: From Refs. [82,88].

where the subscripts *ex* and *est* refer to the exact and estimated parameters, respectively.

The tolerances for the stopping criteria of Techniques I-a and I-b were taken as $\varepsilon_1 = \varepsilon_2 = \varepsilon_3 = 10^{-5}$ in equations (2.2.15a-c), for cases involving errorless measurements, as well as measurements with random errors. For those cases involving errorless measurements, the tolerances for the stopping criterion of Techniques IIa–d were taken as $\varepsilon_1 = 10^{-5}$ in equation (2.3.13). The tolerances for Techniques IIa–d were obtained from equation (2.2.18) based on the discrepancy principle when measurements with random errors were used in the analysis.

Let us consider first in the analysis of Tables 2.6–2.8 the cases involving errorless measurements ($\sigma = 0$). Tables 2.6–2.8 show that all techniques were able to estimate exactly the three different sets of unknown parameters examined, resulting in $e_{RMS} = 0.00$. The highest rates of reduction of the objective function were obtained with the Levenberg-Marquardt method and, for such method, Technique I-b had a better performance than Technique I-a. The rates of reduction of the objective function were of the order of 20 (minimum of 18) or higher for the Levenberg-Marquardt method, while such rates were of the order of 10 (maximum of 14) for the conjugate gradient method.

Tables 2.6–2.8 show a strong decrease in r when measurements with random errors ($\sigma = 0.01Y_{max}$) were used in the analysis, instead of errorless measurements ($\sigma = 0$), as a result of the ill-conditioned inverse problem. Such as for the cases with errorless measurements, Technique I-b generally resulted in the largest rate of reduction of the objective function. For case (*i*), all techniques resulted in $e_{RMS} = 0.01$, when measurements with random errors were used in the analysis. On the other hand, the use of the Levenberg-Marquardt method yielded smaller RMS errors for cases (*ii*) and (*iii*). Large RMS errors can be noticed in Tables 2.7 and 2.8, for cases (*ii*) and (*iii*), respectively, when Techniques II-c and II-d were used for the solution. Such is the case because convergence was not achieved with these techniques for many of the 500 different runs used in the comparison, when measurements with random errors were considered in the analysis of cases (*ii*) and (*iii*).

Table 2.9 was prepared to illustrate the effect of the initial guess for the parameters, i.e., $\mathbf{P}^0$, over the rate of reduction of the objective function. Three different initial guesses were considered for case (*i*) by using errorless measurements in the analysis. Table 2.9 shows that, although the rate of reduction increased to 28 for Technique I-a for the initial guess $\mathbf{P}^0 = [1,1,1]^T$, such rate was relatively insensitive to the three different initial guesses for all techniques examined here. We note that Techniques IIa–d, based on the conjugate gradient method, did not converge to the exact parameters

TABLE 2.9

Effect of the Initial Guess on the Rate of Reduction of the Objective Function

	r		
Technique	$\mathbf{P}^0 = [0.1, 0.1, 0.1]^T$	$\mathbf{P}^0 = [1, 1, 1]^T$	$\mathbf{P}^0 = [3, 3, 3]^T$
I-a	18	28	24
I-b	33	36	36
II-a	14	12	11
II-b	14	13	14
II-c	13	13	14
II-d	13	13	12

Source: From Refs. [82,88].

for initial guesses larger than $\mathbf{P}^0 = [3,3,3]^T$. On the other hand, Techniques I-a and I-b, based on the Levenberg-Marquardt method, were able to converge to the exact parameters even with initial guesses as large as $\mathbf{P}^0 = [10,10,10]^T$.

The foregoing analysis reveals that, besides having the highest rates of reduction of the objective function, the use of the Levenberg-Marquardt method also resulted in the smallest RMS errors of the estimated parameters. Such method was able to converge to the exact parameters even for initial guesses quite far from the exact values.

2.9 TECHNIQUE III: THE CONJUGATE GRADIENT METHOD WITH ADJOINT PROBLEM FOR PARAMETER ESTIMATION

In order to implement the iterative algorithm of the conjugate gradient method, as presented in Section 2.3 for Technique II, the sensitivity matrix needs to be computed at each iteration. For linear problems, it might be quite easy to compute such matrix with an analytical solution. Indeed, the sensitivity matrix being constant for linear problems, it has to be computed only once. On the other hand, for nonlinear inverse problems, the sensitivity matrix needs to be computed most likely by finite differences. This might be very time consuming when the problem involves large numbers of parameters and measurements. For cases involving the estimation of the coefficients of unknown functions parameterized, we present below an alternative implementation of the conjugate gradient method, which does not require the computation of the sensitivity matrix in order to obtain the gradient direction and the search step size. Here, two auxiliary problems, known as the *sensitivity problem* and the *adjoint problem,* are solved in order to compute the search step size β^k and the gradient equation $\nabla S_{ML}(\mathbf{P})$.

We thus revisit the same inverse problem used as an example in Section 2.3 by considering available the transient measurements of one single sensor located at the position x_{meas}, for the estimation of the heat source term $g_p(t)$. Moreover, we assume that the measured data is a function of time, rather than discrete in time. The measurements are supposed to be uncorrelated, with known variance given by the function $\sigma^2(t)$. Thus, the maximum likelihood objective function, equation (2.1.5), is rewritten as:

$$S_{ML}(\mathbf{P}) = \int_{t=0}^{t_f} \frac{[Y(t) - T(x_{meas}, t; \mathbf{P})]^2}{[\sigma(t)]^2} \, dt \tag{2.9.1a}$$

where $Y(t)$ is the measured temperature, $T(x_{meas}, t; \mathbf{P})$ is the estimated temperature at the single measurement location, x_{meas}, and t_f is the duration of the experiment. Therefore, the summation over the number of measurements in equation (2.1.5) is replaced by an integration over the time domain $0 < t \le t_f$, that is, during the period that the measurements are available.

If the variance of the measurements is constant, the minimization of the objective function (2.9.1a) is equivalent to the minimization of:

$$S_{OLS}(\mathbf{P}) = \int_{t=0}^{t_f} [Y(t) - T(x_{meas}, t; \mathbf{P})]^2 \, dt \qquad (2.9.1b)$$

The basic steps for the solution of parameter estimation problems by using the conjugate gradient method with adjoint problem include:

- Direct problem
- Inverse problem
- Analysis of the sensitivity coefficients and design of optimum experiments
- Sensitivity problem
- Adjoint problem
- Gradient equation
- Iterative procedure
- Stopping criterion
- Computational algorithm
- Statistical analysis

The analysis of the sensitivity coefficients, the design of optimum experiments and the statistical analysis were detailed above in Sections 2.4, 2.5 and 2.7, respectively. Therefore, these steps are not repeated here and this section is focused on the other remaining steps of the conjugate gradient method with adjoint problem for parameter estimation.

THE DIRECT PROBLEM

For the test problem considered here, involving the estimation of the strength $g_p(t)$ of a plane heat source, the direct problem is given by equations (2.2.1a-d). This problem is rewritten below in order to facilitate the analysis:

$$\frac{\partial^2 T(x, t)}{\partial x^2} + g_p(t)\delta(x - 0.5) = \frac{\partial T(x,t)}{\partial t} \quad \text{in } 0 < x < 1, \text{ for } t > 0 \qquad (2.9.2a)$$

$$\frac{\partial T(0,t)}{\partial x} = 0 \quad \text{at } x = 0, \text{ for } t > 0 \qquad (2.9.2b)$$

$$\frac{\partial T(1,t)}{\partial x} = 0 \quad \text{at } x = 1, \text{ for } t > 0 \qquad (2.9.2c)$$

$$T(x,0) = 0 \text{ for } t = 0, \text{ in } 0 < x < 1 \qquad (2.9.2d)$$

The direct problem is concerned with the determination of the temperature field $T(x,t)$ in the region $0 < x < 1$, when the strength of the source term $g_p(t)$ is known.

THE INVERSE PROBLEM

The inverse problem, on the other hand, is concerned with the estimation of the unknown strength of the source term by using the readings taken by a sensor located at x_{meas}. Technique III is particularly suitable for problems involving the estimation of the coefficients of basis functions used to approximate an unknown functional form. We consider the unknown function $g_p(t)$ to be parameterized in a general linear form given by:

$$g_p(t) = \sum_{j=1}^{N} P_j C_j(t) \qquad (2.9.3a)$$

where $C_j(t), j = 1, ..., N$ are known basis functions. Thus, the objective of the inverse problem is to estimate the N unknown parameters $P_j, j = 1,..., N$.

THE SENSITIVITY PROBLEM

The *sensitivity function* $\Delta T(x,t)$, solution of the sensitivity problem, is defined as the directional derivative of the temperature $T(x,t)$ in the direction of the perturbation of the unknown function [6,22,92]. The sensitivity function is needed for the computation of the search step size β^k, as will be apparent later in this section.

The *sensitivity problem* can be obtained by assuming that the temperature $T(x,t)$ is perturbed by an amount $\Delta T(x,t)$, when the unknown strength $g_p(t)$ of the source term is perturbed by $\Delta g_p(t)$. Since the strength was parameterized in the form given by equation (2.9.3a), the function $\Delta g_p(t)$ is obtained by perturbing each of the unknown parameters P_j by an amount ΔP_j, that is,

$$\Delta g_p(t) = \sum_{j=1}^{N} \Delta P_j C_j(t) \qquad (2.9.3b)$$

By replacing $T(x,t)$ by $[T(x,t) + \Delta T(x,t)]$ and $g_p(t)$ by $[g_p(t) + \Delta g_p(t)]$ in the direct problem given by equations (2.9.2a-d), and then subtracting the original direct problem from the resulting expressions, we obtain the following *sensitivity problem:*

$$\frac{\partial^2 \Delta T(x,t)}{\partial x^2} + \Delta g_p(t) \, \delta(x - 0.5) = \frac{\partial \Delta T(x,t)}{\partial t} \quad \text{in } 0 < x < 1, \text{for } t > 0 \qquad (2.9.4a)$$

$$\frac{\partial \Delta T(0,t)}{\partial x} = 0 \quad \text{at } x = 0, \text{ for } t > 0 \qquad (2.9.4b)$$

$$\frac{\partial \Delta T(1,t)}{\partial x} = 0 \quad \text{at } x = 1, \text{ for } t > 0 \qquad (2.9.4c)$$

$$\Delta T(x,0) = 0 \text{ for } t = 0, \text{ in } 0 < x < 1 \qquad (2.9.4d)$$

In the sensitivity problem (2.9.4), $\Delta g_p(t)$ given by equation (2.9.3b) is the only forcing function needed for the solution. The computation of $\Delta g_p(t)$ will be addressed later in this section.

THE ADJOINT PROBLEM

A Lagrange multiplier $\lambda(x,t)$ comes into picture in the minimization of the function (2.9.1a) because the temperature $T(x_{\text{meas}},t;\mathbf{P})$ appearing in such function needs to satisfy a constraint, which is the solution of the direct problem. Such Lagrange multiplier, needed for the computation of the gradient equation (as will be apparent below), is obtained through the solution of a problem *adjoint* to the sensitivity problem given by equations (2.9.4a-d). For the definition and properties of adjoint problems, the reader should consult references [6,22,92].

In order to derive the *adjoint problem*, we write the following extended objective function:

$$S_{ML}(\mathbf{P}) = \int\limits_{t=0}^{t_f} \frac{[Y(t) - T(x_{\text{meas}},t;\mathbf{P})]^2}{[\sigma(t)]^2}\,dt + \int\limits_{x=0}^{1}\int\limits_{t=0}^{t_f} \lambda(x,t)\left[\frac{\partial^2 T}{\partial x^2} + g_p(t)\delta(x-0.5) - \frac{\partial T}{\partial t}\right] dt\,dx \qquad (2.9.5)$$

which is obtained by multiplying the partial differential equation of the direct problem, equation (2.9.2a), by the Lagrange multiplier, $\lambda(x,t)$, integrating over the time and space domains and adding the resulting expression to the function $S_{ML}(\mathbf{P})$ given by equation (2.9.1a).

An expression for the variation $\Delta S_{ML}(\mathbf{P})$ of the function $S_{ML}(\mathbf{P})$ can be developed by perturbing $T(x,t)$ by $\Delta T(x,t)$ and $g_p(t)$ by $\Delta g_p(t)$ in equation (2.9.5). We note that $\Delta S_{ML}(\mathbf{P})$ is the directional derivative of $S_{ML}(\mathbf{P})$ in the direction of the perturbation $\Delta \mathbf{P} = [\Delta P_1, \Delta P_2, \ldots, \Delta P_N]^T$ [6,22,92]. Then, by replacing $T(x,t)$ by $[T(x,t) + \Delta T(x,t)]$, $g_p(t)$ by $[g_p(t) + \Delta g_p(t)]$ and $S_{ML}(\mathbf{P})$ by $[S_{ML}(\mathbf{P}) + \Delta S_{ML}(\mathbf{P})]$ in equation (2.9.5), subtracting from the resulting expression the original equation (2.9.5), and neglecting second-order terms, we find:

$$\Delta S_{ML}(\mathbf{P}) = \int\limits_{t=0}^{t_f}\int\limits_{x=0}^{1} 2\frac{[T(x,t;\mathbf{P}) - Y(t)]}{[\sigma(t)]^2}\Delta T(x,t)\delta(x - x_{\text{meas}})\,dx\,dt$$

$$+ \int\limits_{t=0}^{t_f}\int\limits_{x=0}^{1} \lambda(x,t)\left[\frac{\partial^2 \Delta T}{\partial x^2} + \Delta g_p(t)\delta(x-0.5) - \frac{\partial \Delta T}{\partial t}\right]dx\,dt \qquad (2.9.6)$$

where $\delta(.)$ is the Dirac delta function.

The second integral term on the right-hand side of this equation is simplified by integration by parts and by utilizing the boundary and initial conditions of the sensitivity problem. The integration by parts of the term involving the second derivative in the spatial variable yields:

$$\int\limits_{x=0}^{1} \lambda(x,t)\frac{\partial^2 \Delta T}{\partial x^2}\,dx = \left[\lambda\frac{\partial \Delta T}{\partial x} - \frac{\partial \lambda}{\partial x}\Delta T\right]_{x=0}^{1} + \int\limits_{x=0}^{1} \Delta T(x,t)\frac{\partial^2 \lambda}{\partial x^2}\,dx \qquad (2.9.7a)$$

By substituting the boundary conditions (2.9.4b) and (2.9.4c) of the sensitivity problem into equation (2.9.7a), we obtain:

$$\int\limits_{x=0}^{1} \lambda(x,t)\frac{\partial^2 \Delta T}{\partial x^2}\,dx = \frac{\partial \lambda(0,t)}{\partial x}\Delta T(0,t) - \frac{\partial \lambda(1,t)}{\partial x}\Delta T(1,t) + \int\limits_{x=0}^{1} \Delta T(x,t)\frac{\partial^2 \lambda}{\partial x^2}\,dx \qquad (2.9.7b)$$

Similarly, the integration by parts of the term involving the time derivative in equation (2.9.6) gives:

$$\int_{t=0}^{t_f} \lambda(x,t) \frac{\partial \Delta T}{\partial t}\, dt = \left[\lambda(x,t)\, \Delta T(x,t) \right]_{t=0}^{t_f} - \int_{t=0}^{t_f} \Delta T(x,t) \frac{\partial \lambda}{\partial t}\, dt \qquad (2.9.8a)$$

After substituting the initial condition (2.9.4d) of the sensitivity problem, equation (2.9.8a) becomes:

$$\int_{t=0}^{t_f} \lambda(x,t) \frac{\partial \Delta T}{\partial t}\, dt = \lambda(x,t_f)\, \Delta T(x,t_f) - \int_{t=0}^{t_f} \Delta T(x,t) \frac{\partial \lambda}{\partial t}\, dt \qquad (2.9.8b)$$

Equations (2.9.7b) and (2.9.8b) are then substituted into equation (2.9.6) to obtain:

$$\Delta S_{ML}(\mathbf{P}) = \int_{t=0}^{t_f} \int_{x=0}^{1} \left\{ \frac{\partial^2 \lambda(x,t)}{\partial x^2} + \frac{\partial \lambda(x,t)}{\partial t} + 2\frac{[T(x,t;\mathbf{P}) - Y(t)]}{[\sigma(t)]^2} \delta(x - x_{\text{meas}}) \right\} \Delta T(x,t)\, dx\ dt$$

$$+ \int_{t=0}^{t_f} \frac{\partial \lambda(0,t)}{\partial x}\, \Delta T(0,t)\, dt - \int_{t=0}^{t_f} \frac{\partial \lambda(1,t)}{\partial x}\, \Delta T(1,t)\, dt$$

$$- \int_{x=0}^{1} \lambda(x,t_f)\, \Delta T(x,t_f)\, dx + \int_{t=0}^{t_f} \lambda(0.5,t)\, \Delta g_p(t)\, dt \qquad (2.9.9)$$

The boundary value problem for the Lagrange multiplier $\lambda(x,t)$ is obtained by allowing the first four integral terms containing $\Delta T(x,t)$ on the right-hand side of equation (2.9.9) to vanish. This leads to the following *adjoint problem*:

$$\frac{\partial \lambda(x,t)}{\partial t} + \frac{\partial^2 \lambda(x,t)}{\partial x^2} + 2\frac{[T(x,t;\mathbf{P}) - Y(t)]}{[\sigma(t)]^2} \delta(x - x_{\text{meas}}) = 0 \quad \text{in } 0 < x < 1, \text{ for } 0 < t < t_f \qquad (2.9.10a)$$

$$\frac{\partial \lambda(0,t)}{\partial x} = 0 \quad \text{at } x = 0, \text{ for } 0 < t < t_f \qquad (2.9.10b)$$

$$\frac{\partial \lambda(1,t)}{\partial x} = 0 \quad \text{at } x = 1, \text{ for } 0 < t < t_f \qquad (2.9.10c)$$

$$\lambda(x,t_f) = 0 \quad \text{for } t = t_f, \text{ in } 0 < x < 1 \qquad (2.9.10d)$$

We note that in the adjoint problem, the condition (2.9.10d) is the value of the function $\lambda(x,t)$ at the final time $t = t_f$. In the conventional *initial value problem*, the value of the function is specified at time $t = 0$. However, the *final value problem* (2.9.10) can be transformed into an *initial value problem* by defining a new time variable given by $\tau = t_f - t$.

THE GRADIENT EQUATION

After letting the terms containing $\Delta T(x,t)$ vanish, the following integral term remains on the right-hand side of equation (2.9.9):

$$\Delta S_{ML}(\mathbf{P}) = \int_{t=0}^{t_f} \lambda(0.5,t)\Delta g_p(t)\,dt \tag{2.9.11}$$

By substituting $\Delta g_p(t)$ in the parametric form given by equation (2.9.3b) into equation (2.9.11), we obtain:

$$\Delta S_{ML}(\mathbf{P}) = \sum_{j=1}^{N} \int_{t=0}^{t_f} \lambda(0.5,t)C_j(t)\,dt\,\Delta P_j \tag{2.9.12}$$

By definition, the directional derivative of $S_{ML}(\mathbf{P})$ in the direction of the vector $\Delta\mathbf{P}$ is given by the scalar product of the gradient $\nabla S_{ML}(\mathbf{P})$ with an unit vector in the direction $\Delta\mathbf{P}$, that is,

$$\Delta S_{ML}(\mathbf{P}) = \sum_{j=1}^{N} [\nabla S_{ML}(\mathbf{P})]_j\,\Delta P_j \tag{2.9.13}$$

where

$$\Delta\mathbf{P} = [\Delta P_1, \Delta P_2, \ldots, \Delta P_N]^T \tag{2.9.14}$$

We note that the magnitude of the vector $\Delta\mathbf{P}$ was omitted in equation (2.9.13), since it is not relevant for the present analysis.

Therefore, by comparing equations (2.9.12) and (2.9.13), we obtain the jth component of the gradient vector $\nabla S_{ML}(\mathbf{P})$ for the function $S_{ML}(\mathbf{P})$ as:

$$[\nabla S_{ML}(\mathbf{P})]_j = \int_{t=0}^{t_f} \lambda(0.5,t)C_j(t)\,dt \quad \text{for } j = 1,\ldots,N \tag{2.9.15}$$

The use of an adjoint problem for the computation of the gradient vector is most useful for problems involving unknown functions that can be parameterized in a form similar to equation (2.9.3a). With the present approach, *the gradient vector is computed with the solution of a single adjoint problem*. On the other hand, the calculation of the gradient vector in Technique II, as given by equation (2.3.6), may require N additional solutions of the direct problem in order to compute the sensitivity coefficients by forward finite differences (see equation 2.4.11).

THE ITERATIVE PROCEDURE FOR TECHNIQUE III

The iterative procedure of the conjugate gradient method for the computation of the vector of unknown parameters $\mathbf{P}$ is given by equation (2.3.1), with the direction of descent obtained from equation (2.3.5) and the conjugation coefficients from equations (2.3.7–2.3.10). However, the gradient vector components are now computed by using equation (2.9.15), rather than equation (2.3.6).

The search step size β^k is chosen as the one that minimizes the function $S_{ML}(\mathbf{P})$ at each iteration k, that is,

$$\min_{\beta^k} S_{ML}(\mathbf{P}^{k+1}) = \min_{\beta^k} \int_{t=0}^{t_f} \frac{[Y(t) - T(x_{\mathrm{meas}}, t; \mathbf{P}^k + \beta^k \mathbf{d}^k)]^2}{[\sigma(t)]^2} dt \qquad (2.9.16)$$

By linearizing the estimated temperature $T(x_{\mathrm{meas}}, t; \mathbf{P}^k + \beta^k \mathbf{d}^k)$ with a Taylor series expansion and performing the above minimization, we find:

$$\beta^k = \frac{\displaystyle\int_{t=0}^{t_f} \frac{[Y(t) - T(x_{\mathrm{meas}}, t; \mathbf{P}^k)]}{[\sigma(t)]^2} \Delta T(x_{\mathrm{meas}}, t; \mathbf{d}^k)\, dt}{\displaystyle\int_{t=0}^{t_f} \frac{[\Delta T(x_{\mathrm{meas}}, t; \mathbf{d}^k)]^2}{[\sigma(t)]^2} dt} \qquad (2.9.17)$$

where $\Delta T(x_{\mathrm{meas}}, t; \mathbf{d}^k)$ is the solution of the sensitivity problem given by equations (2.9.4a-d), obtained by setting $\Delta P_j = d_j^k, j = 1,..., N$, in the computation of the function $\Delta g_p(t)$ given by equation (2.9.3b). Further details on the derivation of equation (2.9.17) can be found in Note 2 at the end of this chapter.

The reader should note that a single sensitivity problem is solved for the computation of β^k at each iteration, because the unknown function was parameterized in the form given by equation (2.9.3a). Therefore, in the present approach the computation of β^k does not require the computation of the sensitivity matrix as in equation (2.3.4). For problems not involving the estimation of coefficients of basis functions, as in equation (2.9.3a), the use of Techniques I or II is more appropriate.

THE STOPPING CRITERION FOR TECHNIQUE III

As for Technique II, the stopping criterion for Technique III is based on the *discrepancy principle,* when the standard deviation $\sigma(t)$ of the measurements is *a priori* known. It is given by:

$$S_{ML}(\mathbf{P}) < \varepsilon_1 \qquad (2.9.18)$$

where $S_{ML}(\mathbf{P})$ is now computed with equation (2.9.1a). The tolerance ε_1 is then obtained from equation (2.9.1a) by assuming

$$\left| Y(t) - T(x_{\mathrm{meas}}, t; \mathbf{P}) \right| \approx \sigma(t) \qquad (2.9.19)$$

Thus, the tolerance ε_1 is determined as:

$$\varepsilon_1 = t_f \qquad (2.9.20)$$

THE COMPUTATIONAL ALGORITHM FOR TECHNIQUE III

The computational algorithm for the conjugate gradient method with adjoint problem for parameter estimation can be summarized as follows. Suppose that temperature measurements $\mathbf{Y} = \left[Y_1, Y_2, ... , Y_I \right]^T$ are given at times $t_i, i = 1,..., I$, and an initial guess $\mathbf{P}^0$ is available for the vector of unknown parameters $\mathbf{P}$. Set $k = 0$ and then [2,6,7,22,29–43,73,75–79,92]:

Step 1: Compute $g_p(t)$ according to equation (2.9.3a) and then solve the direct problem given by equations (2.9.2a-d) in order to compute $T(x,t)$.

Step 2: Check the stopping criterion given by equation (2.9.18). Continue if not satisfied.

Step 3: Knowing $T(x_{\text{meas}},t;\mathbf{P})$ and the measured temperature $Y(t)$, solve the adjoint problem (2.9.10) to compute $\lambda(0.5,t)$.

Step 4: Knowing $\lambda(0.5,t)$, compute each component of the gradient vector $\nabla S_{ML}(\mathbf{P})$ from equation (2.9.15).

Step 5: Knowing the gradient $\nabla S_{ML}(\mathbf{P})$, compute the conjugation coefficients with one of the pair of equations (2.3.7a, b), (2.3.8a, b), (2.3.9a, b) or (2.3.10a, b). Note that Powell-Beale's version requires restarting if any of the inequalities (2.3.11) or (2.3.12a and b) is satisfied.

Step 6: Compute the direction of descent $\mathbf{d}^k$ by using equation (2.3.5).

Step 7: By setting $\Delta\mathbf{P}^k = \mathbf{d}^k$, compute $\Delta g_p(t)$ from equation (2.9.3b) and then solve the sensitivity problem given by equations (2.9.4a-d) to obtain $\Delta T(x_{\text{meas}},t;\mathbf{d}^k)$.

Step 8: Knowing $\Delta T(x_{\text{meas}},t;\mathbf{d}^k)$, compute the search step size β^k from equation (2.9.17).

Step 9: Knowing β^k and $\mathbf{d}^k$, compute the new estimate $\mathbf{P}^{k+1}$ from equation (2.3.1). Replace k by $k+1$ and return to step 1.

THE USE OF MULTIPLE SENSORS

The above computational algorithm of Technique III can also be applied, with few modifications, to cases where the readings of M sensors are available for the inverse analysis. In such cases, the objective function (2.9.1a) is modified to:

$$S_{ML}(\mathbf{P}) = \sum_{m=1}^{M} \int_{t=0}^{t_f} \frac{[Y_m(t) - T(x_m,t;\mathbf{P})]^2}{[\sigma_m(t)]^2} dt \qquad (2.9.21)$$

where $Y_m(t)$ and $\sigma_m(t)$ are the functions of measurements and standard deviations, respectively, of the sensor located at x_m, for $m = 1,\ldots, M$.

Since the objective function appears in the development of the adjoint problem, such a problem needs also to be modified in order to accommodate the readings of multiple sensors. It can be easily shown that the differential equation for the adjoint problem, equation (2.9.10a), then becomes:

$$\frac{\partial \lambda(x,t)}{\partial t} + \frac{\partial^2 \lambda(x,t)}{\partial x^2} + 2\sum_{m=1}^{M} \frac{[T(x,t;\mathbf{P}) - Y_m(t)]}{[\sigma_m(t)]^2} \delta(x - x_m) = 0 \quad \text{in } 0 < x < 1, \text{ for } 0 < t < t_f \qquad (2.9.22)$$

while the final and boundary conditions, equation (2.9.10b-d), remain unaltered for multiple sensors.

The objective function also appears in the development of the search step size, equation (2.9.17), and of the tolerance for the stopping criterion, equation (2.9.20). For the readings of M sensors, such quantities are respectively obtained from the following expressions:

$$\beta^k = \frac{\displaystyle\sum_{m=1}^{M} \int_{t=0}^{t_f} \frac{[Y_m(t) - T(x_m,t;\mathbf{P}^k)]}{[\sigma_m(t)]^2} \Delta T(x_m,t;\mathbf{d}^k)\, dt}{\displaystyle\sum_{m=1}^{M} \int_{t=0}^{t_f} \frac{[\Delta T(x_m,t;\mathbf{d}^k)]^2}{[\sigma_m(t)]^2} dt} \qquad (2.9.23)$$

and

$$\varepsilon_1 = M t_f \qquad (2.9.24)$$

2.10 ESTIMATION OF A HEAT SOURCE TERM IN A HEAT CONDUCTION PROBLEM

In the previous sections of this chapter, we developed the relevant equations and introduced the computational algorithms for Techniques I, II and III. In this section, we present the results obtained with such techniques, as applied to the solution of the test problem involving the estimation of the strength of a plane heat source term.

As apparent from the analysis of Figure 2.2, the problem of estimating the coefficients of polynomial basis functions used to approximate the unknown source term is quite difficult due to the linear dependence of the sensitivity coefficients. Therefore, we considered here the source term to be approximated by Fourier series, where the basis functions were given by equations (2.4.1e and f). The duration of the experiment was taken as $t_f = 2$, since the rate of increase in $|\mathbf{J}^T \mathbf{J}|$ was strongly reduced for $t > 2$, as shown in Figure 2.4 for a case involving 5 unknown parameters. During the time interval $0 < t \leq 2$, we considered available for the inverse analysis 100 transient measurements of a single sensor located at $x_{meas} = 1$. Techniques I, II and III were applied to the estimation of the coefficients of the basis functions (2.4.1e and f) by using simulated measurements. The simulated measurements were generated with $P_1 = P_2 = P_3 = P_4 = P_5 = 1$, that is,

$$g_p(t) = 1 + \sin \pi t + \cos \pi t + \sin 2\pi t + \cos 2\pi t \qquad (2.10.1)$$

Measurement errors were additive, uncorrelated, Gaussian, with zero mean and constant standard deviation, so that the minimization procedure involved the ordinary least squares norm.

Moré's version of the Levenberg-Marquardt method [74] was used for Technique I. The Fletcher-Reeves version of the conjugate gradient method was used for Techniques II and III, in accordance with the computational algorithms described above. The direct, sensitivity and adjoint problems were solved with finite volumes by using an implicit discretization in time. The spatial domain $0 \leq x \leq 1$ was discretized with 100 volumes, while 100 time steps were used to advance the solutions from $t = 0$ to $t_f = 2$. The sensitivity coefficients, needed for the solutions with Techniques I and II, were evaluated with finite differences by utilizing the forward approximation of equation (2.4.11) with $\varepsilon = 10^{-6}$.

The initial guesses for the unknown parameters were taken as zero, that is, $P_1^0 = P_2^0 = P_3^0 = P_4^0 = P_5^0 = 0$.

Table 2.10 presents the results obtained for the estimated parameters, RMS errors and number of iterations for Techniques I, II and III. Two different levels of measurement errors considered for the analysis included $\sigma = 0$ (errorless) and $\sigma = 0.01Y_{max}$, where Y_{max} is the maximum measured temperature. The RMS error is defined here as:

TABLE 2.10
Results Obtained for the Source Term Function Given by Equation (2.10.1)

Technique	σ	Estimated Parameters $P_1, \quad P_2, \quad P_3, \quad P_4, \quad P_5,$	RMS Error	Iterations
I	0	1.000, 1.000, 1.000, 1.000, 1.000	0.0	2
	$0.01Y_{max}$	0.999, 1.003, 0.997, 1.004, 1.009	0.008	5
II	0	1.000, 1.000, 1.000, 1.000, 1.000	0.0	10
	$0.01Y_{max}$	0.968, 1.020, 0.918, 1.130, 0.894	0.139	5
III	0	1.000, 1.000, 1.000, 1.000, 1.000	0.0	26
	$0.01Y_{max}$	0.981, 1.016, 0.949, 1.065, 0.916	0.087	8

(a)

(b)

(c)

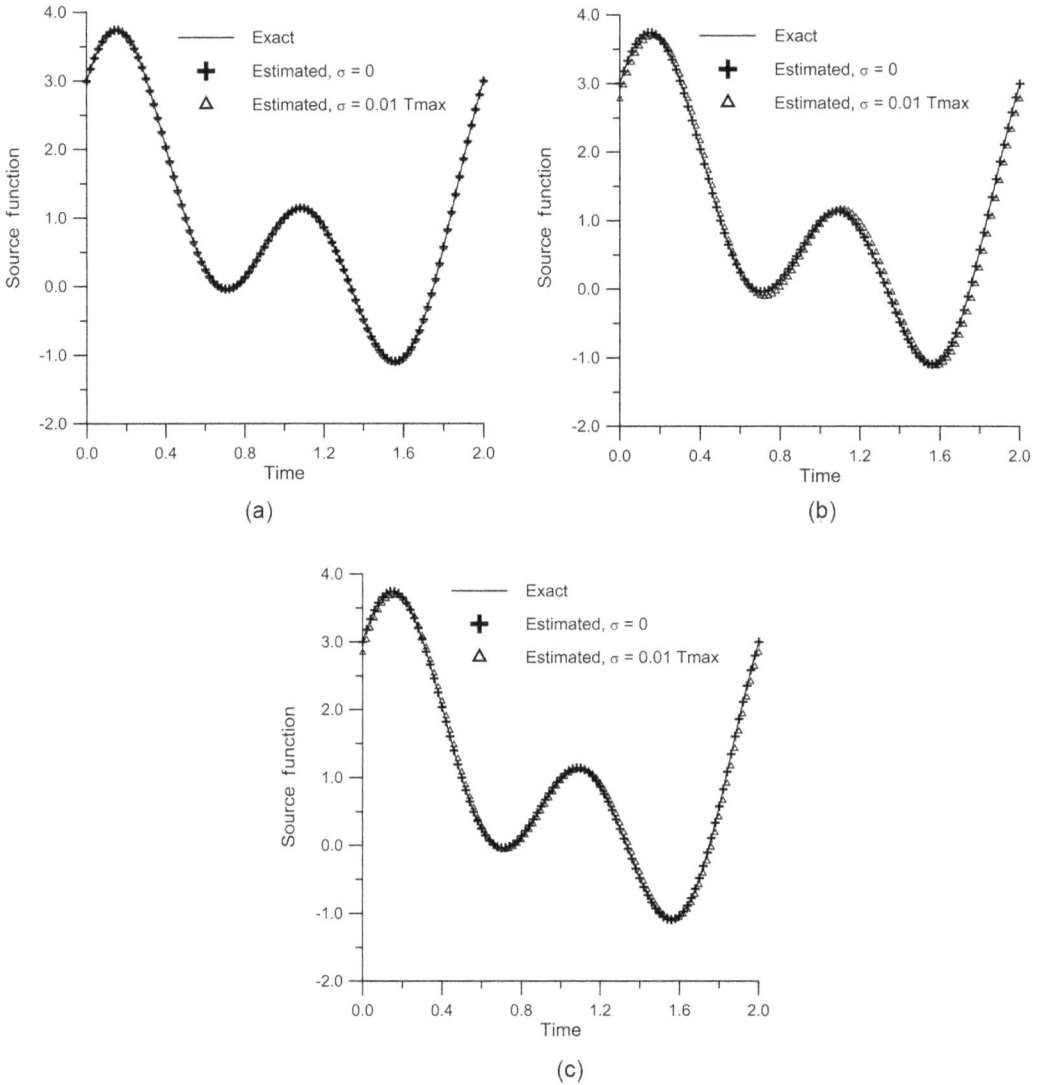

FIGURE 2.13 (a) Estimation of the source term given by equation (2.10.1) with Technique I. (b) Estimation of the source term given by equation (2.10.1) with Technique II. (c) Estimation of the source term given by equation (2.10.1) with Technique III.

$$e_{\text{RMS}} = \sqrt{\frac{1}{I}\sum_{i=1}^{I}\left[g_{\text{est}}(t_i) - g_{\text{ex}}(t_i)\right]^2} \qquad (2.10.2)$$

where $g_{est}(t_i)$ is the estimated source term function at time t_i, $g_{ex}(t_i)$ is the exact source term function (used to generate the simulated measurements) at time t_i and I is the number of measurements.

Table 2.10 shows that the exact values $P_1 = P_2 = P_3 = P_4 = P_5 = 1$ were recovered with these three techniques, when errorless measurements ($\sigma = 0$) were used. In such cases, we had the smallest number of iterations for Technique I. For cases involving measurement errors ($\sigma = 0.01 Y_{max}$), the smallest RMS error was also obtained with Technique I. The foregoing analysis reveals that Technique I, among those examined for parameter estimation, provided the best results in the estimation of the five coefficients of the Fourier series utilized to approximate the source term function. Although this conclusion coincides with that for the example analyzed in Section 2.8, the reader must be aware that it is not general and should not be extended directly to other problems of parameter estimation. The results may depend on the physical character of the problem, number of parameters to be estimated and initial guess [93]. In fact, all the RMS errors shown in Table 2.10 were quite small for all cases considered, as a result of the simplicity of the present test problem.

Figures 2.13a-c present the results for the source term function, obtained with Techniques I, II and III, respectively. These figures clearly show the better results obtained with Technique I when measurements with random errors were used in the analysis, although the results obtained with Techniques II and III were also quite good.

PROBLEMS

2.1 Prove that, for linear parameter estimation problems, the vector of estimated temperatures can be written as $\mathbf{T} = \mathbf{JP}$, where $\mathbf{J}$ is the sensitivity matrix and $\mathbf{P}$ is the vector of parameters.

2.2 Use the relation $\mathbf{T} = \mathbf{JP}$ to derive equation (2.2.10) for linear parameter estimation problems.

2.3 Show that the linearization of the estimated temperatures $\mathbf{T}(\mathbf{P})$ around the vector of parameters $\mathbf{P}^k$ at iteration k can be written in the form given by equation (2.2.11).

2.4 Derive equation (2.2.12).

2.5 Calculate the sensitivity coefficients presented in Figures 2.2 and 2.3 by using forward and central finite difference approximations, instead of using the analytical expression given by equation (2.4.1c). How do the sensitivity coefficients calculated numerically by finite differences compare to those calculated analytically in terms of accuracy and computational time? What is the effect of the factor ε, appearing in equations (2.4.11) and (2.4.12), on the accuracy of the finite difference approximations?

2.6 Derive equation (2.6.3a).

2.7 Derive the sensitivity problem given by equation (2.9.4).

2.8 Derive equation (2.9.6).

2.9 A semi-infinite medium initially at the zero temperature has the temperature at the surface $x = 0$ suddenly changed to a constant value T_0. The formulation of such heat conduction problem is given by:

$$C\frac{\partial T}{\partial t} = k\frac{\partial^2 T}{\partial x^2} \quad \text{for } x > 0 \text{ and } t > 0$$

$$T = T_0 \quad \text{at } x = 0 \text{ for } t > 0$$

$$T = 0 \quad \text{for } t = 0 \text{ and } x > 0$$

Examine the transient variation of the sensitivity coefficients with respect to the volumetric heat capacity $C = \rho c$ and to the thermal conductivity k, for sensors located at different depths below the surface. Is the simultaneous estimation of C and k possible? What is the behavior of $|\mathbf{J}^T\mathbf{J}|$?

2.10 Repeat Problem 2.9 for the surface at $x = 0$ subjected to a constant heat flux q_0, instead of being maintained at the constant temperature T_0. In this case, the formulation of the heat conduction problem is given by:

$$C\frac{\partial T}{\partial t} = k\frac{\partial^2 T}{\partial x^2} \quad \text{for } x > 0 \text{ and } t > 0$$

$$-k\frac{\partial T}{\partial x} = q_0 \text{ at } x = 0 \text{ for } t > 0$$

$$T = 0 \text{ for } t = 0 \text{ and } x > 0$$

2.11 By using the formulation of either Problem 2.9 or 2.10 (whichever is more appropriate) estimate simultaneously k and C with Techniques I and II. Use $C = k = 1$ and $T_0 = 1$ (or $q_0 = 1$), in order to generate the simulated measurements of a single sensor for the analysis. Examine the effects of random measurement errors (additive, uncorrelated, Gaussian, with zero mean and constant standard deviation), initial guess and sensor location on the estimated parameters. Is such parameter estimation problem linear or nonlinear?

2.12 For the physical situation of Problem 2.10, consider k and C known and q_0 unknown. Examine the transient variation of the sensitivity coefficients with respect to q_0 for sensors located at different depths below the surface. Use $C = k = q_0 = 1$ in order to generate the simulated measurements of a single sensor for the analysis. Thus, use such measurements to estimate q_0 by using Techniques I and II. Examine the effects of random measurement errors (additive, uncorrelated, Gaussian, with zero mean and constant standard deviation), initial guess and sensor location on the estimated heat flux. Is it possible to estimate simultaneously k and/or C together with q_0?

2.13 Consider the following heat conduction problem in dimensionless form:

$$\frac{\partial T}{\partial t} = \frac{\partial^2 T}{\partial x^2} \text{ in } 0 < x < 1 \text{ for } t > 0$$

$$\frac{\partial T}{\partial x} = 0 \text{ at } x = 0 \text{ for } t > 0$$

$$\frac{\partial T}{\partial x} = q(t) \text{ at } x = 1 \text{ for } t > 0$$

$$T = 0 \text{ for } t = 0 \text{ in } 0 < x < 1.$$

Formulate all the steps for the solution of the inverse problem of estimating the unknown heat flux $q(t)$ by using Techniques I, II and III. Consider available the transient readings Y_i, $i = 1, ..., I$ of a single sensor located at x_{meas}. Also, assume that $q(t)$ is given in the following general linear parametric form

$$q(t) = \sum_{j=1}^{N} P_j C_j(t)$$

where P_j are the unknown parameters and $C_j(t)$ are known trial functions.

2.14 Is the inverse problem involving the estimation of P_j in Problem 2.13 linear or nonlinear?

2.15 Consider $q(t)$ in Problem 2.13 to be approximated by three trial functions in the form of a polynomial, as given by equation (2.4.1d). Examine the transient variation of the sensitivity coefficients with respect to the parameters P_j, $j = 1, 2, 3$, for a sensor located at $x_{meas} = 0$. Is the estimation of such parameters possible? What is the behavior of $|\mathbf{J}^T \mathbf{J}|$?

2.16 Consider $q(t)$ in Problem 2.13 to be approximated by three trial functions in the form of a Fourier series, as given by equations (2.4.1e and f). Examine the transient variation of the sensitivity coefficients with respect to the parameters P_j, $j = 1, 2, 3$, for a sensor located at $x_{meas} = 0$. Is the estimation of such parameters possible? What is the behavior of $|\mathbf{J}^T\mathbf{J}|$?

2.17 Repeat problem 2–13 by now assuming available the transient readings of M sensors located at $x = x_m$, $m = 1, ..., M$.

NOTE 1: SEARCH STEP-SIZE FOR TECHNIQUE II

The search step size, β^k, for the conjugate gradient method of parameter estimation, is obtained as the one that minimizes the maximum likelihood objective function given by equation (2.1.1) at each iteration, that is,

$$\min_{\beta^k} S_{ML}(\mathbf{P}^{k+1}) = \min_{\beta^k} [\mathbf{Y} - \mathbf{T}(\mathbf{P}^{k+1})]^T \mathbf{W}^{-1} [\mathbf{Y} - \mathbf{T}(\mathbf{P}^{k+1})] \tag{N1.2.1}$$

From the iterative procedure of the conjugate gradient method, equation (2.3.1), we have:

$$\mathbf{P}^{k+1} = \mathbf{P}^k + \beta^k \mathbf{d}^k \tag{N1.2.2}$$

Thus, by substituting $\mathbf{P}^{k+1}$ into equation (N1.2.1), we obtain:

$$\min_{\beta^k} S_{ML}(\mathbf{P}^{k+1}) = \min_{\beta^k} [\mathbf{Y} - \mathbf{T}(\mathbf{P}^k + \beta^k \mathbf{d}^k)]^T \mathbf{W}^{-1} [\mathbf{Y} - \mathbf{T}(\mathbf{P}^k + \beta^k \mathbf{d}^k)] \tag{N1.2.3}$$

We now linearize $T_i(\mathbf{P}^k + \beta^k \mathbf{d}^k)$, for $i = 1,..., I$, by using a Taylor series expansion in the form:

$$T_i(\mathbf{P}^k + \beta^k \mathbf{d}^k) = T_i[(P_1^k + \beta^k d_1^k), (P_2^k + \beta^k d_2^k),...,(P_N^k + \beta^k d_N^k)]$$

$$\approx T_i(P_1^k, P_2^k,..., P_N^k) + \beta^k \frac{\partial T_i}{\partial P_1^k} d_1^k + \beta^k \frac{\partial T_i}{\partial P_2^k} d_2^k + \cdots + \beta^k \frac{\partial T_i}{\partial P_N^k} d_N^k \tag{N1.2.4a}$$

or, in vector form,

$$T_i(\mathbf{P}^k + \beta^k \mathbf{d}^k) \approx T_i(\mathbf{P}^k) + \beta^k \left[\frac{\partial T_i}{\partial \mathbf{P}^k} \right]^T \mathbf{d}^k \tag{N1.2.4b}$$

where

$$\left[\frac{\partial T_i}{\partial \mathbf{P}^k} \right]^T = \left[\frac{\partial T_i}{\partial P_1^k}, \frac{\partial T_i}{\partial P_2^k}, \cdots, \frac{\partial T_i}{\partial P_N^k} \right] \tag{N1.2.5}$$

By writing equation (N1.2.4b) for $i = 1, ..., I$, and using the definition of the sensitivity matrix given by equation (2.2.7), we have:

$$\mathbf{T}(\mathbf{P}^{k+1}) \approx \mathbf{T}(\mathbf{P}^k) + \beta^k \mathbf{J}^k \mathbf{d}^k \tag{N1.2.6}$$

The substitution of equation (N1.2.6) into equation (2.1.1) gives:

$$S_{ML}(\mathbf{P}^{k+1}) = [\mathbf{Y} - \mathbf{T}(\mathbf{P}^k) - \beta^k \mathbf{J}^k \mathbf{d}^k]^T \mathbf{W}^{-1} [\mathbf{Y} - \mathbf{T}(\mathbf{P}^k) - \beta^k \mathbf{J}^k \mathbf{d}^k] \tag{N1.2.7}$$

which can be expanded in the form:

$$S_{ML}(\mathbf{P}^{k+1}) = [\mathbf{Y} - \mathbf{T}(\mathbf{P}^k)]^T \mathbf{W}^{-1} [\mathbf{Y} - \mathbf{T}(\mathbf{P}^k)] - \beta^k [\mathbf{Y} - \mathbf{T}(\mathbf{P}^k)]^T \mathbf{W}^{-1} [\mathbf{J}^k \mathbf{d}^k]$$
$$- \beta^k [\mathbf{J}^k \mathbf{d}^k]^T \mathbf{W}^{-1} [\mathbf{Y} - \mathbf{T}(\mathbf{P}^k)] + (\beta^k)^2 [\mathbf{J}^k \mathbf{d}^k]^T \mathbf{W}^{-1} [\mathbf{J}^k \mathbf{d}^k]$$

(N1.2.8)

By performing the minimization of (N1.2.8) with respect to β^k, we obtain:

$$\frac{\partial S_{ML}(\mathbf{P}^{k+1})}{\partial \beta^k} = 0 = -[\mathbf{Y} - \mathbf{T}(\mathbf{P}^k)]^T \mathbf{W}^{-1} [\mathbf{J}^k \mathbf{d}^k] - [\mathbf{J}^k \mathbf{d}^k]^T \mathbf{W}^{-1} [\mathbf{Y} - \mathbf{T}(\mathbf{P}^k)] + 2\beta^k [\mathbf{J}^k \mathbf{d}^k]^T \mathbf{W}^{-1} [\mathbf{J}^k \mathbf{d}^k]$$

(N1.2.9)

Since $\mathbf{W}^{-1}$ is symmetric we have:

$$[\mathbf{Y} - \mathbf{T}(\mathbf{P}^k)]^T \mathbf{W}^{-1} [\mathbf{J}^k \mathbf{d}^k] = [\mathbf{J}^k \mathbf{d}^k]^T \mathbf{W}^{-1} [\mathbf{Y} - \mathbf{T}(\mathbf{P}^k)]$$

(N1.2.10)

and then the solution of (N1.2.9) for β^k gives:

$$\beta^k = \frac{[\mathbf{J}^k \mathbf{d}^k]^T \mathbf{W}^{-1} [\mathbf{Y} - \mathbf{T}(\mathbf{P}^k)]}{[\mathbf{J}^k \mathbf{d}^k]^T \mathbf{W}^{-1} [\mathbf{J}^k \mathbf{d}^k]}$$

(N1.2.11)

NOTE 2: SEARCH STEP-SIZE FOR TECHNIQUE III

Similarly to Technique II, the search step size for Technique III is obtained as the one that minimizes the objective function given by equation (2.9.1a) at each iteration, that is,

$$\min_{\beta^k} S_{ML}(\mathbf{P}^{k+1}) = \min_{\beta^k} \int_{t=0}^{t_f} \frac{[Y(t) - T(x_{\text{meas}}, t; \mathbf{P}^{k+1})]^2}{[\sigma(t)]^2} \, dt$$

(N2.2.1)

From the iterative procedure of the conjugate gradient method, we have:

$$\mathbf{P}^{k+1} = \mathbf{P}^k + \beta^k \mathbf{d}^k$$

(N2.2.2)

By substituting $\mathbf{P}^{k+1}$ into equation (N2.2.1), we obtain:

$$\min_{\beta^k} S_{ML}(\mathbf{P}^{k+1}) = \min_{\beta^k} \int_{t=0}^{t_f} \frac{[Y(t) - T(x_{\text{meas}}, t; \mathbf{P}^k + \beta^k \mathbf{d}^k)]^2}{[\sigma(t)]^2} \, dt$$

(N2.2.3)

We now linearize $T(\mathbf{P}^k + \beta^k \mathbf{d}^k)$ by using a Taylor series expansion in the form:

$$T(\mathbf{P}^k + \beta^k \mathbf{d}^k) = T[(P_1^k + \beta^k d_1^k), (P_2^k + \beta^k d_2^k), ..., (P_N^k + \beta^k d_N^k)]$$
$$\approx T(P_1^k, P_2^k, ..., P_N^k) + \beta^k \frac{\partial T}{\partial P_1^k} d_1^k + \beta^k \frac{\partial T}{\partial P_2^k} d_2^k + ... + \beta^k \frac{\partial T}{\partial P_N^k} d_N^k$$

(N2.2.4)

By making:

$$d_1^k = \Delta P_1^k, \quad d_2^k = \Delta P_2^k, ... \quad d_N^k = \Delta P_N^k$$

(N2.2.5)

the equation above becomes:

$$T(\mathbf{P}^k + \beta^k \mathbf{d}^k) \approx T(P_1^k, P_2^k, \ldots, P_N^k) + \beta^k \sum_{j=1}^{N} \frac{\partial T}{\partial P_j^k} \Delta P_j^k \qquad (N2.2.6)$$

where N is the number of parameters.

Let

$$\Delta T(\mathbf{d}^k) = \sum_{j=1}^{N} \frac{\partial T}{\partial P_j^k} \Delta P_j^k \qquad (N2.2.7)$$

Then equation (N2.2.6) can be written as:

$$T(\mathbf{P}^k + \beta^k \mathbf{d}^k) \approx T(\mathbf{P}^k) + \beta^k \Delta T(\mathbf{d}^k) \qquad (N2.2.8)$$

By substituting equation (N2.2.8) into equation (N2.2.3), we obtain:

$$\min_{\beta^k} S_{ML}(\mathbf{P}^{k+1}) = \min_{\beta^k} \int_{t=0}^{t_f} \frac{[Y(t) - T(x_{meas}, t; \mathbf{P}^k) - \beta^k \Delta T(x_{meas}, t; \mathbf{d}^k)]^2}{[\sigma(t)]^2} dt \qquad (N2.2.9)$$

which is then minimized with respect to β^k to yield:

$$\beta^k = \frac{\displaystyle\int_{t=0}^{t_f} \frac{[Y(t) - T(x_{meas}, t; \mathbf{P}^k)] \Delta T(x_{meas}, t; \mathbf{d}^k)}{[\sigma(t)]^2} dt}{\displaystyle\int_{t=0}^{t_f} \frac{[\Delta T(x_{meas}, t; \mathbf{d}^k)]^2}{[\sigma(t)]^2} dt} \qquad (N2.2.10)$$

where $\Delta T(x_{meas}, t; \mathbf{d}^k)$ is the solution of the sensitivity problem given by equations (2.9.4a-d), obtained by setting $\Delta P_j = d_j^k, j=1,\ldots, N$, in the computation of:

$$\Delta g_p(t) = \sum_{j=1}^{N} \Delta P_j C_j(t) \qquad (N2.2.11)$$

3 Parameter Estimation
Minimization of an Objective Function with Prior Information about the Unknown Parameters

As a continuation of the previous chapter, this chapter now considers cases where some information is available regarding the values of the unknown parameters before the solution of the inverse problem is attempted. If such information is available, either from someone's practical experience, previous experiments or even literature data, it can be formally taken into account for the solution of the inverse problem. This is accomplished by solving the inverse problem within the Bayesian framework of statistics.

The most complete source for the solution of inverse problems within the Bayesian framework is the book by Kaipio and Somersalo [17]. A very didactical open material on the subject is also available in Ref. [19], while fundamental concepts on Bayesian statistics can be found in the books by Lee [63] and Winkler [67]. Reference [65] is an interesting book, with historical aspects and practical applications of Bayesian statistics in layman's terms.

3.1 OBJECTIVE FUNCTION

For the solution of inverse problems within the Bayesian framework, variables and parameters in the mathematical formulation of the physical problem are considered random. Techniques for the solution of inverse problems within the Bayesian framework can be summarized in the following steps [17]:

1. Based on all information available for the parameters **P** before the measured data **Y** is taken, select a probability distribution function, $\pi(\mathbf{P})$, that represents the *prior* information.
2. Select the *likelihood function*, $\pi(\mathbf{Y}|\mathbf{P})$, that models the measurement errors and involves a relation between the observations and the mathematical model of the physical problem being analyzed.
3. Develop methods to explore the *posterior density function*, which is the conditional probability distribution of the unknown parameters given the measurements, $\pi(\mathbf{P}|\mathbf{Y})$.

Therefore, the solution of the inverse problem within the Bayesian framework is based on statistical inference on the *posterior density function*. In order to obtain the *posterior density function*, the new information (*measurements*) is combined with the previously available information (*prior*) by *Bayes' theorem* [4,17–19,22,48,49,59,60,63,67].

Let **P** and **Y** be continuous random variables. Then, we can write [67]:

$$\pi(\mathbf{P}|\mathbf{Y}) = \frac{\pi(\mathbf{P}, \mathbf{Y})}{\pi(\mathbf{Y})} \tag{3.1.1}$$

that is, the conditional density of the random variable **P** given a value of the random variable **Y** is the joint density of **P** and **Y** divided by the marginal density of **Y**, which is given by:

$$\pi(\mathbf{Y}) = \int_{\mathbb{R}^N} \pi(\mathbf{P}, \mathbf{Y}) d\mathbf{P} \tag{3.1.2}$$

In equation (3.1.2), $\mathbb{R}^N$ is the space of the vector of parameters, **P**, where N is the number of parameters and $\mathbb{R}$ denotes real numbers.

The joint density $\pi(\mathbf{P}, \mathbf{Y})$ is not generally known, but it can be written in terms of the likelihood and the prior as [67]:

$$\pi(\mathbf{P}, \mathbf{Y}) = \pi(\mathbf{Y}|\mathbf{P})\pi(\mathbf{P}) \tag{3.1.3}$$

By substituting (3.1.3) into (3.1.1) we then obtain *Bayes' theorem* as:

$$\pi(\mathbf{P}|\mathbf{Y}) = \frac{\pi(\mathbf{Y}|\mathbf{P})\pi(\mathbf{P})}{\pi(\mathbf{Y})} \tag{3.1.4}$$

In Bayes' theorem, $\pi(\mathbf{P}|\mathbf{Y})$ is the *posterior probability density*, $\pi(\mathbf{P})$ is the *prior density*, $\pi(\mathbf{Y}|\mathbf{P})$ is the likelihood function and $\pi(\mathbf{Y})$ is the marginal probability density of the measurements, which plays the role of a normalizing constant. Since the computation of $\pi(\mathbf{Y})$ with equation (3.1.2) is in general difficult, and usually not needed for practical calculations as will be apparent below, Bayes' theorem is commonly written as:

$$\pi(\mathbf{P}|\mathbf{Y}) \propto \pi(\mathbf{Y}|\mathbf{P})\pi(\mathbf{P}) \tag{3.1.5}$$

Equation (3.1.4) or (3.1.5) clearly shows that the posterior distribution depends not only on the likelihood function, but also on the prior distribution for the parameters. Non-Bayesian approaches for the solution of inverse problems, such as those in the previous chapter, only take into account the information provided by the measurements *via* the likelihood function. While the likelihood is the statistical model for the measurement errors, which is usually obtained from the calibration process of the sensors, the prior distribution is the statistical model of the information available for the parameters before the experiment and the inverse analysis. Therefore, even if the prior information is only qualitatively available, it must be mathematically modeled in terms of a statistical distribution for the solution of the inverse problem within the Bayesian framework. As it will be apparent below, the success of the inverse problem solution strongly depends on the specified prior, which must appropriately reflect the previous knowledge about the parameters.

In this chapter, we consider valid the hypotheses that the measurement errors, ε, are additive Gaussian random variables, with zero means, known covariance matrix **W** and independent of the parameters **P**. Therefore, the likelihood function, which is the conditional probability of the measurements **Y** given the parameters **P**, $\pi(\mathbf{Y}|\mathbf{P})$, is given by (see equation 1.4.5b) [4,17–19,22,48,49,59,60]:

$$\pi(\mathbf{Y}|\mathbf{P}) = (2\pi)^{-D/2} |\mathbf{W}|^{-1/2} \exp\left\{-\frac{1}{2}[\mathbf{Y} - \mathbf{T}(\mathbf{P})]^T \mathbf{W}^{-1}[\mathbf{Y} - \mathbf{T}(\mathbf{P})]\right\} \tag{3.1.6}$$

where $\pi = 4\tan^{-1}(1)$, $D = MI$ is the total number of measurements, M is the number of sensors and I is the number of transient measurements per sensor.

Consider, as an example, that the inverse problem is concerned with the identification of the thermal conductivity, k, of a new material, in an experimental setup constructed for this purpose. Although the thermal conductivity of the material being characterized is not exactly known, let's say that the analyst conducting the experiment and solving the inverse problem is aware that the material is some type of steel. Literature data are abundant on values of thermophysical properties of metallic materials, in particular at low temperatures. The literature data can thus provide minimum and maximum values for the expected thermal conductivity of the new material for the specification of the prior distribution. The interval given by these minimum and maximum values, k_{min} and k_{max}, respectively, may even be enlarged to cope with a possible thermal conductivity outside the range of those of already existing materials.

If all values inside the interval $k_{min} < k < k_{max}$ are equally probable, the prior is then specified in terms of a uniform distribution (see Note 1 in Chapter 1) given by:

$$\pi(k) = \begin{cases} \dfrac{1}{(k_{max} - k_{min})}, & k_{min} < k < k_{max} \\ 0 &, \quad \text{elsewhere} \end{cases} \tag{3.1.7}$$

The uniform prior given by equation (3.1.7) clearly shows that the probability to find a value of k outside the interval $k_{min} < k < k_{max}$ is zero. By substituting the likelihood given by equation (3.1.6) and the prior given by equation (3.1.7) into equation (3.1.5), the posterior distribution for this example is written as:

$$\pi(k|\mathbf{Y}) \propto \begin{cases} \dfrac{(2\pi)^{-D/2} |\mathbf{W}|^{-1/2}}{(k_{max} - k_{min})} \exp\left\{ -\dfrac{1}{2} [\mathbf{Y} - \mathbf{T}(k)]^T \mathbf{W}^{-1} [\mathbf{Y} - \mathbf{T}(k)] \right\}, & k_{min} < k < k_{max} \\ 0 &, \quad \text{elsewhere} \end{cases} \tag{3.1.8}$$

Therefore, the posterior distribution is zero for $k \leq k_{min}$ and for $k \geq k_{max}$, that is, based on the uniform prior given by equation (3.1.7), no solution for this inverse problem is possible outside the interval $k_{min} < k < k_{max}$. Within the interval $k_{min} < k < k_{max}$, a point value can be estimated for k by maximizing $\pi(k|\mathbf{Y})$, which can be recast as a minimization problem involving the following objective function:

$$S_{MAP}(k) = [\mathbf{Y} - \mathbf{T}(k)]^T \mathbf{W}^{-1} [\mathbf{Y} - \mathbf{T}(k)] \quad \text{in} \quad k_{min} < k < k_{max} \tag{3.1.9}$$

Equation (3.1.9) is thus called the *maximum a posteriori (MAP) objective function* for measurement errors that are additive, Gaussian, with zero mean and covariance matrix $\mathbf{W}$, and for a uniform prior for k in the interval $k_{min} < k < k_{max}$. Therefore, for the above uniform prior, the MAP objective function (3.1.9) coincides with the maximum likelihood objective function (equation 2.1.1) within $k_{min} < k < k_{max}$. We note that the constant $(k_{max} - k_{min})$ in the prior (see equation 3.1.7) does not appear in the objective function given by equation (3.1.9).

This simple example shows the importance of the judicious specification of the prior. Imagine if k_{max} is specified much smaller than the thermal conductivity of any metallic material. Then, the correct value of the thermal conductivity of the steel sample will not be retrieved by the inverse analysis because the posterior is zero for $k \geq k_{max}$.

The earlier discussion is now generalized for multiple parameters and for other prior distributions.

MAXIMUM A POSTERIORI OBJECTIVE FUNCTION WITH A UNIFORM PRIOR

As a generalization of the earlier case involving one single parameter (given by the thermal conductivity), consider now that possible values for each model parameter P_j are *a priori* known to be in the intervals $P_{j,min} < P_j < P_{j,max}, j = 1, \dots, N$. Furthermore, if all values inside these intervals are equally probable, the prior for each parameter P_j is given by a uniform distribution, that is, for $j = 1, \dots, N$:

$$\pi(P_j) = \begin{cases} \dfrac{1}{(P_{j,max} - P_{j,min})}, & P_{j,min} < P_j < P_{j,max} \\ 0 &, \quad \text{elsewhere} \end{cases} \tag{3.1.10}$$

With the above prior and the likelihood given by equation (3.1.6), the posterior given by equation (3.1.5) becomes:

$$\pi(\mathbf{P}|\mathbf{Y}) \propto \begin{cases} \dfrac{(2\pi)^{-D/2}|\mathbf{W}|^{-1/2}}{\displaystyle\prod_{j=1,\ldots,N}(P_{j,\max}-P_{j,\min})}\exp\left\{-\dfrac{1}{2}[\mathbf{Y}-\mathbf{T}(\mathbf{P})]^{T}\,\mathbf{W}^{-1}[\mathbf{Y}-\mathbf{T}(\mathbf{P})]\right\}, & \mathbf{P}_{\min}<\mathbf{P}<\mathbf{P}_{\max} \\[4mm] 0 & , \quad \text{elsewhere} \end{cases} \tag{3.1.11}$$

where $\mathbf{P}_{\min}=\left[P_{1,\min},P_{2,\min},\ldots,P_{N,\min}\right]^{T}$ and $\mathbf{P}_{\max}=\left[P_{1,\max},P_{2,\max},\ldots,P_{N,\max}\right]^{T}$. The inequalities in (3.1.11) are considered satisfied if and only if they are individually satisfied for each parameter, that is, $P_{j,\min}<P_{j}<P_{j,\max}$ for all $j=1,\ldots,N$.

A point estimate for $\mathbf{P}$ can be obtained with the solution of a minimization problem involving the following *MAP objective function*:

$$S_{\mathrm{MAP}}(\mathbf{P})=[\mathbf{Y}-\mathbf{T}(\mathbf{P})]^{T}\,\mathbf{W}^{-1}[\mathbf{Y}-\mathbf{T}(\mathbf{P})] \quad \text{in}\quad \mathbf{P}_{\min}<\mathbf{P}<\mathbf{P}_{\max} \tag{3.1.12}$$

which applies for additive and Gaussian measurement errors, with zero mean and covariance matrix $\mathbf{W}$, and for a uniform prior for $\mathbf{P}$ in $\mathbf{P}_{\min}<\mathbf{P}<\mathbf{P}_{\max}$.

Such as for the case dealing with the estimation of one single parameter, the point estimate obtained for $\mathbf{P}$ through the minimization of the MAP objective function (3.1.12) is the same as that obtained through the minimization of the maximum likelihood objective function given by equation (2.1.1), but the parameter space is limited to $\mathbf{P}_{\min}<\mathbf{P}<\mathbf{P}_{\max}$. Therefore, the solution of the inverse problem can be obtained by the minimization of equation (3.1.12) with Techniques I and II described in Chapter 2, as long as the search space is restricted to $\mathbf{P}_{\min}<\mathbf{P}<\mathbf{P}_{\max}$.

MAXIMUM A POSTERIORI OBJECTIVE FUNCTION WITH A GAUSSIAN PRIOR

We now consider a case where each parameter P_{j} can be *a priori* modeled with the largest probability at a point μ_{j} and with symmetric decreasing probabilities at points equidistant from μ_{j}, such as in a Gaussian distribution (see Note 1 in Chapter 1) in the form:

$$\pi(P_{j})=\dfrac{1}{\sigma_{P_{j}}\sqrt{2\pi}}\exp\left[-\dfrac{1}{2}\left(\dfrac{P_{j}-\mu_{j}}{\sigma_{P_{j}}}\right)^{2}\right] \quad \text{for}\quad j=1,\ldots,N \tag{3.1.13}$$

where $\sigma_{P_{j}}$ is the standard deviation of the Gaussian prior for P_{j}. By also *a priori* assuming that the parameters are independent, the posterior distribution is given by:

$$\pi(\mathbf{P}|\mathbf{Y}) \propto (2\pi)^{-D/2}|\mathbf{W}|^{-1/2}\exp\left\{-\dfrac{1}{2}[\mathbf{Y}-\mathbf{T}(\mathbf{P})]^{T}\,\mathbf{W}^{-1}[\mathbf{Y}-\mathbf{T}(\mathbf{P})]\right\}$$

$$\times\left\{\prod_{j=1,\ldots,N}\dfrac{1}{\sigma_{P_{j}}\sqrt{2\pi}}\exp\left[-\dfrac{1}{2}\left(\dfrac{P_{j}-\mu_{j}}{\sigma_{P_{j}}}\right)^{2}\right]\right\} \tag{3.1.14a}$$

which can be rewritten as:

$$\pi(\mathbf{P}|\mathbf{Y}) \propto \dfrac{(2\pi)^{-(D+N)/2}}{|\mathbf{W}|^{1/2}\displaystyle\prod_{j=1,\ldots,N}\sigma_{P_{j}}}\exp\left\{-\dfrac{1}{2}[\mathbf{Y}-\mathbf{T}(\mathbf{P})]^{T}\,\mathbf{W}^{-1}[\mathbf{Y}-\mathbf{T}(\mathbf{P})]-\dfrac{1}{2}\sum_{j=1}^{N}\left(\dfrac{P_{j}-\mu_{j}}{\sigma_{P_{j}}}\right)^{2}\right\} \tag{3.1.14b}$$

Thus, a point estimate for $\mathbf{P}$ can be obtained with the minimization of the following objective function:

$$S_{MAP}(\mathbf{P}) = [\mathbf{Y} - \mathbf{T}(\mathbf{P})]^T \mathbf{W}^{-1}[\mathbf{Y} - \mathbf{T}(\mathbf{P})] + \sum_{j=1}^{N}\left(\frac{P_j - \mu_j}{\sigma_{P_j}}\right)^2 \tag{3.1.15}$$

which is the *MAP objective function* for additive Gaussian measurement errors, with zero mean and covariance matrix $\mathbf{W}$, and Gaussian independent priors given by equation (3.1.13).

Differently from the uniform prior given by equation (3.1.10), which resulted in the same objective function of the previous chapter but with minimum and maximum bounds for each parameter (equation 3.1.12), the use of a Gaussian prior actually modifies the original maximum likelihood objective function (see also equation 2.1.1) by the addition of the *penalization term*:

$$\sum_{j=1}^{N}\left(\frac{P_j - \mu_j}{\sigma_{P_j}}\right)^2$$

If the measurements are uncorrelated and with standard deviation σ_k, where the subscript k is the index that refers to the sensor number and measurement time in accordance with Section 2.6, equation (3.1.15) reduces to:

$$S_{MAP}(\mathbf{P}) = \sum_{k=1}^{D}\left(\frac{Y_k - T_k(\mathbf{P})}{\sigma_k}\right)^2 + \sum_{j=1}^{N}\left(\frac{P_j - \mu_j}{\sigma_{P_j}}\right)^2 \tag{3.1.16}$$

We note that the residuals between the measurements and the solution of the direct problem, $[Y_k - T_k(\mathbf{P})]$, and the differences between the estimated parameters and the Gaussian prior means, $[P_j - \mu_j]$, are weighted by their corresponding standard deviations σ_k and σ_{P_j}, respectively, in the objective function (3.1.16). Therefore, measurements with small uncertainties (small σ_k) and parameters more accurately *a priori* known (small σ_{P_j}) are more significant in the minimization of equation (3.1.16).

Such as for Tikhonov's regularization given by equation (1.4.9a), the above penalization term has a stabilization property in the minimization of the objective functions (3.1.15) or (3.1.16). On the other hand, different from Tikhonov's regularization, this penalization term is now related to the model used for the prior distribution (that must appropriately reflect the information available for the parameters before the solution of the inverse problem). In Tikhonov's regularization, as well as in the Levenberg-Marquardt's iterative procedure, the penalization term is only aimed at smoothing out the solution for the problem and does not reflect the statistical prior information available for the parameters.

Equation (3.1.13) can be written in compact matrix form as:

$$\pi(\mathbf{P}) = (2\pi)^{-N/2}|\mathbf{V}|^{-1/2}\exp\left\{-\frac{1}{2}[\mathbf{P} - \mu]^T\mathbf{V}^{-1}[\mathbf{P} - \mu]\right\} \tag{3.1.17}$$

where μ and $\mathbf{V}$ are the known mean and covariance matrix for $\mathbf{P} \equiv [P_1, P_2, ..., P_N]^T$, respectively. The matrix $\mathbf{V}$ is diagonal if the parameters are independent, such as considered above in equation (3.1.13), that is,

$$\mathbf{V} = \begin{bmatrix} \sigma_{P_1}^2 & & & \\ & \sigma_{P_2}^2 & & \\ & & \ddots & \\ & & & \sigma_{P_N}^2 \end{bmatrix} \tag{3.1.18}$$

However, equation (3.1.17) is general and can be used for Gaussian priors of correlated parameters, where $\mathbf{V}$ is not a diagonal matrix.

By substituting equations (3.1.6) and (3.1.17) into Bayes' theorem given by equation (3.1.5), we obtain:

$$\pi(\mathbf{P}|\mathbf{Y}) \propto (2\pi)^{-(D+N)/2} |\mathbf{W}|^{-1/2} |\mathbf{V}|^{-1/2} \exp\left\{ -\frac{1}{2}[\mathbf{Y} - \mathbf{T}(\mathbf{P})]^T \mathbf{W}^{-1}[\mathbf{Y} - \mathbf{T}(\mathbf{P})]\right.$$

$$\left. -\frac{1}{2}[\mathbf{P} - \mathbf{\mu}]^T \mathbf{V}^{-1}[\mathbf{P} - \mathbf{\mu}]\right\} \tag{3.1.19}$$

By taking the logarithm of equation (3.1.19), the following equation results:

$$\ln\left[\pi(\mathbf{P}|\mathbf{Y})\right] \propto -\frac{1}{2}\left[(D+N)\ln 2\pi + \ln|\mathbf{W}| + \ln|\mathbf{V}| + S_{\mathrm{MAP}}(\mathbf{P})\right] \tag{3.1.20}$$

where

$$S_{\mathrm{MAP}}(\mathbf{P}) = \left[\mathbf{Y} - \mathbf{T}(\mathbf{P})\right]^T \mathbf{W}^{-1}\left[\mathbf{Y} - \mathbf{T}(\mathbf{P})\right] + (\mathbf{P} - \mathbf{\mu})^T \mathbf{V}^{-1}(\mathbf{P} - \mathbf{\mu}) \tag{3.1.21}$$

which is the *MAP objective function* for additive Gaussian measurement errors and Gaussian priors.

Like equations (3.1.15) and (3.1.16), equation (3.1.21) shows the contributions of the likelihood and of the prior distributions in this objective function, given by the first and second terms on the right-hand side, respectively.

It is also now clear that the maximum likelihood objective function (equation 2.1.1) is not a Bayesian estimator, since it does not contain information provided by the prior distribution for the parameters, unless there is a restriction in the parameter space resulting from a uniform prior like in equation (3.1.12). Conspicuously, the least squares norm (equation 2.1.6) and even Tikhonov's regularization are not Bayesian estimators, since they only explore the information provided by the measurements and, eventually, some characteristics of the parameters, like smoothness.

MAXIMUM A POSTERIORI OBJECTIVE FUNCTION WITH A TRUNCATED GAUSSIAN PRIOR

A scalar random variable modeled by a Gaussian distribution has support in $\mathbb{R}$. Hence, it may assume negative values, although this might happen with small probabilities depending on the values of μ_j and σ_j. On the other hand, several physical parameters only allow positive values, such as, for example, thermal conductivity, specific heat and thermal diffusivity. A very simple prior that allows lower and upper bounds for the parameter values is the uniform distribution examined previously, but it does not favor any value in the interval $P_{j,\min} < P_j < P_{j,\max}$. If in this interval, values around a known mean are more likely to occur than elsewhere, like in a Gaussian distribution, but the probability density is zero for $P_j \leq P_{j,\min}$ and $P_j \geq P_{j,\max}$, a prior can be obtained by combining the uniform and the Gaussian distributions, which is called *truncated Gaussian distribution*.

In matrix form, the truncated Gaussian prior can be written as:

$$\pi(\mathbf{P}) = \begin{cases} (2\pi)^{-N/2} |\mathbf{V}|^{-1/2} \exp\left\{ -\frac{1}{2}[\mathbf{P} - \mathbf{\mu}]^T \mathbf{V}^{-1}[\mathbf{P} - \mathbf{\mu}]\right\}, & \mathbf{P}_{\min} < \mathbf{P} < \mathbf{P}_{\max} \\ 0 & , \text{ elsewhere} \end{cases} \tag{3.1.22}$$

and, then, the posterior obtained with equations (3.1.5), (3.1.6) and (3.1.22) is:

$$\pi(\mathbf{P}|\mathbf{Y}) \propto \begin{cases} (2\pi)^{-(D+N)/2} |\mathbf{W}|^{-1/2} |\mathbf{V}|^{-1/2} \exp\left\{ -\frac{1}{2}[\mathbf{Y} - \mathbf{T}(\mathbf{P})]^T \mathbf{W}^{-1}[\mathbf{Y} - \mathbf{T}(\mathbf{P})] \right. \\ \left. -\frac{1}{2}[\mathbf{P} - \mu]^T \mathbf{V}^{-1}[\mathbf{P} - \mu] \right\} , \quad \mathbf{P}_{min} < \mathbf{P} < \mathbf{P}_{max} \\ 0 , \quad \text{elsewhere} \end{cases} \quad (3.1.23)$$

The *MAP objective function* for the truncated Gaussian prior is thus:

$$S_{MAP}(\mathbf{P}) = [\mathbf{Y} - \mathbf{T}(\mathbf{P})]^T \mathbf{W}^{-1}[\mathbf{Y} - \mathbf{T}(\mathbf{P})] + (\mathbf{P} - \mu)^T \mathbf{V}^{-1}(\mathbf{P} - \mu) \quad \text{in} \quad \mathbf{P}_{min} < \mathbf{P} < \mathbf{P}_{max} \quad (3.1.24)$$

Other statistical distributions that can be used for the modeling of the prior information, with different shapes and supports in $\mathbb{R}$ or in a limited interval, can be found in Note 1 of Chapter 1.

3.2   MINIMIZATION OF THE OBJECTIVE FUNCTION

This section is focused on the *MAP objective function* given by equation (3.1.21), which applies to additive Gaussian measurement errors and a Gaussian prior. It also applies to the truncated Gaussian prior in $\mathbf{P}_{min} < \mathbf{P} < \mathbf{P}_{max}$, as given by equation (3.1.24).

The gradient of the MAP objective function (3.1.21) is given by:

$$\nabla S_{MAP}(\mathbf{P}) = -2 \mathbf{J}^T \mathbf{W}^{-1}[\mathbf{Y} - \mathbf{T}(\mathbf{P})] + 2\mathbf{V}^{-1}(\mathbf{P} - \mu) \quad (3.2.1)$$

where $\mathbf{J}$ is the sensitivity matrix defined by equation (2.2.7).

A necessary condition for the minimization of (3.2.1) is that:

$$\nabla S_{MAP}(\mathbf{P}) = 0 \quad (3.2.2)$$

For *linear problems*, where the sensitivity matrix is not a function of $\mathbf{P}$ and

$$\mathbf{T}(\mathbf{P}) = \mathbf{J}\mathbf{P} \quad (3.2.3)$$

the solution of (3.2.2) gives:

$$\mathbf{P} = [\mathbf{J}^T \mathbf{W}^{-1} \mathbf{J} + \mathbf{V}^{-1}]^{-1}[\mathbf{J}^T \mathbf{W}^{-1} \mathbf{Y} + \mathbf{V}^{-1}\mu] \quad (3.2.4)$$

For nonlinear problems, the linearization of $\mathbf{T}(\mathbf{P})$ with a Taylor series expansion around the current solution $\mathbf{P}^k$ at iteration k gives:

$$\mathbf{T}(\mathbf{P}) \approx \mathbf{T}(\mathbf{P}^k) + \mathbf{J}^k(\mathbf{P} - \mathbf{P}^k) \quad (3.2.5)$$

The linearized $\mathbf{T}(\mathbf{P})$ is substituted into equation (3.2.1) and the resulting equation (3.2.2) is solved for $\mathbf{P}$, thus providing the following iterative procedure [4,17,22]:

$$\mathbf{P}^{k+1} = \mathbf{P}^k + [\mathbf{J}^T \mathbf{W}^{-1} \mathbf{J} + \mathbf{V}^{-1}]^{-1} \left\{ \mathbf{J}^T \mathbf{W}^{-1}[\mathbf{Y} - \mathbf{T}(\mathbf{P}^k)] + \mathbf{V}^{-1}(\mu - \mathbf{P}^k) \right\} \quad (3.2.6)$$

where $\mathbf{J} \equiv \mathbf{J}^k$ is the sensitivity matrix evaluated at iteration k. Equation (3.2.6) is the Gauss method for the nonlinear MAP objective function given by equation (3.1.21). The stopping criteria (2.2.15) can be used for the iterative procedure of the Gauss method.

Note in equations (3.2.4) and (3.2.6) that, with the MAP objective function (3.1.21), the conditioning of the matrix $\mathbf{J}^T\mathbf{W}^{-1}\mathbf{J}$ is improved with the matrix $\mathbf{V}^{-1}$, which is the inverse of the covariance matrix of the Gaussian prior. Therefore, the estimation of the parameters can be stabilized if prior information with small variances is available. For this reason, the matrix term $\rho^k\Omega^k$ added to $\mathbf{J}^T\mathbf{W}^{-1}\mathbf{J}$ in Technique I presented in the previous chapter (see equation 2.2.14) might not be needed in equation (3.2.6) for a successful convergence of the iterative procedure of the Gauss method. The equations presented in Chapter 2 for Technique II can be readily applied for the minimization of the objective function (3.1.21) with the gradient obtained from equation (3.2.1).

Despite the desired effect of regularizing the estimation procedure, the MAP estimator based on the minimization of the objective function (3.1.21) is *biased* and the expected value of $\mathbf{P}$ is μ [4]. Such a fact clearly shows the importance of modeling the prior information as accurately as possible for the success of the inverse analysis within the Bayesian framework.

For a linear case, the covariance matrix of the posterior Gaussian distribution is given by [4]:

$$\Phi = \left(\mathbf{J}^T\mathbf{W}^{-1}\mathbf{J} + \mathbf{V}^{-1}\right)^{-1} \tag{3.2.7}$$

which can be used as an approximation for nonlinear cases.

A comparison of equation (3.2.7) with the covariance matrix related to the maximum likelihood objective function (equation 2.7.1a) shows that the covariance of the prior is reduced by solving the inverse problem with the information provided by the measurements, if $\mathbf{J}^T\mathbf{W}^{-1}\mathbf{J}$ is well conditioned. Therefore, the solution of the inverse problem improves the information *a priori* available for the parameters, if the sensitivity coefficients are linearly independent and with large magnitudes, that is, the determinant of $\mathbf{J}^T\mathbf{J}$ is large and $\mathbf{J}^T\mathbf{W}^{-1}\mathbf{J}$ is well conditioned.

Similarly to the classical methods of parameter estimation discussed in Chapter 2, where the sensitivity coefficients directly influence the topology of the objective function based solely on the likelihood and a global minimum might not exist (see Example 2.2 and Figure 2.7), these coefficients also influence the posterior distribution of the parameters in a MAP estimator. Therefore, the sensitivity coefficients need also to be carefully examined if the solution of the inverse parameter estimation problem is to be obtained within the Bayesian framework. In classical methods based on the maximum likelihood objective function, parameters with small and/or linearly dependent sensitivity coefficients are usually deterministically fixed based on values known from previous experience and/or literature. In approaches within the Bayesian framework, uncertainties on such "known" parameters can be appropriately taken into account through their prior distribution functions. Parameters with small and/or linearly dependent sensitivity coefficients require informative prior distributions, with small variances, for the success of the estimation procedure.

Example 3.1

Consider the same physical problem of Example 2.2, involving a semi-infinite medium initially at zero temperature. For times $t>0$, the boundary surface at $x=0$ is subjected to a constant heat flux q_0 W/m². The analytical expression of the temperature at $x=0$ is given by:

$$T(0,t) = \frac{2q_0}{e}\left(\frac{t}{\pi}\right)^{1/2}$$

where $e = \sqrt{k\rho c}$ = thermal effusivity. Plot the MAP objective function given by equation (3.1.21) for $e=12{,}413$ J/(°C m²s$^{1/2}$), $q_0=10{,}000$ W/m², standard deviation of the measurements of 0.1°C, Gaussian-independent priors with means for the parameters e and q_0 given by the above values and with standard deviations of 0.1% of these means. Examine the effects of means and standard deviations of the Gaussian prior on the topology of the objective function, as well as on the parameters estimated with the Gauss method given by equation (3.2.6). Solve the parameter estimation problem with simulated measurements generated with $e = 12{,}413$ J/(°C m^{2}s$^{1/2}$), $q_0 = 10{,}000$ W/m^2 and standard deviation of 0.1°C.

Solution: The parameters q_0 and e are perfectly correlated and the maximum likelihood objective function contains a valley with infinite minima, as shown by Figures 2.7a and b. Therefore, the solution of the inverse problem of estimating q_0 and e is not possible, unless additional information is provided for one of these parameters. Within the Bayesian framework, additional information for the parameters is accounted for by the prior information.

In this example, we consider first a case where the prior means coincide with the exact values of q_0 and e, and the prior standard deviations are 0.1% of these means. In practice, this prior corresponds to a case where the analyst solving the inverse problem a *priori* knows extremely well the values of the parameters q_0 and e. Furthermore, the analyst also has great confidence on the known means, since the standard deviations of the independent Gaussian priors are quite small.

Figure 3.1 shows the contour plot of the objective function (3.1.21) for this case. Differently from the contour plot shown in Figure 2.7b for the maximum likelihood objective function, Figure 3.1 reveals that the MAP objective function has a minimum point at the exact parameter values used to generate the simulated measurements, that is, $e = 12{,}413$ J/(°C m^{2}s$^{1/2}$) and $q_0 = 10{,}000$ W/m^2. Although the ellipses around this minimum value are elongated due to the linear dependence between the parameters, the point of minimum is well defined. In fact, by starting the iterative procedure of the Gauss method given by equation (3.2.6) with initial guesses five times larger than the exact parameters, it converges in only three iterations to $q_0 = 10{,}001$ W/m^2 and $e = 12{,}412$ J/(°C m^{2}s$^{1/2}$), with the following respective uncertainties at the 99% confidence level: ±18 W/m^2 and ±23 J/(°C m^{2}s$^{1/2}$). Although the confidence region is not presented here, it is an ellipse centered at the estimated values and with shape similar to those of the ellipses presented in Figure 3.1.

The case analyzed earlier is very idealistic, since the prior means are very well known and great confidence is attributed to these values. Consider now that, before running the experiment and solving the inverse problem, the analyst is informed by mistake that the parameter values are $e = 24{,}826$ J/(°C m^{2}s$^{1/2}$) and $q_0 = 20{,}000$ W/m^2. The analyst then selects these values as the Gaussian

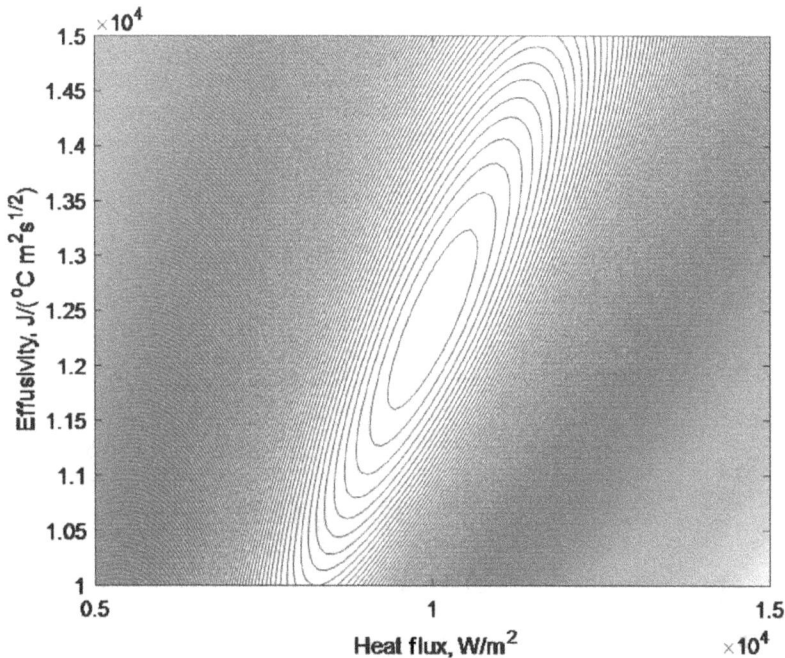

FIGURE 3.1 Contour plot of the objective function (3.1.21) for Example 3.1 with priors centered around the exact parameter values and with small standard deviations.

prior means. Being very confident about these values, the analyst attributes standard deviations of 0.1% of the means to the independent Gaussian priors. The contour plot of the MAP objective function with this prior is presented in Figure 3.2. Although a point of minimum can be identified for the objective function (3.1.21), this point is far from the exact parameter values and centered around the prior means because the MAP estimator is biased. In this case, the minimization of the MAP objective function with the Gauss method results in $e = 24{,}832 \pm 47$ J/(°C m²s$^{1/2}$) and $q_0 = 19{,}996 \pm 38$ W/m². Therefore, the prior must be carefully selected in order to appropriately represent the previous information about the parameters. If there is large uncertainty about the prior information, it must be represented with large variances, differently from the case shown in figure 3.2.

It is also instructive to analyze two other cases of practical interest for this physical problem, namely:

i. Calibration of the apparatus used for the experiment. The objective of the calibration process is to estimate the boundary heat flux q_0 using a reference material with effusivity accurately known. For example, let's assume that the effusivity of the reference material has been *a priori* measured in a metrology institute with a very accurate standard experimental technique.

ii. Thermal characterization of materials with the heat flux obtained in the calibration process. Thus, the heat flux is now accurately known from the calibration process and the inverse problem is aimed at the measurement of the unknown effusivity.

The calibration process is first considered by using as mean and standard deviation for the effusivity 12,413 J/(°C m²s$^{1/2}$) and 1% of the mean, respectively. In order to reflect the lack of knowledge about the heat flux, its prior mean was set as 20,000 W/m² and the prior standard deviation as 20% of this value. Although the minimum of the objective function is not well defined due to the linear dependence of the parameters, as shown by Figure 3.3, the Gauss method converged to $e = 12{,}421 \pm 320$ J/(°C m²s$^{1/2}$) and $q_0 = 10{,}007 \pm 258$ W/m², in three iterations, by starting the iterative procedure with an initial guess five times larger than the exact values. Thus, the means

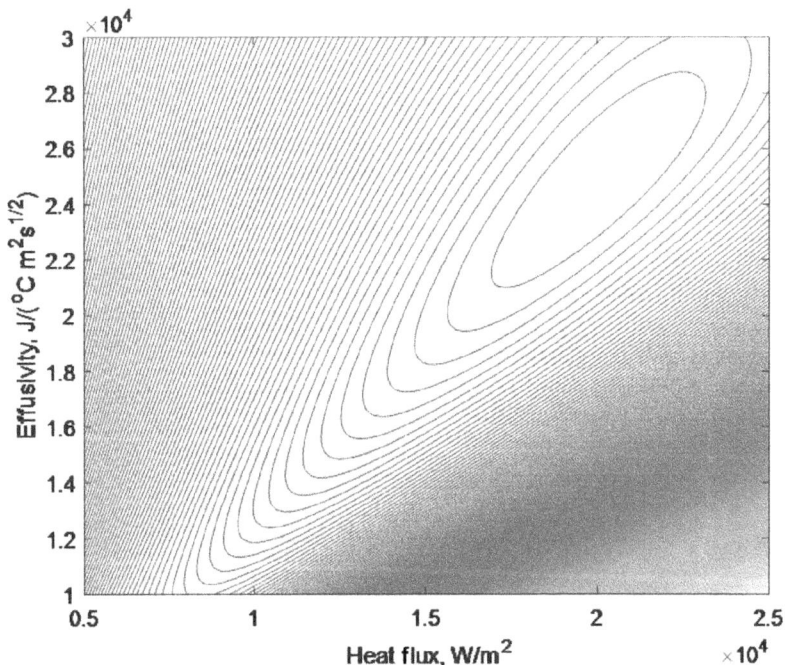

FIGURE 3.2 Contour plot of the objective function (3.1.21) for Example 3.1 with priors centered off the exact parameter values and with small standard deviations.

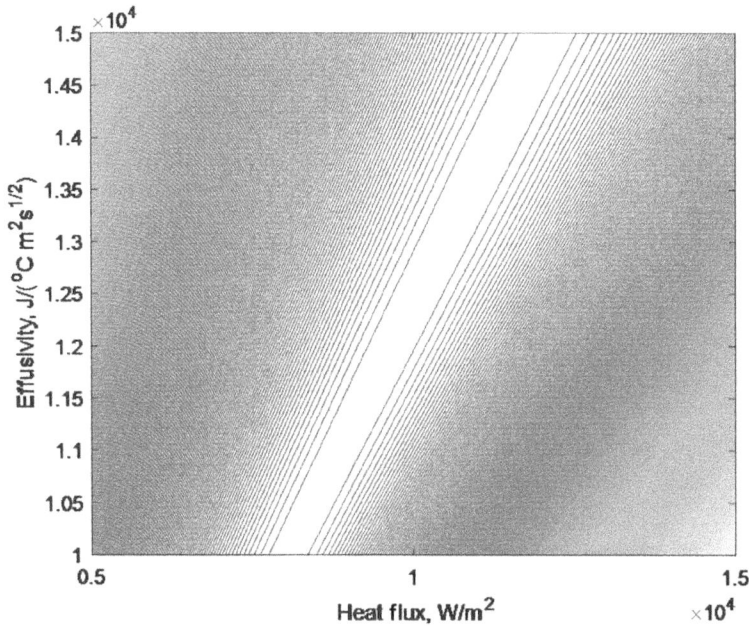

FIGURE 3.3 Contour plot of the objective function (3.1.21) for Example 3.1 in the calibration problem.

of the posterior are in good agreement with the exact parameter values, but the associated 99% confidence intervals are large because the parameters are linearly dependent.

A contour plot similar to that shown in Figure 3.3 was also obtained for the thermal characterization of the material by considering a prior with mean and standard deviation of 10,007 and 100 W/m² , respectively, for the heat flux. The prior for the effusivity was assumed with mean 24,832 J/(°C m²s$^{1/2}$) and standard deviation of 20% of this mean. The estimated parameters obtained with the Gauss method in this case were $e = 12{,}418 \pm 320$ J/(°C m²s$^{1/2}$) and $q_0 = 10{,}006 \pm 258$ W/m².

3.3 IDENTIFICATION OF THE THERMOPHYSICAL PROPERTIES OF SEMI-TRANSPARENT MATERIALS

In this section, we solve the inverse problem of identifying the thermophysical properties of semi-transparent materials in a coupled radiation-conduction three-dimensional problem [94]. Both the conduction and the radiation direct problems are solved with finite volumes. The sensitivity coefficients and the determinant of the information matrix are examined for the experimental design. Parameters are estimated with the Levenberg-Marquardt method (Technique I) of minimization of the least-squares norm (equation 2.1.6b) and with the Gauss method of minimization of the MAP objective function (equation 3.1.21).

The physical problem considered here is similar to the Flash method [95,96]. A thin sample of a material is heated through its front surface by a laser, while the temperature of its back surface is measured with an infrared camera. A furnace is used to heat the sample to high temperatures. After the sample reaches thermal equilibrium with the furnace, it is heated by a laser pulse. The physical problem is shown by Figure 3.4a [97].

THE DIRECT PROBLEM

The formulation for the coupled conduction-radiation problem in Cartesian coordinates is given for a sample in the form of a parallelepiped with sides $2a^*$, $2b^*$ and c^*. The laser heating is supposed to

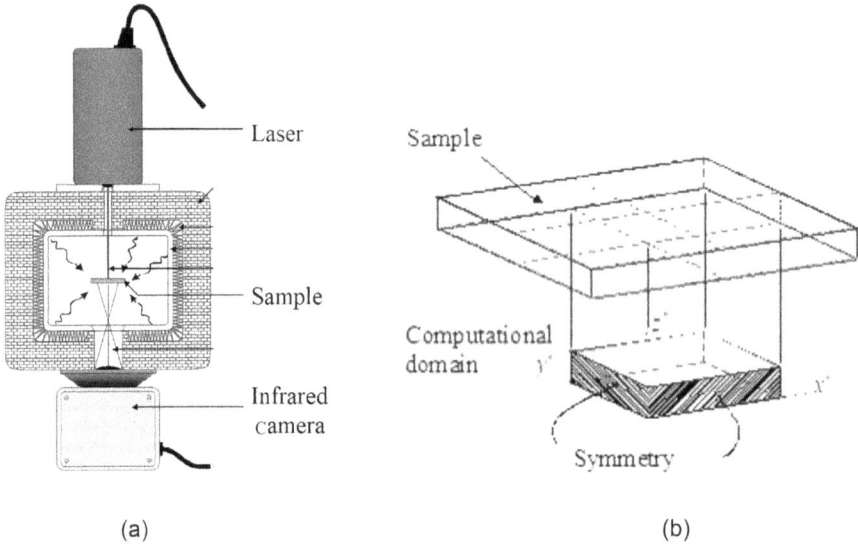

FIGURE 3.4   (a) Physical problem and (b) computational domain. (From Ref. [97].)

be imposed at the center of the sample with a profile that is symmetric with respect to the sample mid-planes (see Figure 3.4b) [97]. The sample is supposed to be initially at the uniform temperature $T^* = T_0^*$ and, for $t^* > 0$, it is heated through its top surface, while it loses heat by radiation and convection through all surfaces, with a combined heat transfer coefficient $h_{rad}^* = h^* + 4n_r^2 \sigma T_0^{*3}$. The solid is assumed to be orthotropic with thermal conductivities k_x^*, k_y^* and k_z^*, along the x^*, y^* and z^* directions, respectively. In addition, the sample is assumed to be a grey-body. The superscript "*" denotes dimensional variables. As traditionally used in the Flash method, the sample is assumed to be coated with graphite paint in order to increase the energy absorbed/emitted. The boundaries are considered opaque [97].

The equation of radiative transfer is used to model the radiation propagation within the sample [98]. It can be written in dimensionless form as:

$$\xi \frac{\partial I^l}{\partial x} + \eta \frac{\partial I^l}{\partial y} + \mu \frac{\partial I^l}{\partial z} = -(\kappa_a + \sigma_s)I^l + S^l \quad \text{in } 0 < x < a, 0 < y < b, 0 < z < c \quad (3.3.1a)$$

where

$$S^l = \kappa_a n_r^2 I_b(T) + \frac{\sigma_s}{4\pi} \int_{\Omega'=4\pi} I^{l'} p(\vec{s}' \to \vec{s}) d\Omega' \quad (3.3.1b)$$

In equation (3.3.1a), ξ, η and μ are the cosine directors along the x, y and z directions, respectively, κ_a is the absorption coefficient and σ_s is the scattering coefficient. In the source term given by equation (3.3.1b), $I_b(T)$ is the blackbody dimensionless intensity and $p(\vec{s}' - \vec{s})$ is the scattering phase-function. Equation (3.3.1a) is subjected to the following boundary conditions:

$$I(\xi,\eta,\mu) = I(-\xi,\eta,\mu) \quad \text{at } \Gamma_1 : \begin{cases} x = 0 \\ 0 < y < b \\ 0 < z < c \end{cases} \quad (3.3.1c)$$

$$I(-\xi,\eta,\mu) = \varepsilon n_r^2 I_b + \frac{1-\varepsilon}{\pi} \int_{\xi'>0} I(\xi',\eta',\mu') \; \xi' d\Omega' \quad \text{at} \quad \Gamma_2 : \begin{cases} x = a \\ 0 < y < b \\ 0 < z < c \end{cases} \tag{3.3.1d}$$

$$I(\xi,\eta,\mu) = I(\xi,-\eta,\mu) \quad \text{at} \quad \Gamma_3 : \begin{cases} 0 < x < a \\ y = 0 \\ 0 < z < c \end{cases} \tag{3.3.1e}$$

$$I(\xi,-\eta,\mu) = \varepsilon n_r^2 I_b + \frac{1-\varepsilon}{\pi} \int_{\eta'>0} I(\xi',\eta',\mu') \; \eta' \; d\Omega' \quad \text{at} \quad \Gamma_4 : \begin{cases} 0 < x < a \\ y = b \\ 0 < z < c \end{cases} \tag{3.3.1f}$$

$$I(\xi,\eta,\mu) = \varepsilon n_r^2 I_b + \frac{1-\varepsilon}{\pi} \int_{\mu'<0} I(\xi',\eta',\mu') \; \mu' \; d\Omega' \quad \text{at} \quad \Gamma_5 : \begin{cases} 0 < x < a \\ 0 < y < b \\ z = 0 \end{cases} \tag{3.3.1g}$$

$$I(\xi,\eta,-\mu) = \varepsilon n_r^2 I_b + \frac{1-\varepsilon}{\pi} \int_{\mu'>0} I(\xi',\eta',\mu') \; \mu' \; d\Omega' \quad \text{at} \quad \Gamma_6 : \begin{cases} 0 < x < a \\ 0 < y < b \\ z = c \end{cases} \tag{3.3.1h}$$

The energy conservation equation can be written in dimensionless form as:

$$C\frac{\partial T}{\partial t} = \frac{\partial}{\partial x}\left(k_x \frac{\partial T}{\partial x}\right) + \frac{\partial}{\partial y}\left(k_y \frac{\partial T}{\partial y}\right) + \frac{\partial}{\partial z}\left(k_z \frac{\partial T}{\partial z}\right) - \nabla \cdot q^{\text{rad}}$$

$$\text{in } 0 < x < a, 0 < y < b, 0 < z < c, \text{ for } t > 0 \tag{3.3.2a}$$

where the divergence of the radiative flux is:

$$\nabla \cdot q^{\text{rad}} = \frac{\kappa_a \tau_0}{N_{pl}}\left[4\pi n_r^2 I_b - \int_{\Omega=4\pi} I^l \, d\Omega\right] \tag{3.3.2b}$$

Equation (3.3.2a) is subjected to the following boundary conditions:

$$\frac{\partial T}{\partial x} = 0 \quad \text{at} \quad \Gamma_1 \quad \text{for} \quad t > 0 \tag{3.3.2c}$$

$$k_x \frac{\partial T}{\partial x} + Bi^{\text{rad}} T = \frac{\varepsilon \tau_0}{N_{pl}}\left[\int_{\xi>0} I^l \xi \cdot \leq d\Omega - n_r^2 \pi I_b\right] + Bi^{\text{rad}} T_\infty \quad \text{at} \quad \Gamma_2 \quad \text{for} \quad t > 0 \tag{3.3.2d}$$

$$\frac{\partial T}{\partial y} = 0 \quad \text{at} \quad \Gamma_3 \quad \text{for} \quad t > 0 \tag{3.3.2e}$$

$$k_y \frac{\partial T}{\partial y} + Bi^{\text{rad}} T = \frac{\varepsilon \tau_0}{N_{pl}} \left[\int_{\eta>0} I^l \eta \cdot d\Omega - n_r^2 \pi I_b \right] + Bi^{\text{rad}} T_{\infty} \quad \text{at } \Gamma_4 \text{ for } t > 0 \qquad (3.3.2\text{f})$$

$$-k_z \frac{\partial T}{\partial z} + Bi^{\text{rad}} T = \frac{\varepsilon \tau_0}{N_{pl}} \left[\int_{\mu<0} I^l \mu \cdot d\Omega - n_r^2 \pi I_b \right] + Bi^{\text{rad}} T_{\infty} \quad \text{at } \Gamma_5 \text{ for } t > 0 \qquad (3.3.2\text{g})$$

$$k_z \frac{\partial T}{\partial z} + Bi^{\text{rad}} T = \frac{\varepsilon \tau_0}{N_{pl}} \left[\int_{\mu>0} I^l \mu \cdot d\Omega - n_r^2 \pi I_b \right] + Bi^{\text{rad}} T_{\infty} + \varepsilon_{10.6\,\mu\text{m}} q_{\text{laser}} (x,y,t) \quad \text{at } \Gamma_6 \text{ for } t > 0$$

$$\qquad (3.3.2\text{h})$$

with initial condition:

$$T = 0 \text{ in } 0 < x < a, 0 < y < b, 0 < z < c, \text{ for } t = 0 \qquad (3.3.2\text{i})$$

For the solution of the coupled conduction-radiation problem given by equations (3.3.1a-h) and (3.3.2a-i), we use the finite volume method [86,97]. The computational program developed for the solution of the direct problem was verified by using several benchmark results available in the literature dealing with different physical situations, such as purely conductive, purely radiative, isotropically and highly directional scattering problems [97].

THE INVERSE PROBLEM

The *inverse problem* of interest here is concerned with the estimation of the following vector of unknown parameters [94]:

$$\mathbf{P}^T = \left[k_x, k_y, k_z, C, Bi^{\text{rad}} \right] \qquad (3.3.3)$$

by using transient temperature measurements taken at the non-heated surface Γ_5 at $z = 0$.

For the solution of the present parameter estimation problem we used minimization techniques presented in Chapter 2 and this chapter. Objective functions were defined involving the differences between measured and estimated temperatures by considering that prior information was either available or not available for the parameters.

The errors in the measured variables were supposed additive, uncorrelated, Gaussian, with zero mean and known constant standard deviation. Thus, in the case where no prior information was assumed available for the parameters, the ordinary least squares norm (equation 2.1.6b) became a minimum variance estimator, as described in Chapter 2. The iterative procedure of the *Levenberg-Marquardt method* (Technique I) given by equation (2.2.14) was used for the minimization of the ordinary least squares in this case. The parameter estimation problem was also solved with independent Gaussian priors for the unknown parameters, with mean μ and covariance matrix $\mathbf{V}$, by minimizing the maximum a posteriori objective function given by equation (3.1.21), with the iterative procedure of the Gauss method (equation 3.2.6).

The Levenberg-Marquardt and the Gauss methods required that the solution of the direct problem be computed several times during the minimization procedure, which was very time consuming, in particular for the present three-dimensional nonlinear conduction-radiation coupled problem. Therefore, in the present work we examined the use of a reduced model for the computation of the sensitivity matrix, in order to reduce the computational cost of the inverse problem solution. The reduced model considered the divergence of the radiative flux in an approximate form in the energy conservation equation (3.3.2a). More specifically, the divergence of the radiative flux was computed only once, with the current estimate for the unknown parameters. Hence, the conduction-radiation problem was decoupled.

In order to generate the simulated measurements, the following physical data were utilized: $C^*=2.5\times10^6\,\mathrm{J/m^3\,K}$, $k_x^*=5\,\mathrm{W/m\,K}$, $k_y^*=5\,\mathrm{W/m\,K}$, $k_z^*=5\,\mathrm{W/m\,K}$, $\kappa_a^*=10\,\mathrm{m^{-1}}$, $\sigma_s^*=10^4\,\mathrm{m^{-1}}$ and $g=0.7$, where g is the asymmetry factor of the Henyey-Greenstein forward phase function. The sample was assumed to be a parallelepiped with dimensions $2a^* = 2b^*=0.01\,\mathrm{m}$ and $c^* = 0.001\,\mathrm{m}$, with a black coating $(\varepsilon_w = 1)$ and heated by a laser with a power of 0.25 W and a Gaussian profile. For the heat flux imposed by the laser, 99% of its power was assumed to be delivered within a circle with radius of 2 mm centered on the sample. The sample was assumed to be initially at the uniform temperature $T_0^*=1800\mathrm{K}$, which was the same temperature of the surrounding environment, that is, $T_\infty^*=1800\mathrm{K}$. The combined heat transfer coefficient, which takes into account convective $(h^*=50\,\mathrm{W/m^2\,K})$ and radiative heat losses, was evaluated as $h_{\mathrm{rad}}^*=1373\,\mathrm{W/m^2\,K}$.

ANALYSIS OF THE SENSITIVITY COEFFICIENTS AND DESIGN OF OPTIMUM EXPERIMENTS

Figures 3.5a-c present the transient variations of the dimensionless temperature and of the reduced sensitivity coefficients at the following (x^*,y^*,z^*) positions: A=(0, 0, 0) m; B=(0.002, 0, 0) m and C=(0, 0.002, 0) m, respectively. We note in these figures that, at the three locations examined, the reduced sensitivity coefficients were of the same order of magnitude of the temperature. Therefore, the temperature measurements were relatively sensitive to variations in the unknown parameters. In Figure 3.5a, where the measurement point was located below the center of the applied heat flux, the sensitivity coefficients with respect to the transversal conductivity components (k_x and k_y) were identical. Therefore, temperature measurements taken off the center of the sample were required for the estimation of such conductivity components. At this location, the sensitivity coefficient with respect to the heat transfer coefficient (Bi^{rad}) was linearly dependent with respect to those for the transversal conductivity components, while the sensitivity coefficient with respect to the volumetric heat capacity (C) was not linearly dependent to the others. The largest magnitude of the sensitivity coefficient for the longitudinal conductivity component (k_z) took place at location A (see also Figures 3.5b and c). Figures 3.5b and c show that the sensitivity coefficients with respect to k_x and k_y at location B are identical to those for k_y and k_x at location C, respectively. Such was the case because locations B and C are 0.002 m off the origin on the x and y axes, respectively, and the transversal conductivity components are identical. Despite the fact that some of the sensitivity coefficients were linearly dependent at each position, the constant of proportionality between each pair was different at different positions. Therefore, if the measurements of more than one sensor are used for the inverse analysis, the columns of the sensitivity matrix are not linearly dependent and all the parameters can be simultaneously estimated.

For the results presented below, we assumed that the temperature measurements were taken with an infrared camera, so that each recorded image permitted the measurements over a grid fixed by the camera resolution. We now examine the determinant of $\mathbf{J}^T\mathbf{J}$, aiming at the optimum design of the experiment with respect to the heating and final times and with respect to the number of transient measurements (camera images). Figure 3.6a presents $|\mathbf{J}^T\mathbf{J}|$ for heating times of $t_{\mathrm{laser}}=20, 40, 60, 80$ and 100, in dimensionless form. For this figure, we assumed that the camera was capable of recording 500 images during the duration of the experiment with a spatial resolution of 64×64 pixels. We notice in Figure 3.6a, a very small increase in $|\mathbf{J}^T\mathbf{J}|$ at the moment that the heating was stopped. Also, such a determinant became constant afterward and smaller than that obtained with continuous heating. Figure 3.6a shows that continuous heating should be used for the experimental design in order to maximize $|\mathbf{J}^T\mathbf{J}|$ and to reduce the confidence region of the estimated parameters. However, care must be taken regarding the maximum temperature variation within the sample, which should not exceed the limit for the assumption of constant properties. The effects of the number of transient measurements (camera images) on $|\mathbf{J}^T\mathbf{J}|$ are presented in Figure 3.6b for continuous heating. This figure shows that $|\mathbf{J}^T\mathbf{J}|$ can be substantially increased by increasing the number of images (or the frequency of measurements). Therefore, the maximum measurement frequency allowing uncorrelated measurements should be used.

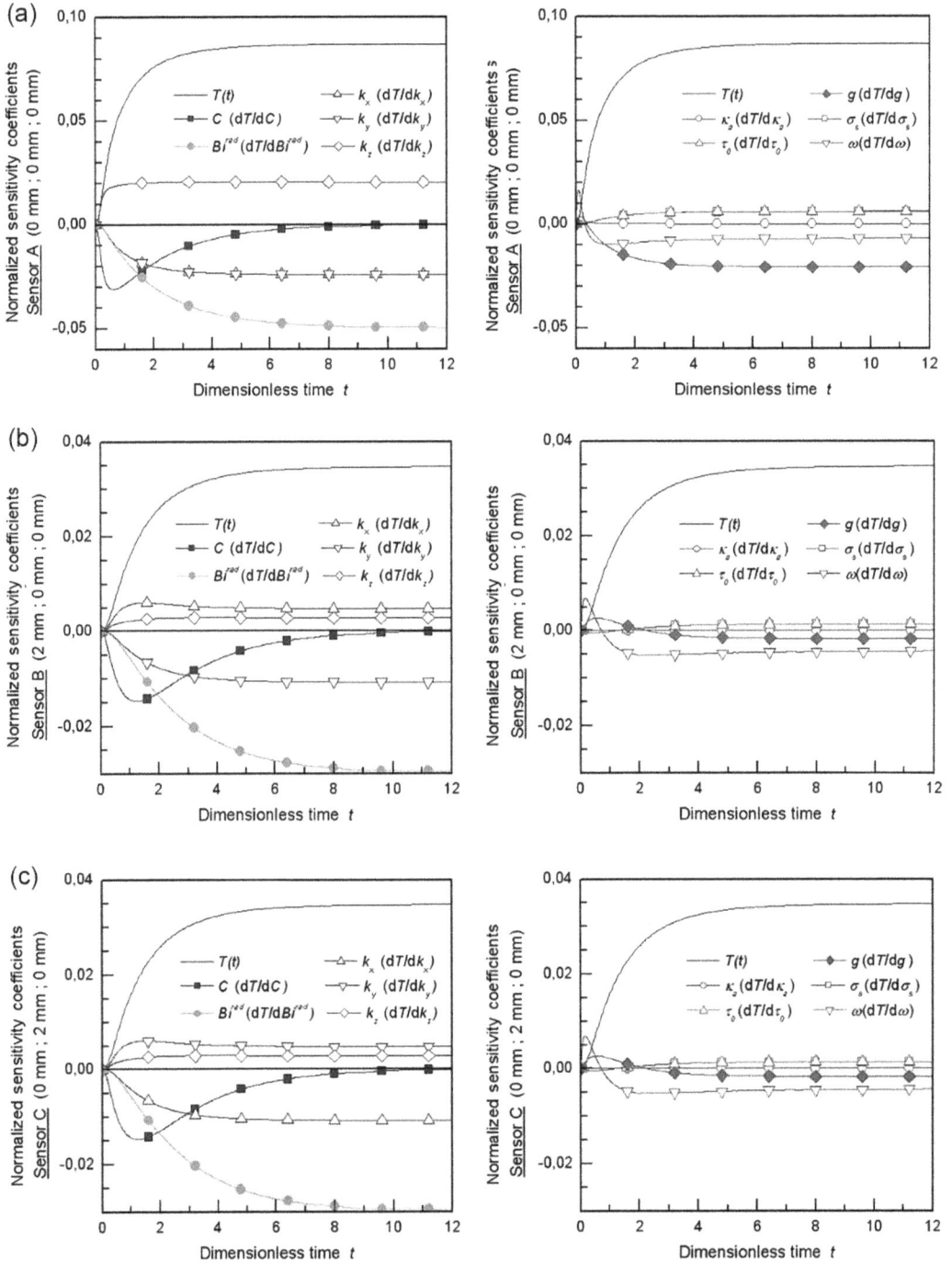

FIGURE 3.5 Transient variation of temperature and reduced sensitivity coefficients at: (a) position A: (0, 0, 0) m; (b) position B: (0.002, 0, 0) m; (c) position C: (0, 0.002, 0) m. (From Ref. [94].)

PARAMETER ESTIMATION AND STATISTICAL ANALYSIS

We now consider the simultaneous estimation of the five unknown parameters by using transient measurements of the temperature at the surface opposite to the heating imposed by the laser. The duration of the experiment was taken as 20 s, which corresponded to a dimensionless time of 40.

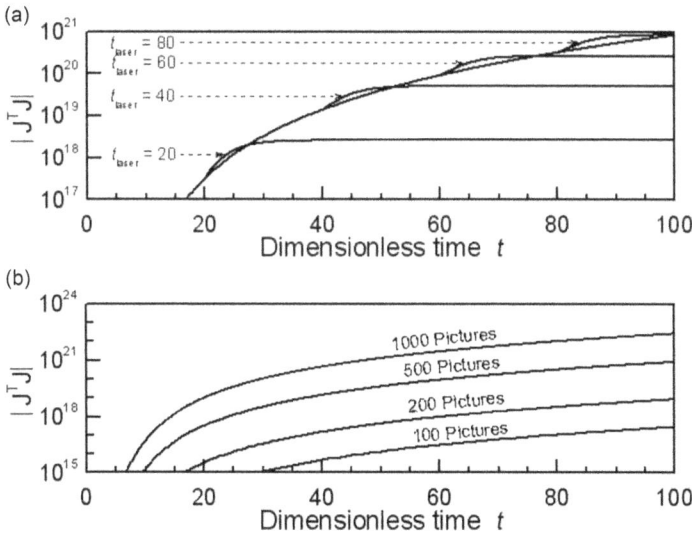

FIGURE 3.6   Effects on $\left| \mathbf{J}^T \mathbf{J} \right|$ of: (a) heating time and (b) number of images. (From Ref. [94].)

TABLE 3.1

Cases Examined

Case	Objective Function	Method	Model for the Direct Problem	Model for the Gradient
1	Least squares	Levenberg-Marquardt	Complete	Reduced
2			Complete	Complete
3	Maximum a Posteriori	Gauss	Complete	Reduced
4			Complete	Complete

Source: From Ref. [94].

With the prescribed heating time the maximum temperature increase in the sample was 5 K. The measurement frequency was considered as 50 Hz, so that 1000 images were taken during the duration of the experiment. The spatial resolution assumed for the camera was a grid of 64×64 pixels. For the results presented below, the measurements were assumed to contain additive, uncorrelated, Gaussian random errors, with zero mean and constant standard deviation of 0.8 K.

Table 3.1 summarizes the four cases compared for the estimation of $\mathbf{P}^T = [k_x, k_y, k_z, C, Bi^{\text{rad}}]$. These cases involved the methods described earlier, as well as the complete and reduced models for the computation of the sensitivity matrix. The initial guesses used for the unknown parameters were: $C^{*0} = 2.8 \times 10^6 \, \text{J/m}^3 \, \text{K}$, $k_x^{*0} = k_y^{*0} = k_z^{*0} = 8 \, \text{W/m K}$ and $h_{rad}^* = 800 \, \text{W/m}^2 \, \text{K}$. For the Gaussian-independent priors, the means were taken equal to these initial values. The variances were calculated so that the widths of the 99% confidence intervals were the double of the initial guesses used for the parameters.

Table 3.2 presents a comparison of the results obtained for the simultaneous estimation of the volumetric heat capacity, the three thermal conductivity components and the heat transfer coefficient, in dimensional form. In addition to the estimates obtained for the parameters (means and 99% confidence intervals), Table 3.2 also shows the numbers of iterations for each case. This table reveals that quite accurate estimates were obtained for the unknown parameters for all cases. A comparison of the results obtained with cases 1 and 2 shows no significant changes in the number of iterations and no change on the accuracy of the estimates, when the reduced model was used for

TABLE 3.2
Results Obtained

				Estimates		
Case	Number of Iterations	$C^* \times 10^{-6}$ (J/m³ K)	k_x^* (W/m K)	k_y^* (W/m K)	k_z^* (W/m K)	h_{rad}^* (W/m² K)
1	16	2.51 ± 0.03	4.99 ± 0.07	5.01 ± 0.07	5.0 ± 0.2	1373 ± 5
2	16	2.51 ± 0.03	5.00 ± 0.07	5.01 ± 0.07	5.0 ± 0.2	1373 ± 5
3	13	2.51 ± 0.03	5.00 ± 0.07	5.01 ± 0.07	5.0 ± 0.2	1373 ± 5
4	6	2.51 ± 0.03	5.00 ± 0.07	5.01 ± 0.07	5.0 ± 0.2	1373 ± 5

Source: From Ref. [94].

the calculation of the sensitivity matrix for the Levenberg-Marquardt method. On the other hand, the use of the reduced model to calculate the sensitivity matrix for the Gauss method increased the number of iterations required for convergence and, consequently, resulted in larger computational times. Therefore, for the present inverse problem, the use of the reduced model for the sensitivity matrix did not cause loss of accuracy on the estimated parameters, but increased the computational time due to the larger number of iterations required for convergence, because of the low precision of the sensitivity coefficients.

Reduced models for the solution of inverse problems, not only for the computation of the sensitivity matrix like in this example but also for the direct problem solution, must be used with care. Of course, large errors between reduced and complete models may significantly affect the inverse problem solution. However, the model errors can be formally accommodated in the inverse analysis within the Bayesian framework of statistics, as it will be apparent in the next chapter.

PROBLEMS

3.1 Derive equation (3.2.1).

3.2 Derive the iterative procedure of the Gauss method given by equation (3.2.6).

3.3 A semi-infinite medium initially at the zero temperature has the temperature at the surface $x=0$ suddenly changed to a constant value T_0. The formulation of such heat conduction problem is given by:

$$\frac{1}{\alpha} \frac{\partial T}{\partial t} = \frac{\partial^2 T}{\partial x^2} \text{ for } x > 0 \text{ and } t > 0$$

$$T = T_0 \quad \text{at } x = 0 \text{ for } t > 0$$

$$T = 0 \quad \text{for } t = 0 \text{ and } x > 0$$

Examine the transient variation of the sensitivity coefficients with respect to the thermal diffusivity α and to the initial temperature T_0 for sensors located at different depths below the surface. Is the simultaneous estimation of α and T_0 possible? What is the behavior of $|\mathbf{J}^T\mathbf{J}|$?

3.4 Solve the inverse problem of estimating α and T_0 with Gaussian independent priors by applying the Gauss method given by equation (3.2.6). Examine the effects of the prior means and standard deviations on the inverse problem solution. Use simulated measurements in the inverse analysis, which contain additive, uncorrelated, Gaussian errors with zero mean and constant standard deviation. The standard deviation is 0.2°C. Use $T_0 = 20°C$ and $\alpha = 10^{-6} m^2/s$ to generate the simulated measurements.

3.5 Consider the following heat conduction problem in dimensionless form:

$$\frac{\partial T}{\partial t} = \frac{\partial^2 T}{\partial x^2} \text{ in } 0 < x < 1 \text{ for } t > 0$$

$$\frac{\partial T}{\partial x} = 0 \text{ at } x = 0 \text{ for } t > 0$$

$$\frac{\partial T}{\partial x} = q^* \text{ at } x = 1 \text{ for } t > 0$$

$$T = 0 \text{ for } t = 0 \text{ in } 0 < x < 1$$

Consider available the transient readings Y_i, $i=1, \ldots, I$ of a single sensor located at $x_{meas} = 0.5$ for the estimation of q^*. Generate simulated measurements by using $q^* = 1$. The simulated measurements contain additive, uncorrelated, Gaussian errors with zero mean and constant standard deviation of 1% of the maximum measured temperature. Solve the inverse problem with a Gaussian prior by applying the Gauss method given by equation (3.2.6). Examine the effects of the prior mean and standard deviation on the inverse problem solution.

4 Parameter Estimation

Stochastic Simulation with Prior Information about the Unknown Parameters

The Gaussian likelihood and the priors examined in Chapter 3 resulted in posterior distributions from which a *maximum a posteriori* point estimate could be obtained for the parameters, provided that the minimum of the objective functions existed. For a Gaussian (or a truncated Gaussian) prior and a Gaussian likelihood, the prior is *conjugate* to the likelihood in a linear problem [17,59,63,67]. For a likelihood function, a class Ψ of prior distributions is said to form a *conjugate family* if the posterior density is in the same class Ψ for all **P**, whenever the prior density is in Ψ [63]. Although this property is valid for many cases that involve continuous distributions, in special those that belong to the exponential family [59,63], the posterior probability distribution may not be analytical if non-conjugate prior probability densities are assumed for the parameters.

We also note that the computation of the *maximum a posteriori* point estimate presented in Chapter 3 involved an optimization problem, that is,

$$\mathbf{P}_{MAP} = \arg\max_{\mathbf{P} \in \mathbb{R}^N} \pi(\mathbf{P} \mid \mathbf{Y})$$

which was reformulated as the minimization of a *maximum a posteriori objective function*. For Gaussian likelihood and Gaussian prior, this objective function is given by equation (3.1.21) and, for the linear case, the resulting posterior distribution, $\pi(\mathbf{P} \mid \mathbf{Y})$, is also Gaussian, with mean $\mathbf{\mu}$ and covariance matrix $\mathbf{\Phi} = (\mathbf{J}^T \mathbf{W}^{-1} \mathbf{J} + \mathbf{V}^{-1})^{-1}$ (see equation 3.2.7).

The solution of the inverse problem is expected to be reported not only in terms of a point estimate (the *mode* of the posterior distribution in the case of the maximum a posteriori estimate), but also of other statistics of the posterior distribution that permit the assessment of the related uncertainties. In general, statistics of the posterior distribution typically require numerical integration, such as the *conditional mean* defined by [17,59]:

$$\mathbf{P}_{CM} = E(\mathbf{P}) = \int_{\mathbb{R}^N} \mathbf{P}\pi(\mathbf{P} \mid \mathbf{Y})d\mathbf{P}$$

where $E(.)$ denotes the expected value (see Note 1 in Chapter 1) and the posterior distribution is given by Bayes' theorem, that is,

$$\pi(\mathbf{P}|\mathbf{Y}) = \frac{\pi(\mathbf{Y}|\mathbf{P})\pi(\mathbf{P})}{\pi(\mathbf{Y})}$$

The integration in $\mathbf{P}_{CM}$ is challenging because the normalizing constant, $\pi(\mathbf{Y})$, in the denominator of $\pi(\mathbf{P}|\mathbf{Y})$ is not explicitly known for most practical cases. Actually, the calculation of $\pi(\mathbf{Y})$ from its definition

$$\pi(\mathbf{Y}) = \int_{\mathbb{R}^N} \pi(\mathbf{Y}|\mathbf{P})\pi(\mathbf{P})d\mathbf{P}$$

includes, itself, a numerical integration in the parameter space with dimension N. Inverse problems may involve parameter spaces with large dimensions or prior distributions with unbounded variances. As a consequence, the above numerical integrations might not feasible.

For those cases that the posterior does not allow an analytical treatment or numerical integrations required for estimates are not possible, *stochastic simulation* with *Markov Chain Monte Carlo (MCMC) methods* can provide a solution for the inverse problem. In this approach, samples of the posterior distribution are generated with stochastic simulation and inference on the posterior probability is performed through inference on these samples [17–22,48,49,59,60,63,67]. For example, the conditional mean can be approximately calculated by Monte Carlo integration with:

$$\mathbf{P}_{CM} = \int_{\mathbb{R}^N} \mathbf{P}\,\pi(\mathbf{P}|\mathbf{Y})\,d\mathbf{P} \approx \frac{1}{n}\sum_{t=1}^{n}\mathbf{P}^{(t)}$$

where $\mathbf{P}^{(t)}$, for $t = 1,\ldots, n$, are samples of $\pi(\mathbf{P}|\mathbf{Y})$ obtained with the stochastic simulation.

Due to their inherent simplicity, MCMC methods have recently become quite popular for the solution of inverse problems, being applied even for cases where a *maximum a posteriori* estimate would be possible. One clear disadvantage on the application of Monte Carlo methods is the large required computational time. On the other hand, the use of computationally fast reduced or surrogate models for the physical problem can be formally accommodated within the Bayesian framework, so that the application of MCMC methods to many practical inverse problems is nowadays possible.

The MCMC method will be referred to in this book as **Technique IV**. The steps for the implementation of this method can be summarized as follows:

- Direct problem
- Inverse problem
- Analysis of the sensitivity coefficients and design of optimum experiments
- Stochastic simulation
- Analysis of the Markov chains

This chapter is focused on the last two steps. Concepts and properties of Markov chains are presented in the next section, which is followed by a powerful, simple and popular implementation of stochastic simulation via MCMC methods, namely, the *Metropolis-Hastings algorithm* [17–22,48,49,59,60,63,67,99,100]. Some practical aspects and speedup techniques for the implementation of MCMC methods are addressed in this chapter, which also includes elements for the analysis of the Markov chains. Deeper details about MCMC methods can be found in the book by Gamerman and Lopes [59] and in the book edited by Brooks et al. [60].

4.1 MARKOV CHAINS

A stochastic process is a collection of random variables, say $\{\mathbf{P}^{(t)} : t \in T\}$, which is indexed by a set T and where each variable in the collection is uniquely associated with an element in the indexing set. For the solution of inverse problems, the state space S of these random variables is a subset of $\mathbb{R}^N$, that is, the support of the parameter vector, while T is the set of natural numbers [59].

A Markov chain is a stochastic process that satisfies the following condition [59]:

$$q\left(\mathbf{P}^{(t+1)} = \mathbf{y} \mid \mathbf{P}^{(t)} = \mathbf{x}, \mathbf{P}^{(t-1)} = \mathbf{x}^{(t-1)}, \ldots, \mathbf{P}^{(0)} = \mathbf{x}^{(0)}\right) = q\left(\mathbf{P}^{(t+1)} = \mathbf{y} \mid \mathbf{P}^{(t)} = \mathbf{x}\right)$$

$$\text{for all } \mathbf{y}, \mathbf{x}, \mathbf{x}^{(t-1)}, \ldots, \mathbf{x}^{(0)} \in S$$

(4.1.1)

where $q(\mathbf{y} \mid \mathbf{x})$ is a transition probability from $\mathbf{x}$ to $\mathbf{y}$. Therefore, in a Markov chain, given the present state, $\mathbf{P}^{(t)} = \mathbf{x}$, the next state $\left(\mathbf{P}^{(t+1)} = \mathbf{y}\right)$ is independent of the previous states $\left(\mathbf{P}^{(t-1)} = \mathbf{x}^{(t-1)}, \ldots, \mathbf{P}^{(0)} = \mathbf{x}^{(0)}\right)$.

Some concepts regarding Markov chains are now briefly presented. The reader shall consult references [17–22,48,49,59,60,63,67] for further details on Markov chains.

If the transition probability does not depend on t, that is, if

$$q\left(\mathbf{P}^{(t+m+1)} = \mathbf{y} \mid \mathbf{P}^{(t+m)} = \mathbf{x}\right) = q\left(\mathbf{P}^{(t+1)} = \mathbf{y} \mid \mathbf{P}^{(t)} = \mathbf{x}\right) \text{ for all } m \in T \qquad (4.1.2)$$

the Markov chain is said to be *homogenous*.

A distribution π^* is said to be a *stationary distribution* of a chain if, once the chain is in π^*, it stays in this distribution. Suppose now that $\pi^{(t)} \to \pi^*$ as $t \to \infty$ for any $\pi^{(0)}$, where $\pi^{(t)}$ is the distribution at state t of the chain. Then, π^* is the *equilibrium distribution* of the Markov chain and the chain is said to be *ergodic*.

Consider the sequence $\mathbf{x} \to \mathbf{k}_1 \to \mathbf{k}_2 \to \cdots \mathbf{k}_t \to \mathbf{y}$ so that the transition probabilities $q(\mathbf{k}_1 \mid \mathbf{x}) \neq 0$, $q(\mathbf{k}_2 \mid \mathbf{k}_1) \neq 0$, $\ldots$, $q(\mathbf{y} \mid \mathbf{k}_t) \neq 0$. Then, there is a sequence from $\mathbf{x}$ to $\mathbf{y}$ with a nonzero probability of occurring in the Markov chain. It is said that $\mathbf{x}$ and $\mathbf{y}$ communicate. If $\mathbf{y}$ and $\mathbf{x}$ also communicate through nonzero transition probabilities, it is said that these two states intercommunicate. If all states in S intercommunicate, then the state space is said to be *irreducible* under the transition probability $q(. \mid .)$.

A Markov chain is *reversible* if $\pi(\mathbf{x})q(\mathbf{y} \mid \mathbf{x}) = \pi(\mathbf{y})q(\mathbf{x} \mid \mathbf{y})$, that is, the probability of being at a point $\mathbf{y}$ multiplied by the probability of moving from $\mathbf{y}$ to $\mathbf{x}$ is equal to the probability of being at point $\mathbf{x}$ multiplied by the probability of moving from $\mathbf{x}$ to $\mathbf{y}$.

The period of $\mathbf{x}$ in the Markov chain, denoted by $d_\mathbf{x}$, is the largest common divisor of the set $\{m \geq 1 : q^{(m)}(\mathbf{x} \mid \mathbf{x}) > 0\}$, where m indicates the number of states required for the point $\mathbf{x}$ be reached with transition probability $q(. \mid .)$, by starting from this same point $\mathbf{x}$. A state $\mathbf{x}$ is aperiodic if $d_\mathbf{x} = 1$, that is, $\mathbf{x}$ can be reached in one single transition from the point $\mathbf{x}$ itself. A chain is *aperiodic* if all of its states are aperiodic.

4.2 TECHNIQUE IV: MARKOV CHAIN MONTE CARLO (MCMC) METHOD

The most common implementations of MCMC methods are the Gibbs Sampler and the Metropolis-Hastings algorithm [17–22,48,49,59,60,63,67,99,100]. The Gibbs Sampler is not presented here. The Metropolis-Hastings algorithm was first published by Metropolis et al. [99] in 1953, in a work aimed at the calculation of properties of substances composed of interacting molecules. It was, therefore, a statistical mechanics work not focused on the solution of inverse problems. Although the paper has five co-authors [99], only the name of the first author became popular to designate the developed algorithm, which was later generalized by Hastings in 1970 [100]. The Metropolis-Hastings algorithm was devised to generate reversible Markov chains.

The following statement applies to the Metropolis-Hastings algorithm [19,60]: *Let π be a given probability distribution. The Markov chain simulated by the Metropolis-Hastings algorithm is reversible with respect to π. If it is also irreducible and aperiodic, then it defines an ergodic Markov chain with unique equilibrium distribution π^*.*

Unfortunately, it might not be possible to prove that the chain is irreducible and/or aperiodic for practical cases. In fact, linearly dependent parameters generally result in periodic and correlated chains, so that an equilibrium distribution is not clearly reached. Such as for the classical methods presented in Chapter 2, where correlated parameters directly influence the topology of the objective function based on the likelihood and a global minimum might not exist (see Example 2.2), the posterior distribution, which is now sought with the Markov chain, will be affected by parameters with sensitivity coefficients of small magnitude and/or linearly dependent.

The Metropolis-Hastings algorithm draws samples from a candidate density, like in the acceptance-rejection method [59,63]. The acceptance-rejection method can be used to generate samples from a density $p(\mathbf{P}) = \tilde{p}(\mathbf{P})/K$, where the normalizing constant K might be unknown, such as the posterior distribution given by Bayes´ theorem. Instead of sampling from $p(\mathbf{P})$, assume that there exists a candidate density $h(\mathbf{P})$ that is easy to simulate samples from, where $\tilde{p}(\mathbf{P}) \le c\,h(\mathbf{P})$ and c is a constant. The following steps are then used to obtain a random variable $\hat{\mathbf{P}}$ with density $p(\mathbf{P})$ by using the acceptance-rejection method [63]:

1. Generate a random variable $\mathbf{P}^*$ from the density $h(\mathbf{P})$;
2. Generate a random value $U \sim U(0,1)$, which is uniformly distributed in $(0,1)$;
3. If $U \le \dfrac{\tilde{p}(\mathbf{P}^*)}{c\,h(\mathbf{P}^*)}$, let $\hat{\mathbf{P}} = \mathbf{P}^*$. Otherwise, return to step 1.

Example 4.1

Use the acceptance-rejection method to generate samples of a Gaussian distribution:

$$p(x) = \frac{1}{\sigma\sqrt{2\pi}} \exp\left[-\frac{1}{2}\left(\frac{x-\mu}{\sigma} \right)^2 \right]$$

with mean $\mu = 0.5$ and standard deviation $\sigma = 0.5$, in the interval $-1 < x < 3$, from the samples of a uniform distribution in this interval, that is,

$$h(x) = \begin{cases} \dfrac{1}{(b-a)}, & a < x < b \\ 0, & \text{elsewhere} \end{cases}$$

where $a = -1$ and $b = 3$.

Solution: Figure 4.1 shows the uniform distribution $h(x)$, which was sampled in order to generate random numbers with the Gaussian distribution $p(x)$ (also shown in this figure). For the

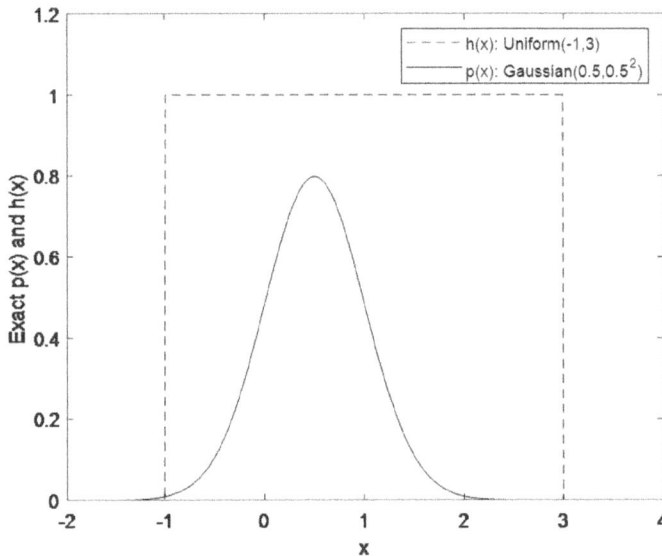

FIGURE 4.1   Uniform and Gaussian distributions of Example 4.1. Samples of the uniform distribution are used to generate samples of the Gaussian distribution with the acceptance-rejection method.

acceptance-rejection method, we used $c = (b - a)$ and generated 20,000 samples from the uniform distribution. The histogram of these samples is presented in Figure 4.2a. Figure 4.2b shows the histogram of the samples of the Gaussian distribution $p(x)$, which were obtained with the acceptance-rejection method. The sought Gaussian distribution is also presented in Figure 4.2b. Therefore, the samples generated with the acceptance-rejection method are distributed in accordance with the Gaussian distribution $p(x)$. We note that around 25% of the samples generated from the uniform distribution were accepted as belonging to the Gaussian distribution.

The implementation of the Metropolis-Hastings algorithm starts with the selection of a proposal distribution $q(\mathbf{P}^* \mid \mathbf{P}^{(t)})$, which is used to draw a new candidate sample $\mathbf{P}^*$ given the current sample $\mathbf{P}^{(t)}$ of the Markov chain. For the solution of the inverse problem within the Bayesian framework, one aims at simulating the posterior distribution $\pi_{\text{posterior}}(\mathbf{P}) \propto \pi(\mathbf{Y}|\mathbf{P})\pi(\mathbf{P})$ (see equation 3.1.5). Hence, the balance (reversibility) condition of the Markov chain of interest is given by:

$$\pi_{\text{posterior}}(\mathbf{P}^{(t)})q(\mathbf{P}^* \mid \mathbf{P}^{(t)}) = \pi_{\text{posterior}}(\mathbf{P}^*)q(\mathbf{P}^{(t)} \mid \mathbf{P}^*) \tag{4.2.1}$$

In order to avoid eventual cases that $\pi_{\text{posterior}}(\mathbf{P}^{(t)})q(\mathbf{P}^* \mid \mathbf{P}^{(t)}) > \pi_{\text{posterior}}(\mathbf{P}^*)q(\mathbf{P}^{(t)} \mid \mathbf{P}^*)$, that is, the process moves from $\mathbf{P}^{(t)}$ to $\mathbf{P}^*$ more often than the reverse, a probability $\alpha(\mathbf{P}^* \mid \mathbf{P}^{(t)})$ is introduced in equation (4.2.1), so that [63]:

$$\pi_{\text{posterior}}(\mathbf{P}^{(t)})q(\mathbf{P}^* \mid \mathbf{P}^{(t)})\alpha(\mathbf{P}^* \mid \mathbf{P}^{(t)}) = \pi_{\text{posterior}}(\mathbf{P}^*)q(\mathbf{P}^{(t)} \mid \mathbf{P}^*) \tag{4.2.2}$$

Therefore,

$$\alpha(\mathbf{P}^* \mid \mathbf{P}^{(t)}) = \min\left[1, \frac{\pi_{\text{posterior}}(\mathbf{P}^*)q(\mathbf{P}^{(t)} \mid \mathbf{P}^*)}{\pi_{\text{posterior}}(\mathbf{P}^{(t)})q(\mathbf{P}^* \mid \mathbf{P}^{(t)})}\right] \tag{4.2.3}$$

where $\alpha(\mathbf{P}^* \mid \mathbf{P}^{(t)}) = 1$ when the reversibility condition is satisfied.

Equation (4.2.3) is also called the Metropolis-Hastings ratio. Notice that, for the computation of equation (4.2.3), there is no need to know the normalizing constant that appears in the definition of the posterior distribution (see equation 3.1.4). Equation (4.2.2) shows that the probability of moving from the sample at the current state, $\mathbf{P}^{(t)}$, to $\mathbf{P}^*$ is now given by $[q(\mathbf{P}^* \mid \mathbf{P}^{(t)})\alpha(\mathbf{P}^* \mid \mathbf{P}^{(t)})]$. The Metropolis-Hastings algorithm follows a procedure similar to the acceptance-rejection method described above.

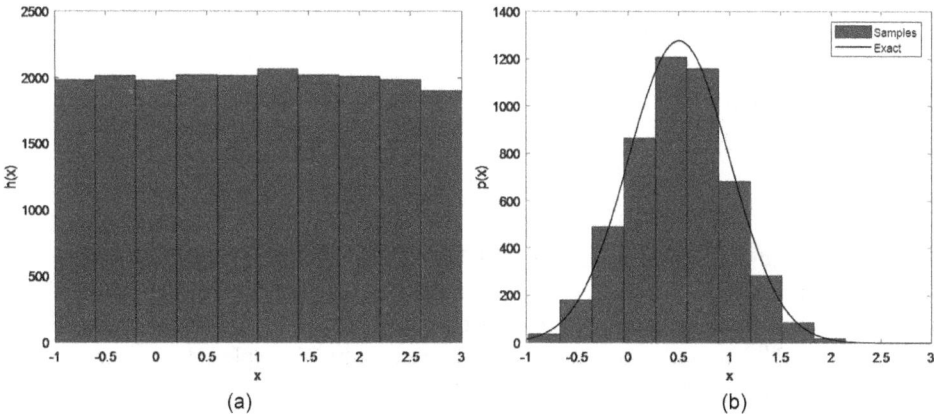

FIGURE 4.2   (a) Histogram of the samples obtained from the uniform distribution; (b) histogram of the samples of the Gaussian distribution generated with the acceptance-rejection method.

The Metropolis-Hastings algorithm can then be summarized in the following steps [17–22,48,49,59,60,63,67,99,100–102]:

1. Let $t = 0$ and start the Markov chain with sample $\mathbf{P}^{(0)}$ at the initial state.
2. Sample a candidate point $\mathbf{P}^*$ from a proposal distribution $q(\mathbf{P}^* \mid \mathbf{P}^{(t)})$.
3. Calculate the probability $\alpha(\mathbf{P}^* \mid \mathbf{P}^{(t)})$ with equation (4.2.3).
4. Generate a random value $U \sim U(0,1)$, which is uniformly distributed in $(0,1)$.
5. If $U \leq \alpha(\mathbf{P}^* \mid \mathbf{P}^{(t)})$, set $\mathbf{P}^{(t+1)} = \mathbf{P}^*$. Otherwise, set $\mathbf{P}^{(t+1)} = \mathbf{P}^{(t)}$.
6. Make $t = t + 1$ and return to step 2 in order to generate the sequence $\{\mathbf{P}^{(1)}, \mathbf{P}^{(2)}, \ldots, \mathbf{P}^{(n)}\}$.

Hopefully, the sequence $\{\mathbf{P}^{(1)}, \mathbf{P}^{(2)}, \ldots, \mathbf{P}^{(n)}\}$ can be used to represent the posterior distribution and inference on this distribution can be obtained from inference on $\{\mathbf{P}^{(1)}, \mathbf{P}^{(2)}, \ldots, \mathbf{P}^{(n)}\}$. We note that values of $\mathbf{P}^{(t)}$ must be ignored until the chain has not converged to equilibrium. The number of states from the beginning of the chain until it reaches equilibrium is called the *burn-in period*.

For the computational implementation of the Metropolis-Hastings algorithm, the test in step 5 is performed by taking the logarithm of both sides, that is, $\ln[U] \leq \ln[\alpha(\mathbf{P}^* \mid \mathbf{P}^{(t)})]$. This is required in order to avoid numerical errors, since $\pi_{\text{posterior}}(\mathbf{P})$ commonly involve exponentials and the ratio in $\alpha(\mathbf{P}^* \mid \mathbf{P}^{(t)})$ may become a number that cannot be represented within the computer numerical limits, if $\pi_{\text{posterior}}(\mathbf{P}^{(t)}) q(\mathbf{P}^* \mid \mathbf{P}^{(t)}) \ll \pi_{\text{posterior}}(\mathbf{P}^*) q(\mathbf{P}^{(t)} \mid \mathbf{P}^*)$.

We note that different versions of the Metropolis-Hastings algorithm have been proposed [60]. One of these versions, in which the candidates are generated by blocks of parameters, is presented in Note 1 at the end of this chapter. Sampling by blocks of parameters is convenient for cases involving sets of linearly dependent parameters.

PROPOSAL DISTRIBUTION

The proposal distribution plays a fundamental role for the success of the Metropolis-Hastings algorithm. Typical choices for $q(\mathbf{P}^* \mid \mathbf{P}^{(t)})$ are presented below.

Random Walk

In this case $\mathbf{P}^* = \mathbf{P}^{(t)} + \mathbf{\Psi}$, where $\mathbf{\Psi}$ is a vector of random variables with distribution $q_1(\mathbf{\Psi})$. Therefore, $q(\mathbf{P}^* \mid \mathbf{P}^{(t)}) = q_1(\mathbf{\Psi})$. If the proposal distribution is symmetric $q(\mathbf{P}^* \mid \mathbf{P}^{(t)}) = q(\mathbf{P}^{(t)} \mid \mathbf{P}^*)$, that is, $q_1(\mathbf{\Psi}) = q_1(-\mathbf{\Psi})$, and equation (4.2.3) reduces to:

$$\alpha(\mathbf{P}^* \mid \mathbf{P}^{(t)}) = \min\left[1, \frac{\pi_{\text{posterior}}(\mathbf{P}^*)}{\pi_{\text{posterior}}(\mathbf{P}^{(t)})}\right] \qquad (4.2.4a)$$

or, by using equation (3.1.5),

$$\alpha(\mathbf{P}^* \mid \mathbf{P}^{(t)}) = \min\left[1, \frac{\pi(\mathbf{Y} \mid \mathbf{P}^*)\pi(\mathbf{P}^*)}{\pi(\mathbf{Y} \mid \mathbf{P}^{(t)})\pi(\mathbf{P}^{(t)})}\right] \qquad (4.2.4b)$$

Thus, with the choice of a symmetrical proposal density, the candidate point $\mathbf{P}^*$ is always accepted if it is located in a region of higher posterior probability, in step 5 of the Metropolis-Hastings algorithm. However, the candidate point can also be accepted if $\pi_{\text{posterior}}(\mathbf{P}^*) < \pi_{\text{posterior}}(\mathbf{P}^{(t)})$ with probability $\alpha(\mathbf{P}^* \mid \mathbf{P}^{(t)})$, thus allowing that the state space be highly explored.

Uniform and Gaussian distributions are commonly used as the symmetrical density $q_1(\mathbf{\Psi})$. Although uniform and Gaussian priors can also be used for the parameters, as discussed in Chapter 3, we note that the random walk proposal is totally independent of the prior. Therefore, Gaussian random walk proposals can be used with uniform priors and *vice versa*. While the prior must reflect the information *a priori* known about the parameters, the proposal distribution simply governs how the candidates are generated in the Markov chain.

Consider one single component P_j ($j = 1,..., N$) of the parameter vector **P**. For a *random walk proposal with a uniform distribution*, one can write:

$$P_j^* = P_j^{(t)} + w_j(2u_j - 1) \tag{4.2.5}$$

where u_j is a random number with uniform distribution in (0,1), that is, $u_j \sim U(0,1)$, while w_j is the maximum variation to generate the candidate parameter at each state of the Markov chain.

For a *random walk proposal with a Gaussian distribution*, we have:

$$P_j^* = P_j^{(t)} + r_j \tag{4.2.6}$$

where r_j is a Gaussian random number with zero mean and standard deviation ξ_j.

The probability of accepting the candidate P_j^* increases with small variations w_j or with small standard deviations $\xi_j, j = 1,..., N$, in the random walk proposal with uniform and Gaussian distributions, respectively. Such is the case because it is more likely to move to regions of high posterior around $P_j^{(t)}$ with small w_j or ξ_j. With candidates generated from small perturbations of $P_j^{(t)}$, the number of accepted states can thus be large and the resulting Markov chains may take too long to reach an equilibrium distribution for the parameters. On the other hand, large perturbations of $P_j^{(t)}$ may lead to small acceptance rates, meaning that the parameter values at the current state may be repeated at many successive states in the Markov chain, in accordance with step 5 of the Metropolis-Hastings algorithm. Although large perturbations of $P_j^{(t)}$ may fast lead to an equilibrium distribution, long chains may still be needed to generate enough samples with different (and independent) values that can be used to represent the posterior distribution. The selection of w_j or ξ_j will be addressed later in this chapter.

Independent Move

This choice for the proposal density is of the kind $q(\mathbf{P}^* \mid \mathbf{P}^{(t)}) = q_2(\mathbf{P}^*)$, that is, it does not depend on the parameter values at the current state, $\mathbf{P}^{(t)}$. The natural choice for the independent move is to select the proposal density $q(\mathbf{P}^* \mid \mathbf{P}^{(t)})$ as the prior density, $\pi(\mathbf{P}^*)$. Then, with $\pi_{\text{posterior}}(\mathbf{P}) \propto \pi(\mathbf{Y}|\mathbf{P})\pi(\mathbf{P})$ and $q(\mathbf{P}^* \mid \mathbf{P}^{(t)}) = \pi(\mathbf{P}^*)$, equation (4.2.3) is rewritten as:

$$\alpha(\mathbf{P}^* \mid \mathbf{P}^{(t)}) = \min\left[1, \frac{\pi(\mathbf{Y} \mid \mathbf{P}^*)\pi(\mathbf{P}^*)}{\pi(\mathbf{Y} \mid \mathbf{P}^{(t)})\pi(\mathbf{P}^{(t)})} \frac{\pi(\mathbf{P}^{(t)})}{\pi(\mathbf{P}^*)}\right] \tag{4.2.7a}$$

that is, the Metropolis-Hastings ratio for the independent move is given by the ratio of the likelihoods:

$$\alpha(\mathbf{P}^* \mid \mathbf{P}^{(t)}) = \min\left[1, \frac{\pi(\mathbf{Y} \mid \mathbf{P}^*)}{\pi(\mathbf{Y} \mid \mathbf{P}^{(t)})}\right] \tag{4.2.7b}$$

Such as for the random walk proposal, candidates moving to regions of higher probability (in this case, the likelihood) are always accepted. Candidates in regions of lower likelihoods can be accepted with probability $\alpha(\mathbf{P}^* \mid \mathbf{P}^{(t)})$. Although the probability $\alpha(\mathbf{P}^* \mid \mathbf{P}^{(t)})$ given by equation (4.2.7b) does not involve the prior distribution, the Markov chain still depends on the prior since it is used to generate the candidates. This kind of proposal can be very effective for parameters with small prior variances. On the other hand, it may result in very small acceptance rates if the prior have large variances. Also, it cannot be applied with an improper prior due to its unlimited variance.

The application of the Metropolis-Hastings algorithm to the solution of inverse heat conduction problems is illustrated by examples presented in the next three sections.

4.3   MCMC ESTIMATION OF THERMAL CONDUCTIVITY COMPONENTS OF AN ORTHOTROPIC HEAT CONDUCTING MEDIUM

In this section we revisit the inverse problem discussed in Section 2.8 but estimate the thermal conductivity components with the Metropolis-Hastings algorithm presented above [103].

THE DIRECT PROBLEM

The mathematical formulation of the physical problem of interest is given in *dimensionless form* by:

$$k_1 \frac{\partial^2 T}{\partial x^2} + k_2 \frac{\partial^2 T}{\partial y^2} + k_3 \frac{\partial^2 T}{\partial z^2} = \frac{\partial T}{\partial t} \quad \text{in } 0 < x < a, 0 < y < b, 0 < z < c \, ; t > 0 \qquad (4.3.1\text{a})$$

$$T = 0 \text{ at } x = 0; \quad k_1 \frac{\partial T}{\partial x} = q_1(t) \text{ at } x = a, \quad \text{for } t > 0 \qquad (4.3.1\text{b,c})$$

$$T = 0 \text{ at } y = 0; \quad k_2 \frac{\partial T}{\partial y} = q_2(t) \text{ at } y = b, \quad \text{for } t > 0 \qquad (4.3.1\text{d,e})$$

$$T = 0 \text{ at } z = 0; \quad k_3 \frac{\partial T}{\partial z} = q_3(t) \text{ at } z = c, \quad \text{for } t > 0 \qquad (4.3.1\text{f,g})$$

$$T = 0 \text{ for } t = 0; \text{ in } 0 < x < a, \, 0 < y < b, \, 0 < z < c \qquad (4.3.1\text{h})$$

THE INVERSE PROBLEM

Simulated temperature measurements taken at the centers of the heated surfaces were used in the inverse analysis. The measurement errors were additive, uncorrelated, Gaussian, and with constant standard deviations of 0.01 or 0.05. The experimental variables optimally selected in [104] were used here, for $k_1 = 1$, $k_2 = 15$ and $k_3 = 15$, where $b/a = c/a = q_2/q_1 = q_3/q_1 = 3.87$ and the heating and final times were $t_h = t_f = 1$. Uniform priors were used for the thermal conductivity components in the interval (0.1,50). The results presented below were obtained with 20,000 samples in the Markov chain.

STOCHASTIC SIMULATION

Figures 4.3a and b present the samples resulting from the Metropolis-Hastings algorithm for standard deviations of the measurement errors of $\sigma = 0.01$ and $\sigma = 0.05$, respectively. The samples of the posterior distributions after the Markov chains reached equilibrium resemble ellipsoids, like for linear estimation problems, even though the present estimation problem is nonlinear. Such is probably the case because of the optimized experimental variables used in the estimation [104]. Figures 4.3a and b also show the burn-in period required to reach an equilibrium distribution, when the samples start from their initial values and tend to the posterior distribution for the parameters around the exact values used to generate the simulated measurements. As expected, the posterior distribution is more spread for larger measurement errors. Such conclusions are also apparent from the analysis of Figures 4.4a-c and 4.5a-c, which present the states of the Markov chains of each parameter, for $\sigma = 0.01$ and $\sigma = 0.05$, respectively. The burn-in period for $\sigma = 0.01$ was of the order of 4000 states, while for $\sigma = 0.05$ the burn-in period was of the order of 2000 states, as shown by Figures 4.4a-c and 4.5a-c, respectively.

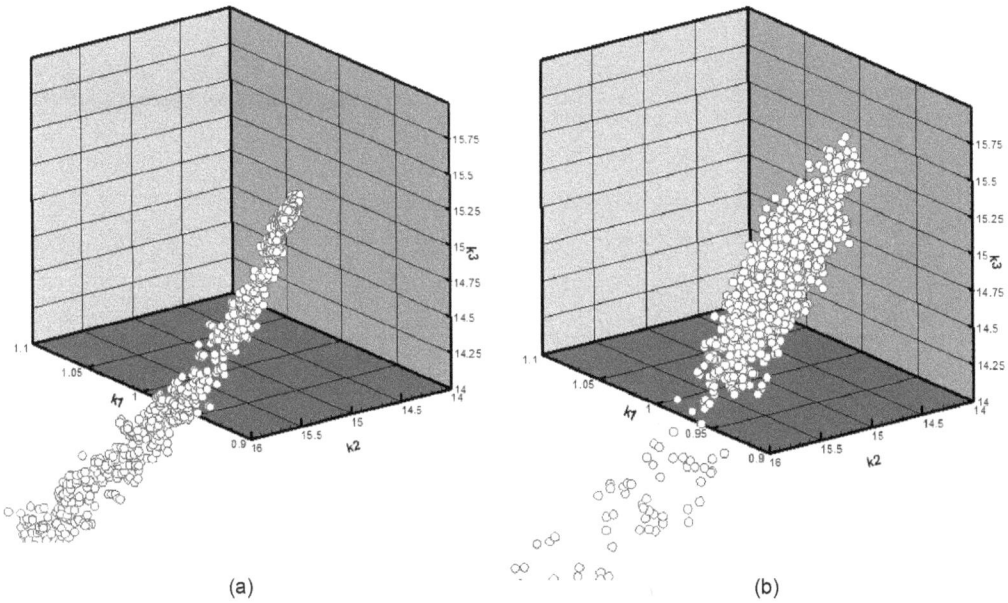

FIGURE 4.3 Samples of the posterior distribution for: (a) $\sigma = 0.01$ and (b) $\sigma = 0.05$. (From Ref. [103].)

The means and the standard deviations of the parameter values in the Markov chains after the burn-in period are presented in Table 4.1. This table shows that the MCMC Bayesian approach with the Metropolis-Hastings algorithm provided very accurate estimates for the unknown thermal conductivity components, even for very large magnitudes of the measurement errors, such as $\sigma = 0.05$. As expected, the standard deviations of the estimated parameters increased with the magnitude of the measurement errors.

4.4 MCMC ESTIMATION OF THERMAL CONDUCTIVITY AND VOLUMETRIC HEAT CAPACITY OF VISCOUS LIQUIDS WITH THE LINE HEAT SOURCE PROBE

The classical transient line heat source probe method involves a constant heat generation by Joule effect within a thin cylindrical needle, which is placed inside the material to be thermally characterized. A sensor inside the probe measures its temperature variation, as the probe and the surrounding material are heated. It is a popular method for the measurement of thermal conductivity [105,106], particularly for viscous liquids, where the effects of natural convection can be avoided. The transient line heat source probe method, in its classical form, is based on an asymptotic mathematical solution for an ideal system, composed of an infinite linear heat source of zero mass [107]. Such assumptions may, in some practical situations, lead to measurements errors.

Inverse parameter estimation techniques, which involve accurate modeling of the physical problem, have been successfully applied to the transient line heat source probe method, allowing the simultaneous estimation of more than one parameter, besides the thermal conductivity [108–110]. A common feature of these works is that the solution of the inverse problem was retrieved by assuming several model parameters as deterministically known. Therefore, these solutions provided no means to quantify the uncertainties associated with the supposedly "known" parameters. On the other hand, the Bayesian inference approach to the solution of inverse problems offers a way to cope with the uncertainties of all parameters appearing in the formulation, by using a complete probabilistic description via prior modeling.

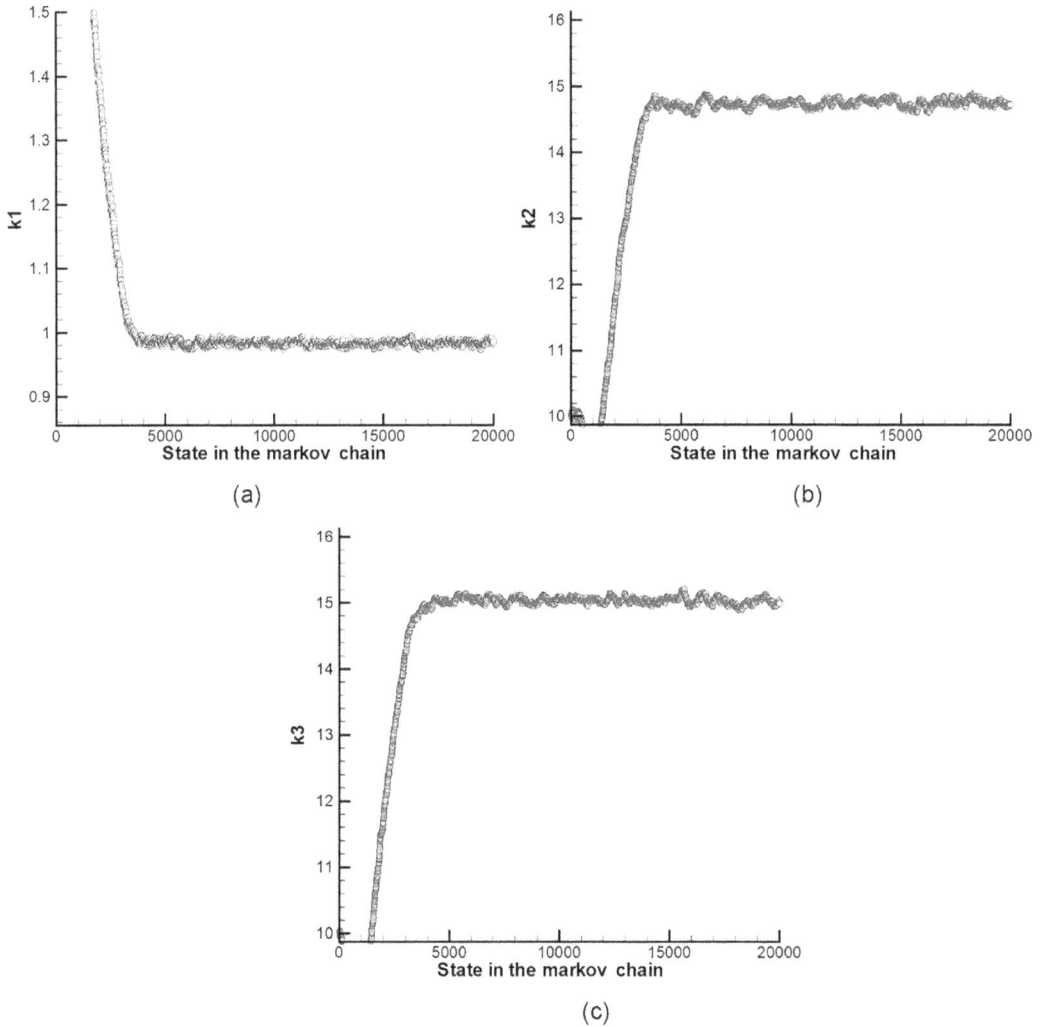

FIGURE 4.4 Markov chains obtained with simulated measurements with $\sigma = 0.01$ for: (a) k_1, (b) k_2, and (c) k_3. (From Ref. [103].)

THE DIRECT PROBLEM

In this section, we follow reference [111] for the mathematical model that takes into account the probe and the surrounding material. The physical problem under analysis consists of a long and thin cylinder (the probe) of radius r_s, modeled as a lumped system. The probe is inserted into the sample of a material with unknown properties, which is contained in a cylindrical cell. The cell internal and external radiuses are r_{int} and r_{ext}, respectively. We assume heat transfer by conduction within the material and the cell, which are considered as homogeneous and isotropic media with constant thermophysical properties. Furthermore, a perfect thermal contact is assumed at all interfaces, while the surface at $r = r_{ext}$ exchanges heat by convection with a surrounding fluid, such as a liquid in a thermostatic bath, with a heat transfer coefficient h. The system is considered to be initially in thermal equilibrium at the temperature T_0, which is also assumed to be the constant temperature of the surrounding fluid. For the time scale of interest, the probe can be assumed as uniformly heated with a constant volumetric heat source g. By neglecting end-effects, the physical problem under picture can be formulated as one dimensional. The proposed mathematical formulation in dimensionless form is given by:

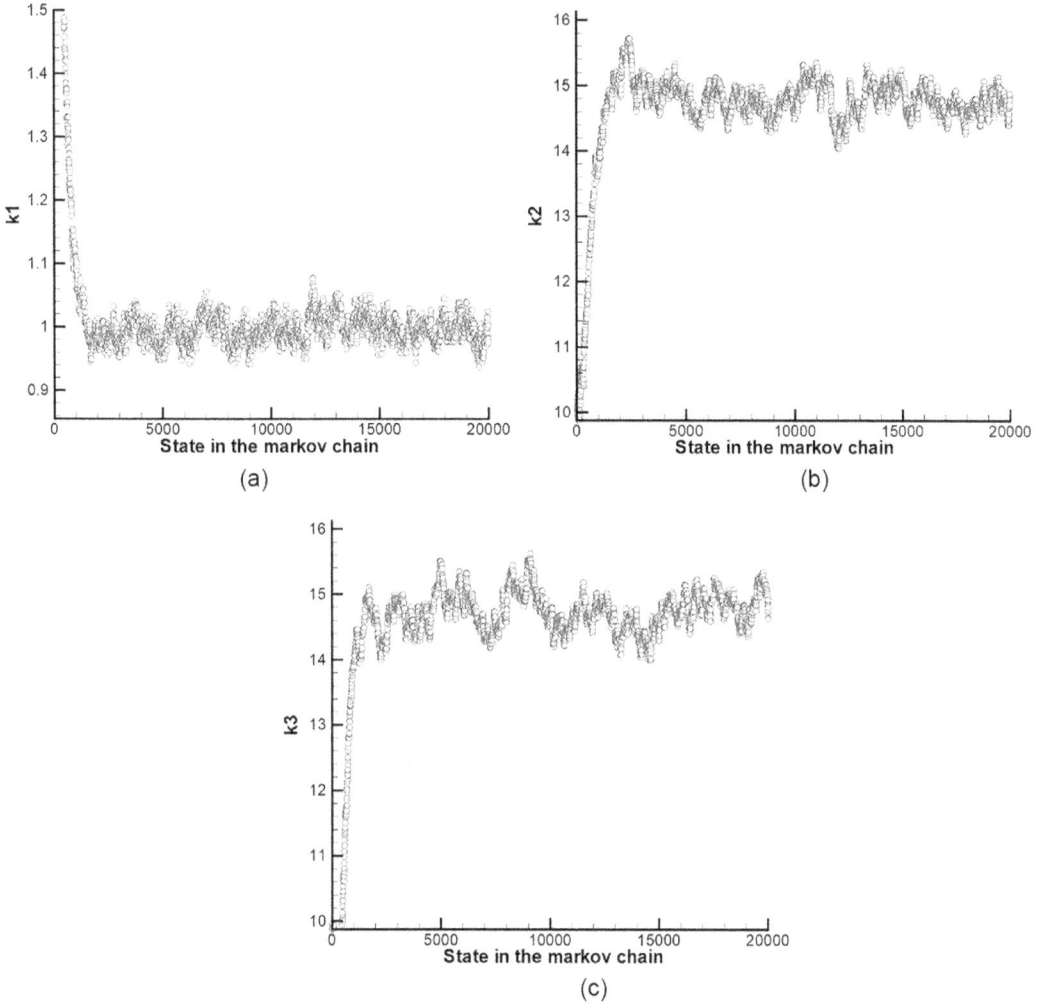

FIGURE 4.5 Markov chains obtained with simulated measurements with $\sigma = 0.05$ for: (a) k_1, (b) k_2 and (c) k_3. (From Ref. [103].)

$$C_p^* \frac{d\Theta_p(\tau)}{d\tau} = G^* + 2K_m^* \left. \frac{\partial \Theta_m(R,\tau)}{\partial R} \right|_{R=1} , \text{ for } \tau > 0 \qquad (4.4.1a)$$

$$C_m^* \frac{\partial \Theta_m(R,\tau)}{\partial \tau} = \frac{K_m^*}{R} \frac{\partial}{\partial R}\left(R \frac{\partial \Theta_m(R,\tau)}{\partial R} \right), \text{ in } 1 < R < R_{int}, \text{for } \tau > 0 \qquad (4.4.1b)$$

$$C_c^* \frac{\partial \Theta_c(R,\tau)}{\partial \tau} = \frac{K_c^*}{R} \frac{\partial}{\partial R}\left(R \frac{\partial \Theta_c(R,\tau)}{\partial R} \right), \text{ in } R_{int} < R < R_{ext}, \text{for } \tau > 0 \qquad (4.4.1c)$$

$$\Theta_p(\tau) = \Theta_m(R,\tau) \quad \text{at } R = 1, \ \tau > 0 \qquad (4.4.1d)$$

$$\Theta_m(R,\tau) = \Theta_c(R,\tau) \text{ at } R = R_{int}, \tau > 0 \qquad (4.4.1e)$$

$$K_m^* \frac{\partial \Theta_m(R,\tau)}{\partial R} = K_c^* \frac{\partial \Theta_c(R,\tau)}{\partial R} \text{ at } R = R_{int}, \tau > 0 \qquad (4.4.1f)$$

TABLE 4.1

Results Obtained with the MCMC Method

Standard Deviation for the Measurements	Parameter	Mean	Standard Deviation
$\sigma = 0.01$	k_1	0.983	0.004
	k_2	14.74	0.06
	k_3	15.04	0.05
$\sigma = 0.05$	k_1	0.99	0.02
	k_2	14.8	0.2
	k_3	14.7	0.3

Source: From Ref. [103].

$$K_c^* \frac{\partial \Theta_c(R,\tau)}{\partial R} + Bi\Theta_c(R,\tau) = 0 \ \ \text{at} \ \ R = R_{\text{ext}}, \tau > 0 \tag{4.4.1g}$$

$$\Theta_p(\tau) = \Theta_m(R,\tau) = \Theta_c(R,\tau) = 0 \ \ \text{in} \ 1 < R < R_{\text{ext}}, \text{at} \ \tau = 0 \tag{4.4.1h}$$

where the subscripts p, m and c denote the probe, the sample of the material and the cell, respectively. The following dimensionless parameters were introduced [111]:

$$\Theta(R,\tau) = \frac{T(r,t) - T_0}{\frac{g_{\text{ref}} r_s^2}{k_{\text{ref}}}}; \ \tau = \frac{k_{\text{ref}} t}{C_{\text{ref}} r_s^2}; \ R = \frac{r}{r_s}; \ R_{\text{int}} = \frac{r_{\text{int}}}{r_s};$$

$$(4.4.2\text{a-i})$$

$$R_{\text{ext}} = \frac{r_{\text{ext}}}{r_s}; \ K^* = \frac{k}{k_{\text{ref}}}; C^* = \frac{C}{C_{\text{ref}}}; \ Bi = \frac{h r_s}{k_{\text{ref}}}; G^* = \frac{g}{g_{\text{ref}}}$$

where the subscript 'ref' denotes a reference value.

THE INVERSE PROBLEM

The *inverse problem* aims at the simultaneous estimation of the material's thermal conductivity and volumetric heat capacity, K_m^* and C_m^*, respectively. However, the probe's volumetric heat capacity (C_p^*), the cell's thermophysical properties (K_c^*, C_c^*), the probe source term (G^*) and the Biot number (Bi) at the surface of the cell, are not deterministically known. Therefore, the vector of model parameters is given by $\mathbf{P} = \left[K_m^*, C_m^*, Bi, K_c^*, C_c^*, C_p^*, G^* \right]^T$. For the solution of the inverse problem, transient temperature measurements taken within the probe are assumed available.

For the results presented below, the heat source probe consisted of a needle of stainless steel with 70 mm length and 1.2 mm of external diameter [111]. Inside the needle, there exists a heating wire, as well as a K-type thermocouple. The probe is inserted into the sample of the liquid to be characterized, which is contained in a beaker made of Pyrex, with an internal radius of 40 mm and a wall thickness of 2 mm. In order to minimize the effects of natural convection in the liquid during the experiments, the temperature increase of the probe was kept below 6°C. The focus of this study was on the characterization of viscous liquids, in which the effects of natural convection resulting from the probe heating can be minimized and the heat conduction model given by equations (4.4.1a-d) is appropriate. Therefore, results are presented below for glycerin, which has a viscosity two orders of magnitude larger than that of water at room temperature.

ANALYSIS OF THE SENSITIVITY COEFFICIENTS AND DESIGN OF OPTIMUM EXPERIMENTS

Figure 4.6 presents the transient behavior of the reduced sensitivity coefficients with respect to the parameters appearing in the formulation. The heating duration was taken as 60 s. The variation of the dimensionless temperature of the probe was also included in this figure. The reduced sensitivity coefficients for the parameters K_m^* and C_m^* have the same order of magnitude of the dimensionless temperatures. It must also be noticed that the sensitivity coefficients of the heat source term and of the thermal conductivity are linearly dependent; as a consequence, these two parameters cannot be estimated simultaneously. The sensitivity coefficient of the probe's volumetric heat capacity is very close to zero. Similarly, it can be noticed that the sensitivity coefficients with respect to the properties of the cell, as well as to the Biot number, are practically null. Consequently, the estimation of such parameters is difficult and not accurate. Anyhow, the parameters with small sensitivity coefficients, that is, Bi, K_c^*, C_c^* and C_p^*, do not have a significant effect on the measured temperatures.

Based on the analysis of the sensitivity coefficients, the solution of the inverse problem was considered in two steps, as follows. First, we consider a liquid with known thermophysical properties, where the objective of the inverse problem is to estimate the heat source and the volumetric heat capacity of the probe. The second step aims effectively at the characterization of a liquid with unknown thermophysical properties. In the second step, the estimates of the heat source and the volumetric heat capacity of the probe, which were obtained in the first step, are then used to estimate the thermophysical properties of the liquid. As mentioned above, accurate estimates of the cell properties and of the Biot number cannot be obtained in the inverse analysis. Thus, these quantities are assumed as known for both steps.

STOCHASTIC SIMULATION

The prior distributions for the parameters considered as known for the inverse analysis were assumed as Gaussian. The means and standard deviations for these distributions were taken from measurements performed with other experiments and from a correlation available in the literature for the external heat transfer coefficient. Uniform distributions were used for the other parameters to be estimated in each of the steps described above. For the first step, a uniform distribution centered on the value obtained from the measured electrical power, with lower and upper bounds of ±20% around this value, was assumed for the heat source term. A uniform distribution was assumed for the probe's volumetric heat capacity with upper bound given by the volumetric heat capacity of steel. For the liquid thermophysical properties in the second step, uniform distributions centered on

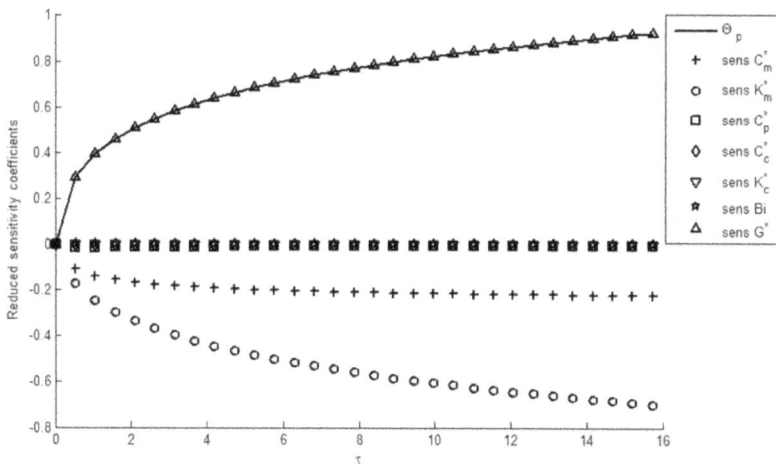

FIGURE 4.6   Sensitivity coefficients: glycerin. (From Ref. [111].)

literature values and with bounds of ±20% of these values were used. The same fluid was used in both steps, for the sake of validation of the proposed methodology.

Figure 4.7 presents the states of the Markov chains for each parameter during the application of the MCMC method for the first step. This figure shows that the burn-in period, which is required for the Markov chain corresponding to the probe's heat source to reach equilibrium, is roughly 20,000 states. It must be noticed in Figure 4.7 the stationary behavior of the Markov chains corresponding to the liquid's thermal conductivity and volumetric heat capacity. Also, one can notice that the Markov chains corresponding to the volumetric heat capacity of the probe, cell properties and Biot number, present an oscillatory behavior within their assigned prior distributions, due to their small sensitivity coefficients. From the Markov chains, we computed the means and standard deviations of the marginal distributions for each parameter by discarding the first 20,000 samples.

The estimated parameter means were used to calculate the theoretical temperature variation, which is compared to the experimental temperatures in Figure 4.8. The calculated and experimental temperatures are in excellent agreement, as also revealed by Figure 4.9 that presents the temperature residuals. This figure shows that the residuals have small magnitude and are only slightly correlated, thus indicating that the estimated means and the proposed mathematical model appropriately represent the physics of the problem. The relative large magnitudes of the residuals observed at small times are due to a small lag between the beginning of heating and the beginning of the experimental data acquisition.

The probe's volumetric heat source term and volumetric heat capacity estimated in the first step were then used in the second step with informative normal priors. The second step was concerned with the estimation of the thermal properties of the liquid and, thus, uniform priors were assumed for these parameters. Figure 4.10 presents the states of the Markov chains for each parameter, obtained in the second step of the inverse analysis. The burn-in period was taken as 15,000 states based on the analysis of these chains. Table 4.2 presents the means and standard deviations of the samples in the Markov chains for each parameter. This table shows that accurate estimates were obtained for the thermal conductivity and volumetric heat capacity of glycerin, which are in excellent agreement with the reported literature values [111].

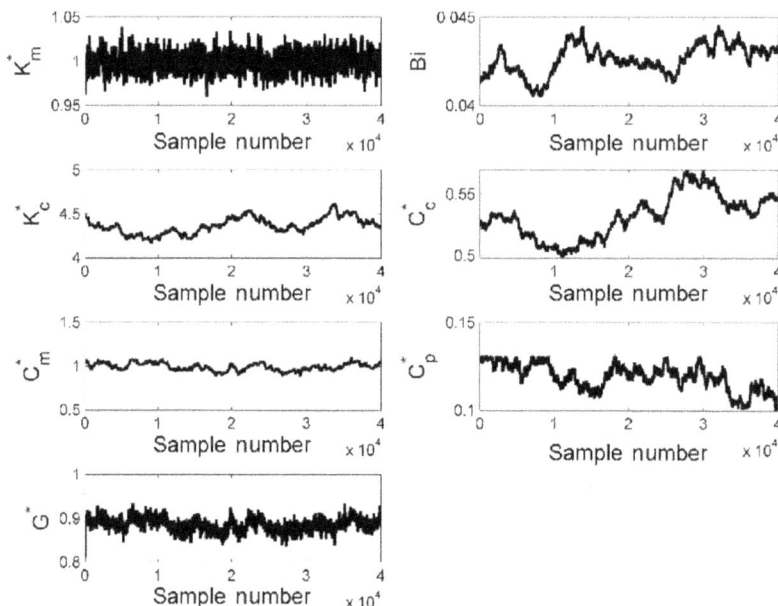

FIGURE 4.7   Markov Chains—first step. (From Ref. [111].)

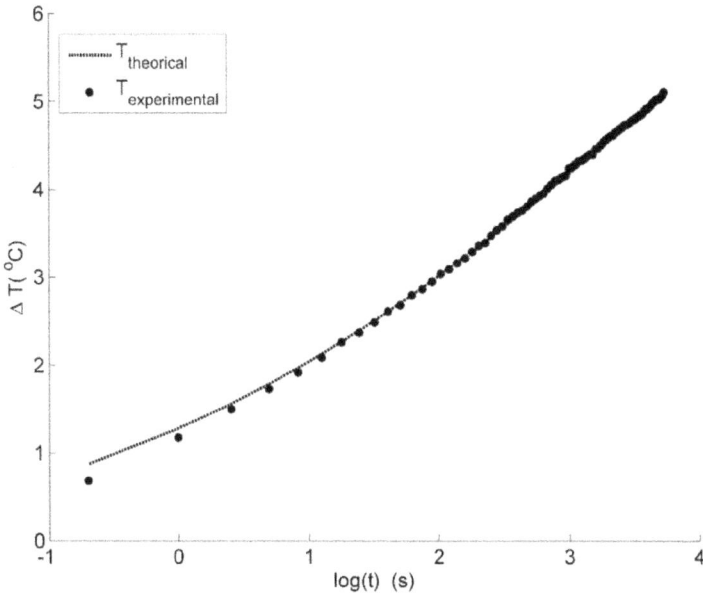

FIGURE 4.8   Comparison of measured and estimated temperatures—first step. (From Ref. [111].)

A comparison of measured and estimated temperatures for the second step, and their associated residuals, are presented in Figures 4.11 and 4.12, respectively. As for the first step (see Figures 4.8 and 4.9), these figures show an excellent agreement between the estimated and the measured temperatures, which resulted in small and slightly correlated residuals, except at small times. This clearly demonstrates the accuracy of the proposed technique for the simultaneous estimation of thermal conductivity and volumetric heat capacity with the linear heat source probe.

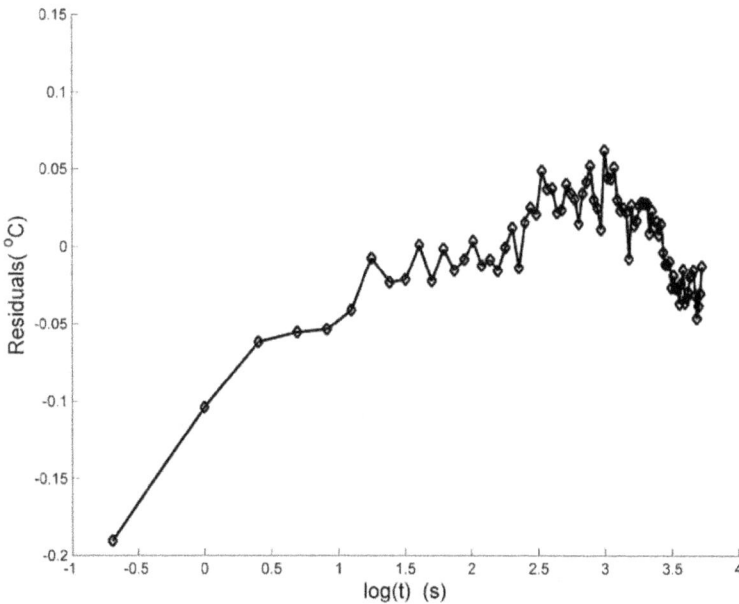

FIGURE 4.9   Temperature residuals—first step. (From Ref. [111].)

TABLE 4.2

Posterior Statistics—Second Step

Parameters	Means	Standard Deviations
k_m (W/m K)	0.2	0.013
h (W/m² K)	20.6	0.6
k_c (W/m K)	1.34	0.05
C_c (kJ/m³ K)	1596	46
C_m (kJ/m³ K)	2959	339
C_p (kJ/m³ K)	344	12
g (W)	0.386	0.007

Source: From Ref. [111].

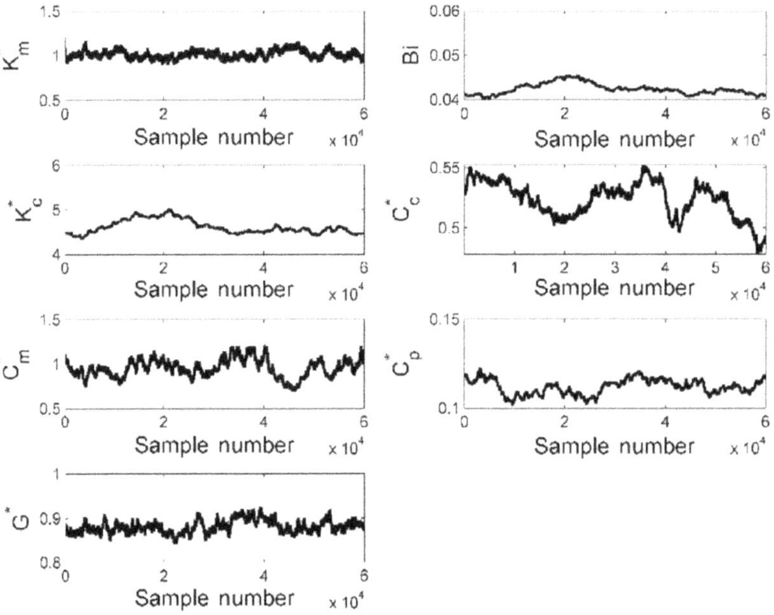

FIGURE 4.10   Markov chains—second step. (From Ref. [111].)

4.5   MCMC ESTIMATION OF THERMOPHYSICAL PARAMETERS OF THIN METAL FILMS HEATED BY FAST LASER PULSES

The thermal nonequilibrium between electrons and lattice is an important phenomenon in the study of heat transfer in thin metal films subjected to fast laser pulses. The photon energy of the laser pulse absorbed by the electrons gives rise to a hot free-electron gas, which diffuses through the metal and heats up the lattice by electron-phonon collisions. For laser pulses of duration longer than the electron-phonon thermalization time, the electrons have enough time to establish equilibrium with the lattice, so that they have the same temperature. On the other hand, for laser pulses of the order of femtoseconds and in a time scale of up to few picoseconds, the variation of the lattice temperature is small as compared to the electron temperature rise and thermal nonequilibrium can be experimentally observed [112–119].

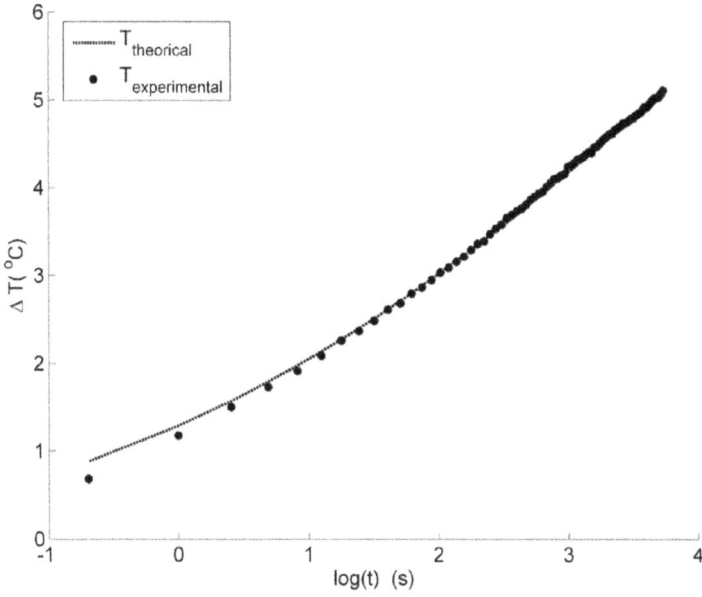

FIGURE 4.11   Comparison of measured and estimated temperatures—second step. (From Ref. [111].)

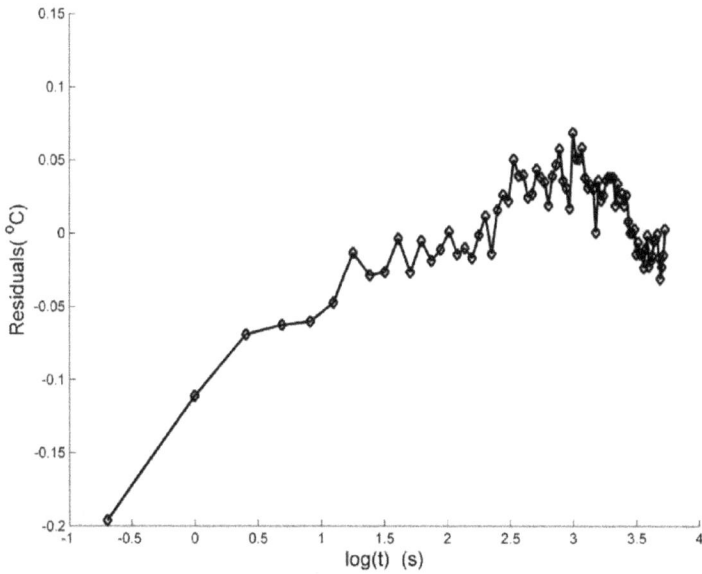

FIGURE 4.12   Temperature residuals—second step. (From Ref. [111].)

THE DIRECT PROBLEM

For the current range of laser pulse durations used for the fast heating of thin metal films, the transient nonequilibrium temperatures of electrons, T_e, and lattice, T_l, can be described by the following model, which is written for a one-dimensional problem [112–119]:

$$C_e(T_e)\frac{\partial T_e(x,t)}{\partial t} = \frac{\partial}{\partial x}\left(K\frac{\partial T_e}{\partial x}\right) - G(T_e - T_l) + Q(x,t) \qquad (4.5.1a)$$

$$C_l \frac{\partial T_l(x,t)}{\partial t} = G(T_e - T_l) \qquad\qquad (4.5.1b)$$

In equations (4.5.1a and b), C_l and C_e are the lattice and the electron volumetric heat capacities, respectively, K is the thermal conductivity of the electron gas, $Q(x, t)$ is the source term resulting from the laser heating and G is the electron-phonon coupling factor, which controls heat transfer between electrons and lattice. Diffusion can be neglected in equation (4.5.1b), since heat is mainly carried by free electrons in metals during the nonequilibrium state duration. The electron-phonon coupling factor can be theoretically predicted, but inverse analysis techniques, such as the Levenberg-Marquardt method, can also be used for its estimation [116]. The thermal conductivity, the lattice volumetric heat capacity, and the electron-phonon coupling factor can be assumed as constant. For electron temperatures like those observed in experiments, the electron heat capacity is known to vary linearly with temperature in the form [116]:

$$C_e(T_e) = \gamma T_e \qquad\qquad (4.5.2)$$

The metal film is assumed to be initially at uniform temperature and in thermal equilibrium, that is,

$$T_l(t_0) = T_e(t_0) = T_0 \qquad\qquad (4.5.3)$$

where t_0 is the initial time. Heat losses at the film surfaces are neglected due to the short duration of the related experiment, that is,

$$\frac{\partial T_e}{\partial x} = 0 \text{ at } x = 0 \text{ and at } x = L = \text{thickness of the medium} \qquad (4.5.4a,b)$$

The source term $Q(x, t)$ resultant from the laser heating has the form Ref. [116]:

$$Q(x,t) = (1-R)I\alpha e^{-\alpha x} e^{-(t/t_p)^2} \qquad\qquad (4.5.5)$$

where I is the maximum laser power flux, R is the surface reflectivity and t_p is the laser pulse duration. The absorption coefficient α is determined from the relation:

$$\alpha = \frac{4\pi n'}{\lambda} \qquad\qquad (4.5.6)$$

where λ is the wavelength of the heating laser and n' is the extinction coefficient, i.e., the coefficient of the imaginary part of the complex refractive index at the same wavelength.

For the *direct problem*, all the parameters appearing in the formulation of the physical problem are considered known and the electron and lattice temperature fields can then be obtained. The solution of the direct problem was obtained by finite differences [119]. The calculations were started at time $t_0 = -3t_p$, when the heat source term is four orders of magnitude smaller than its maximum value, so that its effects can be neglected for previous times. For the calculation of the source term, the optical properties of the metal were assumed independent of light intensity, laser pulse duration and temperature [119].

THE INVERSE PROBLEM

The *inverse problem* considered here deals with the estimation of parameters appearing in the two-temperature model, given by equations (4.5.1–4.5.6), by using transient measurements of the temperature of the electron gas at the surface heated by the laser pulse. Such measurements can be obtained with a pump-probe setup. In such an arrangement, a laser beam (pump) is used to heat the sample, while another laser of much smaller intensity (probe) is used to measure changes in the metal's reflectivity [112–119]. The sample reflectivity changes with variations in the electron and lattice temperatures. However, in a timescale of up to few picoseconds the effects due to variations of the lattice temperature can be neglected, since the electron temperature rise is much larger than that of the lattice. Therefore, the measured data is taken in the form of the normalized temperature variation, which is given by:

$$Y_i = \frac{\Delta R(t_i)}{\Delta R_{max}}$$
(4.5.7)

where $\Delta R(t_i) = R(t_i) - R(t_0)$, for $i = 1,\ldots, I$, is the surface reflectivity variation at time t_i and ΔR_{max} indicates the maximum reflectivity variation.

In order to estimate the parameters within the Bayesian framework, the likelihood and the prior distributions must be modeled. In this work, the measurement errors are considered to be additive, uncorrelated, Gaussian, with zero mean and a constant standard deviation, σ_{meas}. In such a case, the likelihood function can be written as (see equation 3.1.6):

$$\pi(\mathbf{Y}|\mathbf{P}) \propto \exp\left\{-\frac{[\mathbf{Y}-\mathbf{F}(\mathbf{P})]^T[\mathbf{Y}-\mathbf{F}(\mathbf{P})]}{2\sigma_{meas}^2}\right\}$$
(4.5.8)

where

$$\mathbf{F}^T(\mathbf{P}) = [F_1(\mathbf{P}), F_2(\mathbf{P}),\ldots, F_I(\mathbf{P})]$$
(4.5.9a)

$$F_i(\mathbf{P}) = \frac{T_e(0,t_i;\mathbf{P}) - T_0}{T_{e,max}(0;\mathbf{P}) - T_0}$$
(4.5.9b)

In equation (4.5.9b), $T_e(0,t_i;\mathbf{P})$ is the solution of the direct problem for the electron temperature at time t_i and $T_{e,max}(0;\mathbf{P})$ is the maximum electron temperature, both evaluated at the surface of the metal film where the laser heating is imposed, that is, at $x = 0$.

The parameters appearing in the formulation of the two-temperature model (see equations 4.5.1–4.5.6) are:

$$\mathbf{P}^T = [G, K, C_l, \gamma, L, t_p, I, \lambda, R, n', T_0]$$
(4.5.10)

The parameters $\mathbf{P}_{known}^T = [L, t_p, I, \lambda, R, n', T_0]$ are fairly known from other experiments, as well as from the settings of the experimental setup. For those parameters, the priors were then defined as normal distributions with means given by their nominal values and with constant standard deviations of 1% of these values. The modeling of the prior information for the other parameters is discussed below.

The solution of the present inverse problem was obtained by using simulated measured data. Three different levels of random errors were examined for the simulated measurements, with standard deviations $\sigma_{meas} = 0.01, 0.02$ or 0.05. The parameters used to generate the simulated measurements were $L = 10^{-7}$m, $t_p = 96$ fs, $I = 1.04 \times 10^{14}$ W/m^2, $\lambda = 630$ nm, $R = 0.89$, $n' = 3.183$, $T_0 = 300$ K, $G = 2.6 \times 10^{16}$ W/m^3 K, $K = 315$ W/m K, $C_l = 2.5 \times 10^6$ J/m^3 K and $\gamma = 70$ J/m^3 K^2, where the sample material was considered as gold [119].

ANALYSIS OF THE SENSITIVITY COEFFICIENTS AND DESIGN OF OPTIMUM EXPERIMENTS

The transient variations of the electron and lattice temperatures at the surface of the heated film are presented in Figure 4.13. This figure clearly shows that thermal equilibrium between electrons and lattice is only reached after 5 ps. During the period of thermal nonequilibrium, the electron

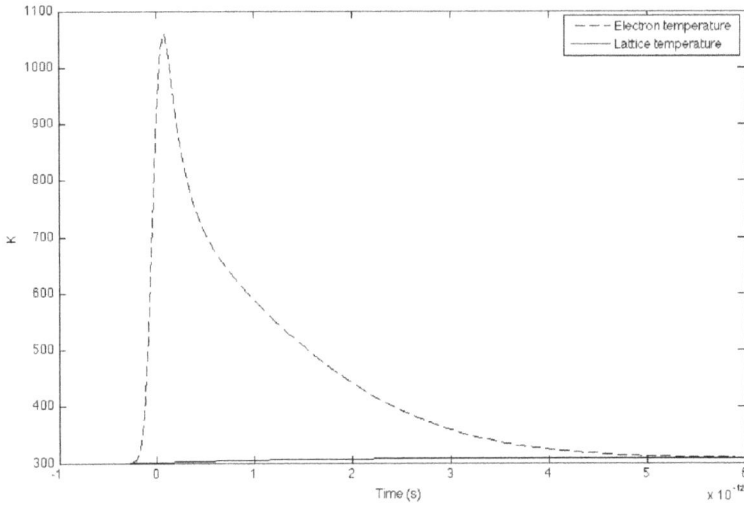

FIGURE 4.13 Electron and lattice temperatures. (From Ref. [119].)

temperature rise is much larger than that for the lattice. Therefore, the transient electron temperature rise is used as the measured data for the inverse analysis.

Figure 4.14 presents the reduced sensitivity coefficients with respect to the parameters G, K, C_l
and γ. This figure shows that the order of magnitude of the sensitivity coefficient with respect to
C_l is much smaller than those for the other parameters, being practically null during the supposed
duration of the experiment, when the thermal nonequilibrium between electrons and lattice can
be observed. Therefore, an inverse problem involving the estimation of the lattice volumetric heat
capacity during this time period is likely to be very ill-conditioned if a non-informative prior is
used for this parameter. However, the lattice heat capacity can be identified through other experimental techniques at much larger experimental times, when thermal equilibrium exists between
electrons and lattice. Hence, for the results presented below the prior distribution for C_l was taken as
Gaussian, with mean given by the nominal value used to generate the simulated measurements, and
with a constant standard deviation of 1% of this value. Anyhow, the effects of this parameter on the
measured variable are very small due to its small sensitivity coefficient.

It is also noticeable in Figure 4.14 that the relative sensitivity coefficients for G, K and γ are quite
small until around $t = 0.1$ ps and practically linearly dependent until around $t = 0.3$ ps. An analysis
of the sensitivity coefficients with respect to such parameters also reveals that they tend to zero for
large times ($t > 5$ ps), when electrons and lattice approach thermal equilibrium. Furthermore, they
tend to be linearly dependent for times greater than $t = 3$ ps. The foregoing analysis of the sensitivity coefficients reveals the possibility of simultaneously estimating G, K and γ, but it also suggests
that the experimental data should be taken within $0.1 < t < 3$ ps, when the sensitivity coefficients are
relatively large and linearly independent.

In order to optimize the time range when the temperature measurements are used for the estimation
of such three parameters, the D-optimum approach described in Chapter 2 was used. The maximi-
zation of the determinant of $\mathbf{J}^T\mathbf{J}$ was obtained for measurements taken within $0.1 < t < 2.5$ ps. Hence,
$I = 20$ equally spaced measurements in this time range were used for the estimation of G, K and
γ in the test cases examined below, where the prior information for such parameters was coded as
uniform distributions, with equally probable states within the following intervals:

$2.0 \times 10^{16} < G < 3.0 \times 10^{16}$ W/m³ K
$300 < K < 400$ W/m K
$65 < \gamma < 90$ J/m³ K²

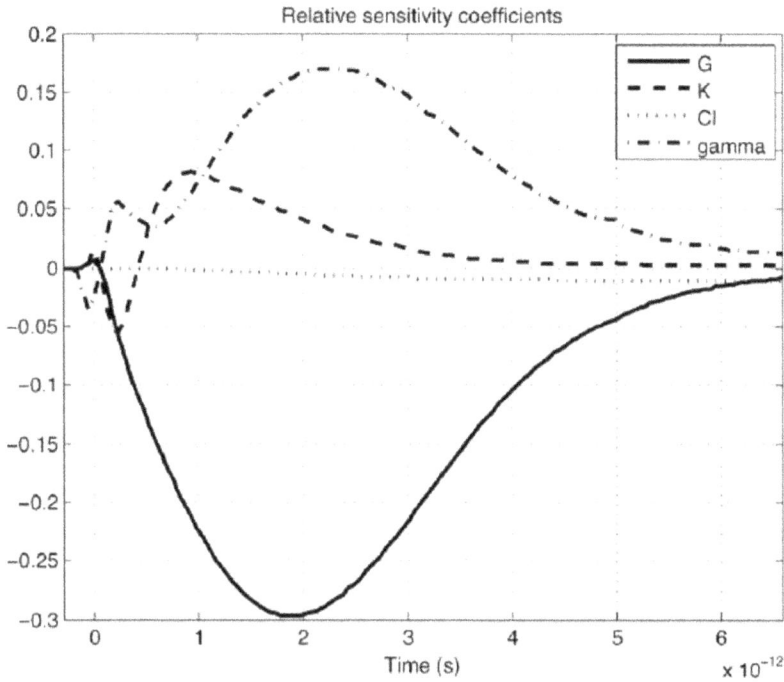

FIGURE 4.14 Reduced sensitivity coefficients. (From Ref. [119].)

STOCHASTIC SIMULATION

Table 4.3 summarizes the results obtained with such prior distributions for G, K and γ by using 20,000 states in the Markov chain and by neglecting the first 3000 states (burn-in period) for the computation of the means and standard deviations of the marginal posterior distributions of each parameter. This table shows that, for the three standard deviations of the measurement errors examined here, the estimated means are in excellent agreement with the exact parameter values used to generate the simulated measurements. In addition, the estimated standard deviations are relatively small, being at most 9% for the cases involving the largest standard deviations of the measurement errors.

TABLE 4.3
Estimated Means and Standard Deviations—MCMC

		σ_{meas}		
Estimates		**0.01**	**0.02**	**0.05**
Means	G (10^{16} W/m³ K)	2.58	2.62	2.59
	K (W/m K)	322	359	344
	γ (J/m³ K²)	70	73	76
Standard Deviations	G (10^{16} W/m³ K)	0.07	0.15	0.12
	K (W/m K)	17	24	30
	γ (J/m³ K²)	3	6	7
Relative Standard	G	3%	6%	5%
Deviations	K	5%	7%	9%
	γ	4%	8%	9%

Source: From Ref. [119].

Figures 4.15a-c present the histograms of the marginal posterior distributions for G, K and γ, respectively, for the case involving $\sigma_{meas} = 0.01$. The histogram for G resembles a Gaussian distribution. The histograms for K and γ are also similar to a Gaussian distribution, but truncated at the lower bounds of the intervals used in the uniform prior distributions for these parameters. This is due to the fact that the exact values of these two parameters are quite close to their lower bounds. However, note in Figures 4.15a-c that the exact parameter values used to generate the simulated measurements are in the regions of most probable states in their corresponding marginal posterior distributions.

4.6   ANALYSIS OF MARKOV CHAINS

In the examples presented in the previous three sections, the burn-in period was established by visual inspection of the Markov chain of each parameter. Also, the calculated statistics of the posterior distribution were only the means and standard deviations of the samples of the Markov chains after the burn-in period. Additional statistics and other concepts regarding the convergence and analysis of the Markov chains are now presented [19,59]. For simplicity in the analysis, we consider one single component P_j of the vector of parameters $\mathbf{P}$.

STATISTICS

Let $\left\{ P_j^{(1)}, P_j^{(2)}, \ldots, P_j^{(n)} \right\}$ be a homogeneous and reversible Markov chain for P_j. A real-valued function $f(P_j^{(n)})$ of the random sequence $\left\{ P_j^{(1)}, P_j^{(2)}, \ldots, P_j^{(n)} \right\}$ is called a *statistic* if it does not depend on any other unknown parameters [63,72].

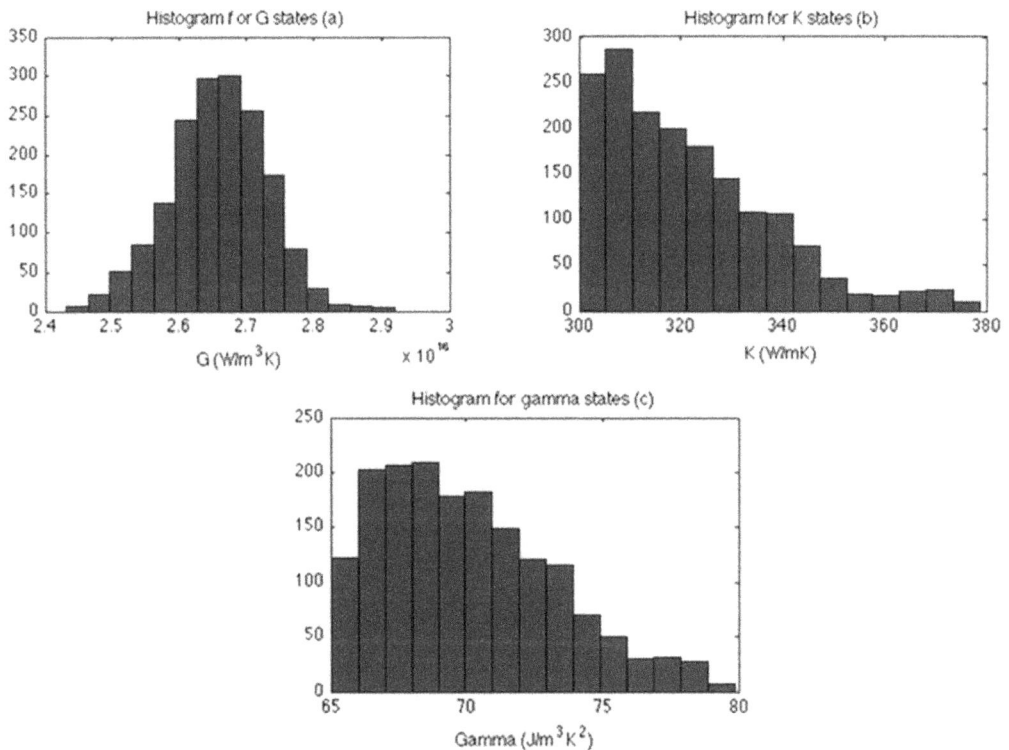

FIGURE 4.15   Histograms of the estimated parameters for measurement errors with $\sigma_{meas} = 0.01$ (From Ref. [119].)

Some useful statistics are:

$$\text{Minimum Value}: f\left(P_j^{(n)}\right) = P_{j,\min}^{(n)} = \min\left\{P_j^{(1)}, P_j^{(2)}, \ldots, P_j^{(n)}\right\} \tag{4.6.1a}$$

$$\text{Maximum Value}: f\left(P_j^{(n)}\right) = P_{j,\max}^{(n)} = \max\left\{P_j^{(1)}, P_j^{(2)}, \ldots, P_j^{(n)}\right\} \tag{4.6.1b}$$

$$\text{Median}: f\left(P_j^{(n)}\right) = \tilde{P}_j^{(n)} = \text{med}\left\{P_j^{(1)}, P_j^{(2)}, \ldots, P_j^{(n)}\right\} \tag{4.6.1c}$$

$$\text{Mean}: f\left(P_j^{(n)}\right) = \overline{P}_j^{(n)} = \frac{1}{n}\sum_{t=1}^{n} P_j^{(t)} \tag{4.6.1d}$$

$$\text{Variance}: f\left(P_j^{(n)}\right) = \text{var}\left(P_j^{(n)}\right) = \frac{1}{n-1}\sum_{t=1}^{n}\left(P_j^{(t)} - \overline{P}_j^{(n)}\right)^2 \tag{4.6.1e}$$

The variability of the samples in the Markov chain can also be reported by percentiles. A percentile x is the value of the sample for which x percent of the samples fall below or coincide with the value itself. Hence, the median is the 50th or 50% percentile [72].

Since $\left\{P_j^{(1)}, P_j^{(2)}, \ldots, P_j^{(n)}\right\}$ are realizations of a random variable, a statistic is a random variable, as well. A statistic of the samples will be a good representation of a statistic of the population if the samples are a good representation of the population. This certainly depends on the size n of the set $\left\{P_j^{(1)}, P_j^{(2)}, \ldots, P_j^{(n)}\right\}$ and on the independence of the samples. Furthermore, since the samples are obtained from a Markov chain, the chain should already have reached equilibrium before statistics can be computed to represent the solution of the inverse problem. For this reason, samples of the Markov chain are discarded before the chain reaches equilibrium.

If z states are needed for the chain to reach equilibrium, that is, the size of the burn-in period is z, the samples used for the computation of the statistics are $\left\{P_j^{(z+1)}, P_j^{(z+2)}, \ldots, P_j^{(n)}\right\}$. The index of this set is changed from $t = z+1, \ldots, n$ to $r = 1, \ldots, s$ for simplicity in the notation hereafter, where $s = n - z$ is the number of samples used for the computation of the statistics.

The *mean* of the set $P_j^{(r)} \equiv \left\{P_j^{(1)}, P_j^{(2)}, \ldots, P_j^{(s)}\right\}$ is

$$\overline{P}_j^s = \frac{1}{s}\sum_{r=1}^{s} P_j^{(r)} \tag{4.6.2}$$

If the chain is ergodic, this mean provides a strongly consistent estimate of the mean of the limiting distribution, that is,

$$\overline{P}_j^s \to E\left[P_j\right] \quad \text{as} \quad s \to \infty \tag{4.6.3}$$

This result is the *law of large numbers* for a Markov chain.

If $\left\{P_j^{(1)}, P_j^{(2)}, \ldots, P_j^{(s)}\right\}$ are independent samples, then the *variance of the mean* $\overline{P}_j^s$ is [19,59]:

$$\text{var}[\overline{P}_j^s] = \frac{\text{var}[P_j^{(r)}]}{s} \tag{4.6.4a}$$

where $\text{var}[P_j^{(r)}]$ is the variance of the set $\left\{P_j^{(1)}, P_j^{(2)}, \ldots, P_j^{(s)}\right\}$.

Unfortunately, the samples in the Markov chain usually have some degree of correlation. Hence, equation (4.6.4a) needs to be rewritten as:

$$\text{var}[\bar{P}_j^s] = \frac{\tau_j \, \text{var}[P_j^{(r)}]}{s} \tag{4.6.4b}$$

where τ_j is the *integrated autocorrelation time* (IACT) of parameter P_j, which represents the number of correlated samples between independent samples in the chain $\{P_j^{(1)}, P_j^{(2)}, ..., P_j^{(s)}\}$. Therefore, the *effective chain size*, which gives the number of independent samples in the chain, is $s_{\text{eff},j} = s/\tau_j$.

The statistical efficiency of the sampling algorithm can be assessed by examining τ_j for each parameter P_j, $j = 1,..., N$. Algorithms that result in small values of τ_j promote better sampling. The analysis of τ_j for each parameter P_j can be difficult when the problem involves many parameters. For such cases, the statistical efficiency can be examined with the IACT of the posterior distribution $\pi(\mathbf{P}^{(r)} \mid \mathbf{Y})$, $r = 1,...,s$ [60,102].

The IACT is calculated by [19,59]:

$$\tau_j = 1 + 2 \sum_{k=1}^{\infty} \rho_j(k) \tag{4.6.5}$$

where $\rho_j(k)$ is the *normalized autocovariance function of lag k*, that is,

$$\rho_j(k) = \frac{C_j(k)}{C_j(0)} \tag{4.6.6}$$

In the previous equation, $C_j(k)$ is the *autocovariance function of lag k* of the chain for the parameter P_j, which is defined by [19,59]:

$$C_j(k) = \text{cov}[P_j^{(r)}, P_j^{(r+k)}] \tag{4.6.7}$$

Clearly, the variance of $P_j^{(r)}$ is $C_j(0)$. Besides that, $\rho_j(0) = 1$, which means that the sample $P_j^{(r)}$ is perfectly correlated with itself.

The calculation of the normalized autocovariance function is straightforward, since several computational packages have functions available for this purpose. For the calculation of τ_j, the summation in equation (4.6.5) needs to be truncated at a finite number of terms $s^* \le s$. In fact, $\rho_j(k)$ is expected to tend to zero as k increases, because the samples tend to be less correlated when they are farther apart. Besides that, $\rho_j(k)$ is dominated by noise for large k. Therefore, s^* can be selected by increasing k until $\rho_j(k)$ approaches zero, thus avoiding terms that are dominated by noise in the summation of equation (4.6.5).

For s sufficiently large and for an ergodic chain, the distribution of $\dfrac{\bar{P}_j^s - E[P_j]}{\sqrt{\text{var}[\bar{P}_j^s]}}$, where $\text{var}[\bar{P}_j^s]$ is given by equation (4.6.4b), tends to a standard Gaussian distribution, with zero mean and unitary standard deviation. Thus [19,59]:

$$\frac{\bar{P}_j^s - E[P_j]}{\sqrt{\text{var}[\bar{P}_j^s]}} \to N(0,1) \quad \text{as} \quad s \to \infty \tag{4.6.8}$$

Equation (4.6.8) is a statement of the *central limit theorem* of the distribution for the mean $\bar{P}_j^s$.

Therefore, the mean of the samples in the Markov chain can be reported with its related Gaussian uncertainties as $\bar{P}_j^s \pm \eta \sqrt{\text{var}[\bar{P}_j^s]}$, where η is a constant that defines the confidence interval of $\bar{P}_j^s$ (for example, $\eta = 2.576$ for a 99% confidence interval—see Note 1 in Chapter 1).

CONVERGENCE OF THE MARKOV CHAIN

Quantitative techniques are available for the analysis of the convergence of a Markov chain to an equilibrium distribution, from which the samples can be used for the solution of the inverse problem. In the technique proposed by Geweke [120], the means calculated with samples of two different ranges of the Markov chain are compared.

Let:

$$\bar{P}_j^a = \frac{1}{s_a} \sum_{r=1}^{s_a} P_j^{(r)} \quad \text{and} \quad \bar{P}_j^b = \frac{1}{(s - s_b + 1)} \sum_{r=s_b}^{s} P_j^{(r)} \qquad (4.6.9\text{a,b})$$

be the means calculated with s_a and $(s - s_b + 1)$ states, respectively.

Geweke [120] recommended:

$$s_a = 0.1s \quad \text{and} \quad s_b = 0.5s + 1 \qquad (4.6.9\text{c,d})$$

that is, the means of the samples of the first 10% and of the last 50% of the states in the Markov chain are compared. If an equilibrium distribution is reached, $\left| \bar{P}_j^a - \bar{P}_j^b \right| \approx 0$.

For the convergence analysis, it is also recommended to repeat the sampling procedure by starting the Markov chains from different initial values. Gelman and Rubin [121] developed a method for inference on multiple chains, based on two steps: (i) An estimate is obtained for the posterior distribution with an initial Markov chain, which is then used to start new independent chains. The initial states for these new multiple chains must have a dispersion larger than that of the initial chain; (ii) the new multiple chains are then used for inference with analyses inter chains and within each chain. The posterior distribution simulated with the multiple chains exhibits a variability larger than that of the initial chain.

The multiple chains also allow a convergence analysis to an equilibrium distribution that represents the sought posterior. We consider the case of a parameter P_j, $j = 1,..., N$. The variance of the means of m chains, each one with s states, is given by [121]:

$$B_j = \frac{1}{(m-1)} \sum_{k=1}^{m} \left(\bar{P}_j^k - \bar{P}_j \right)^2 \qquad (4.6.10)$$

where $\bar{P}_j^k$ is the mean of the chain k, $k = 1,..., m$, and $\bar{P}_j$ is the mean of these means.

The mean of the m variances of the chains $k = 1,..., m$, is given by [121]:

$$W_j = \frac{1}{m(s-1)} \sum_{k=1}^{m} \sum_{r=1}^{s} \left(P_j^{(r),k} - \bar{P}_j^k \right)^2 \qquad (4.6.11)$$

where $P_j^{(r),k}$ is the sample for P_j at state r, $r = 1,..., s$, of chain k, $k = 1,..., m$.

The variance of the posterior distribution simulated with the multiple chains for P_j is thus obtained as [121]:

$$\hat{\sigma}_j^2 = \left(1 - \frac{1}{s} \right) W_j + \frac{1}{s} B_j \qquad (4.6.12)$$

The variance of the total number of samples of the multiple chains, $\hat{\sigma}_j^2$, overestimate the variance of the actual posterior if the equilibrium distribution has not been reached. On the other hand, W_j underestimates the variance of the actual posterior if each chain has not reached equilibrium. Gelman and Rubin [121] thus proposed a parameter to indicate convergence based on $\hat{\sigma}_j^2$ and W_j, called the *scale reduction coefficient*, which was simplified by Gamerman and Lopes [59] and is given by:

$$\hat{R}_j = \sqrt{\frac{\hat{\sigma}_j^2}{W_j}} \qquad (4.6.13)$$

Note that $\hat{R}_j > 1$, but $\hat{R}_j \to 1$ when $s \to \infty$. Gelman and Shirley [122] have suggested the empirical test $\hat{R}_j < 1.1$ for convergence of the multiple chains, but larger threshold values have also been proposed [59].

Example 4.2

We now consider a dimensionless version of the same problem of Examples 2.2 and 3.1. Let a semi-infinite medium, initially at temperature T_0, be heated with constant heat flux q_0 at $x = 0$, for times $t > 0$. The mathematical formulation of this problem is given by:

$$\frac{1}{\alpha}\frac{\partial T(x,t)}{\partial t} = \frac{\partial^2 T}{\partial x^2} \quad x > 0, t > 0 \qquad (4.6.14a)$$

$$-k\frac{\partial T}{\partial x} = q_0 \quad x = 0, t > 0 \qquad (4.6.14b)$$

$$T = T_0 \quad x > 0, t = 0 \qquad (4.6.14c)$$

By defining the following dimensionless groups:

$$X = \frac{x}{L} \quad \tau = \frac{\alpha_{ref} t}{L^2} \quad \theta(X,\tau) = \frac{T(x,t) - T_0}{\dfrac{q_{ref} L}{k_{ref}}} \quad \alpha^* = \frac{\alpha}{\alpha_{ref}} \qquad (4.6.15\text{a-d})$$

$$k^* = \frac{k}{k_{ref}}, \ C^* = \frac{C}{C_{ref}}, \ \alpha_{ref} = \frac{k_{ref}}{C_{ref}}, \ \phi = \frac{q_0}{q_{ref}} \qquad (4.6.15\text{e-h})$$

the analytical solution of the dimensionless version of problem (4.6.14) at the heated boundary is given by:

$$\theta(X = 0,\tau) = 2\frac{\phi}{e^*}\sqrt{\frac{\tau}{\pi}} \qquad (4.6.16)$$

where the subscript "ref" in equations (4.6.15) denotes reference values and $e^* = \sqrt{k^* C^*}$ is the dimensionless thermal effusivity. Similarly to the dimensional problem of Examples 2.2 and 3.1, equation (4.6.16) shows a perfect correlation between the parameters ϕ and e^*. Use the Metropolis-Hastings algorithm to solve the inverse problem for calibration of the test apparatus (see Example 3.1), that is, use an informative prior for the effusivity and estimate the heat flux. For the inverse analysis, use 50 simulated measurements of the dimensionless temperature $\theta(X = 0,\tau)$ up to time $\tau = 5$. Generate the simulated measurements with $\phi = 1$ and $e^* = 1$, by using additive, Gaussian and uncorrelated errors, with zero mean and constant standard deviation (σ_{meas}) equal to 1% of the maximum value of $\theta(X = 0,\tau)$.

Solution: The Metropolis-Hastings algorithm was applied here with a Gaussian proposal distribution for the parameters, with standard deviation of 0.5% of the parameter value at the current state of the Markov chain, that is,

$$P_j^* = P_j^{(t)} + \xi_j \omega_j \tag{4.6.17}$$

where ($\xi_j = 0.005 P_j^{(t)}$ see equation 4.2.6) and ω_j is a Gaussian random number with zero mean and unitary standard deviation. The Markov chains were simulated with 50,000 states. Gaussian priors with a positivity constraint were used for the parameters. A material with known effusivity was used for the calibration procedure with mean 1 and standard deviation 0.01. For the unknown heat flux, the mean and standard deviation were assigned as 2 and 0.2, respectively, such as in Example 3.1.

Figures 4.16a and b present the Markov chains for the parameters ϕ and e^*, respectively, and Figure 4.17 shows the simulated posterior distribution. Figure 4.16 shows that the Markov chains reached equilibrium after about 1000 states, so that the burn-in period was safely taken as 10,000 states. The convergence of the Markov chains to equilibrium distributions was verified with Geweke's technique described above, which revealed a relative difference between the means at the beginning and end of the chains, $\left|\bar{P}_j^a - \bar{P}_j^b\right|$, of 0.2%, where $\bar{P}_j^a$ and $\bar{P}_j^b$ are given by equations (4.6.9a and b), respectively. It is interesting to notice in Figure 4.17 the evolution of the samples in the Markov chain to eventually reach an equilibrium distribution that resembles a very stretched ellipse because of the strong correlation between the parameters. A similar result was obtained with the maximum a posteriori objective function in Example 3.1 (Figure 3.3).

The correlation between the parameters is also evident from the analysis of Figure 4.18, which presents the normalized autocovariance functions (equation 4.6.6) for both parameters. Due to the strong parameter correlation, the normalized autocovariance functions of both parameters are practically the same and the curves are superposed. The IACT given by equation (4.6.5) corresponds to 89 states for ϕ and e^*. Figure 4.18 shows that the normalized autocovariance functions decrease to zero as the lag increases, and then oscillate around zero due to the stochastic character of the Markov chains.

The histograms of the samples in the Markov chains after the burn-in period are shown by Figures 4.19a and b for the heat flux and the effusivity, respectively. These histograms resemble Gaussian distributions and are centered around the exact values used to generate the simulated measurements. The corresponding 99% credible intervals, obtained from the 0.5% and 99.5% percentiles are also shown in these figures. Note that these intervals are practically the same for the heat flux and the effusivity because of the strong correlation between the two parameters.

The variation of the posterior distribution in the Markov chain is shown by Figure 4.20. With the above hypotheses regarding the measurement errors and the Gaussian priors, the posterior distribution is given by equation (3.1.23) for positive parameters, that is, by imposing the positivity

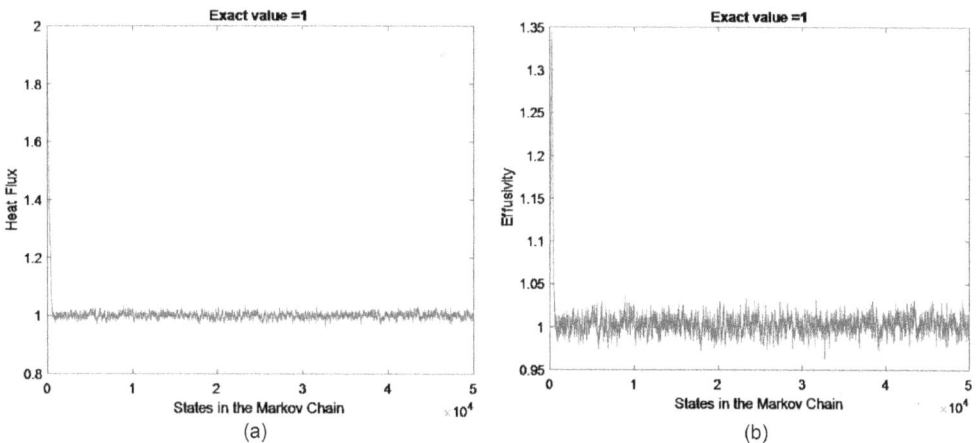

FIGURE 4.16   Markov chains for: (a) ϕ and (b) e^*.

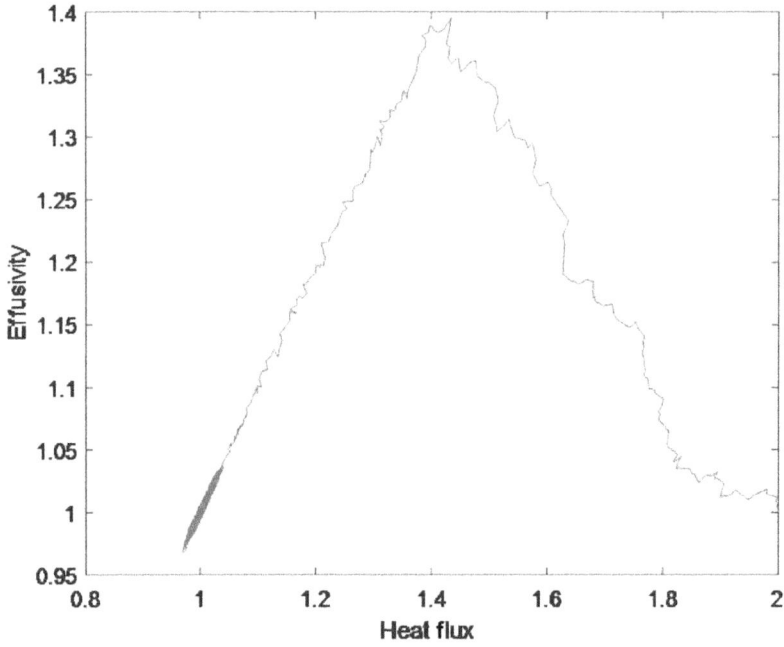

FIGURE 4.17   Simulated posterior distribution.

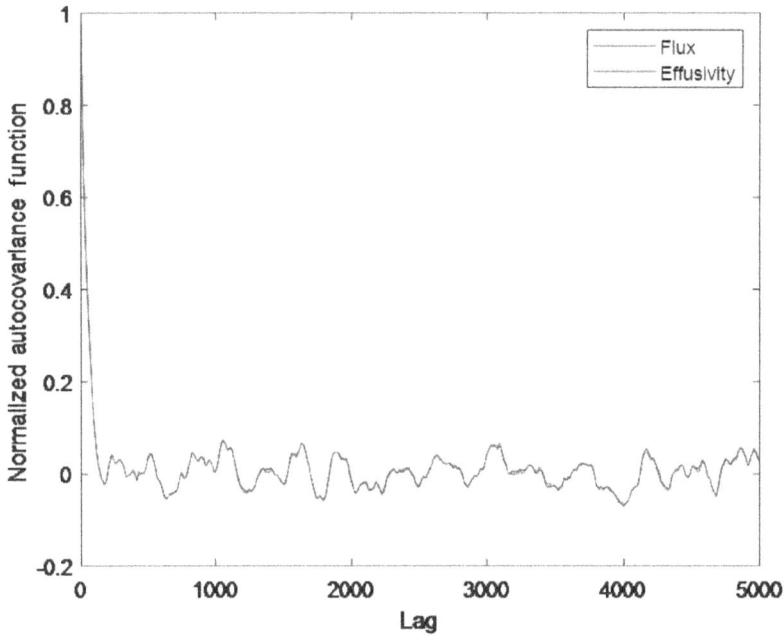

FIGURE 4.18   Normalized autocovariance function.

constraint, $\pi(\mathbf{P}|\mathbf{Y}) = 0$ for $\mathbf{P} < 0$. Figure 4.20 presents the variation of $-\ln[\pi(\mathbf{P}|\mathbf{Y})]$, which coincides with the maximum a posteriori objective function given by equation (3.1.16) for this case, that is,

$$-\ln[\pi(\mathbf{P}|\mathbf{Y})] = \sum_{i=1}^{I}\left(\frac{Y_i - \theta(X=0, \tau_i; \mathbf{P})}{\sigma_{\text{meas}}}\right)^2 + \sum_{j=1}^{N}\left(\frac{P_j - \mu_j}{\sigma_{P_j}}\right)^2 \qquad (4.6.18)$$

Exact value =1 Mean =1.0011 99% Credible interval = (0.97527,1.0273)

Exact value =1 Mean =1.0028 99% Credible interval = (0.97765,1.0281)

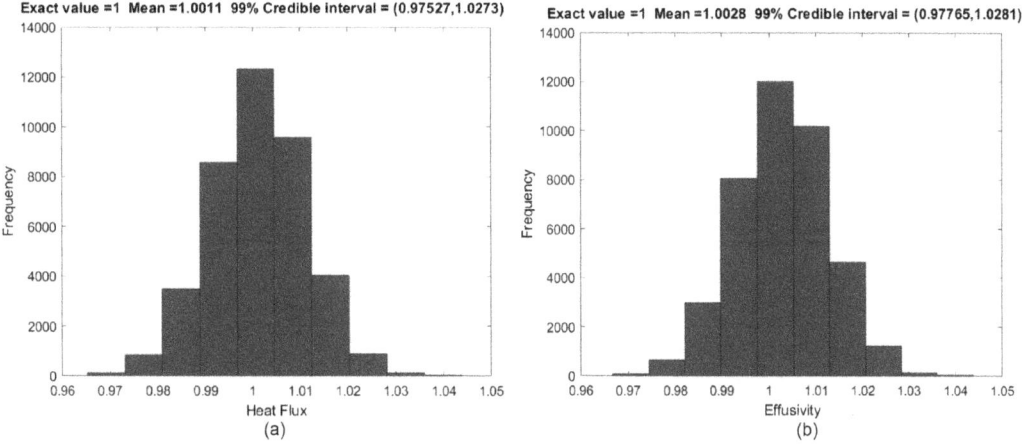

FIGURE 4.19 Histograms of the samples after the burn-in for: (a) ϕ and (b) e^{*}.

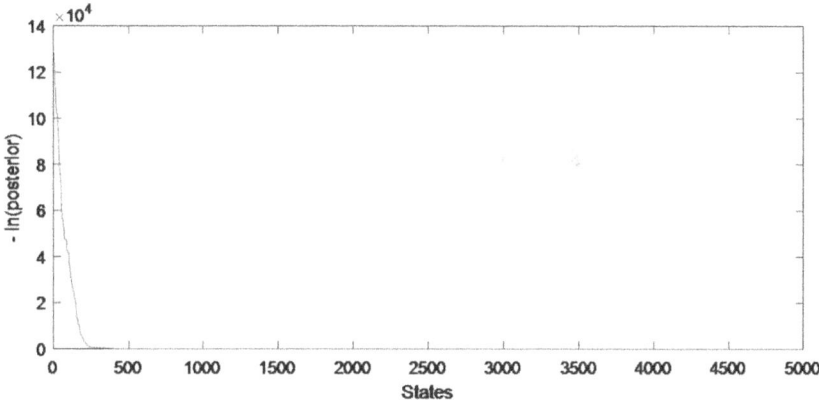

FIGURE 4.20 Variation of the posterior distribution.

Figure 4.20 shows that $-\ln[\pi(\mathbf{P}|\mathbf{Y})]$ drops very fast in the beginning of the Markov chain and then becomes practically constant with values close to zero. An enlargement of Figure 4.20 reveals that actually $-\ln[\pi(\mathbf{P}|\mathbf{Y})]$ exhibits a stochastic behavior after the Markov chain reaches equilibrium, as presented in Figure 4.21. It is interesting to notice in this figure that $-\ln[\pi(\mathbf{P}|\mathbf{Y})]$ may increase along the Markov chain because the Metropolis-Hastings algorithm may accept candidates that lead to regions of smaller posteriors in order to effectively explore the parameter space (see step 5 of the Metropolis-Hastings algorithm in Section 4.2).

The number of candidates generated with the proposal distribution that are accepted in step 5 of the Metropolis-Hastings algorithm is presented in Figure 4.22. The acceptance rate is about 30% in this case.

It is instructive to examine the effects of the standard deviation of the Gaussian proposal distribution on the Markov chains. Let us first increase ξ_j to $0.5P_j^{(t)}$ in equation (4.6.17). With such a large standard deviation, the probability of generating candidates quite far from $P_j^{(t)}$, in regions of low posteriors, increased. Hence, the acceptance rate decreased significantly and only 14 states were accepted out of the 50,000 candidates generated. Due to the low acceptance rate, many samples are repeated in the Markov chain, in accordance with step 5 of the Metropolis-Hastings algorithm. The repeated samples are correlated, so that the IACT increased from 89 for $\xi_j = 0.005P_j^{(t)}$ to 7440 for $\xi_j = 0.5P_j^{(t)}$. Therefore, only few samples in the Markov chain are statistically independent and the posterior distribution is not adequately represented by the samples generated with the Metropolis-Hastings algorithm. Such is the fact, despite that the samples in the Markov chains tended to the exact parameter values, as shown in Figures 4.23a and b. The repeated (and

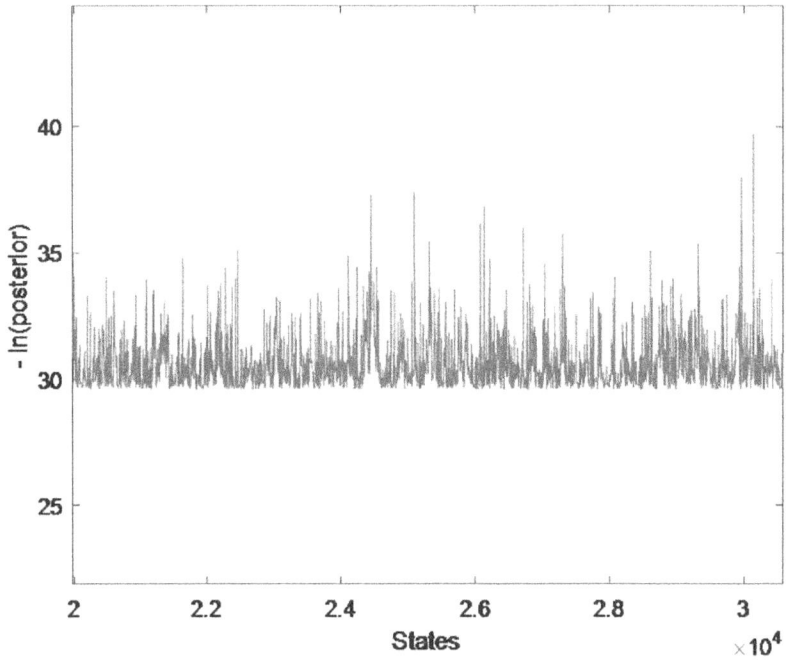

FIGURE 4.21   Enlargement of the variation of the posterior distribution.

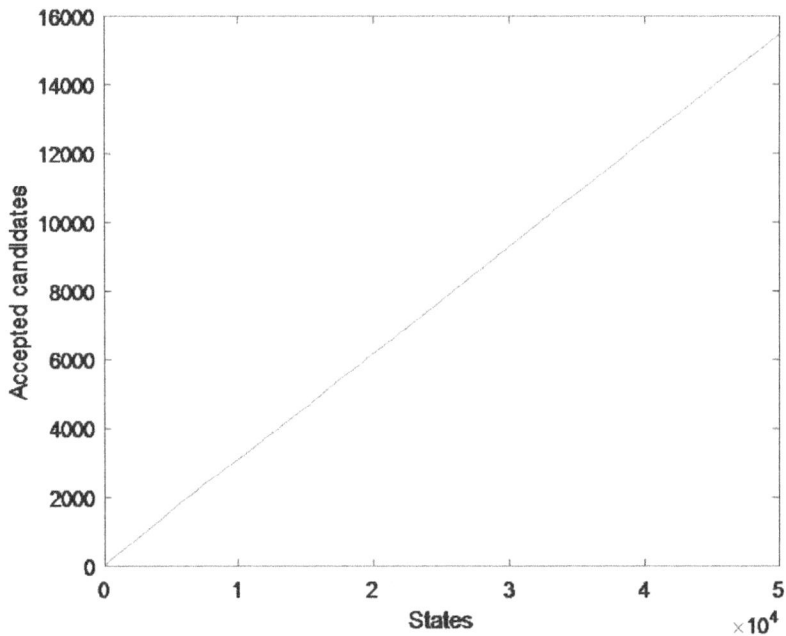

FIGURE 4.22   Number of candidates accepted.

correlated) samples in the Markov chains are clearly observed in these figures. If a Markov chain exhibits behavior similar to those of Figures 4.23a and b, a new chain needs to be generated with smaller standard deviation ξ_j of the Gaussian proposal distribution, or with a smaller maximum step size w_j in the case of a uniform proposal distribution (equation 4.2.5).

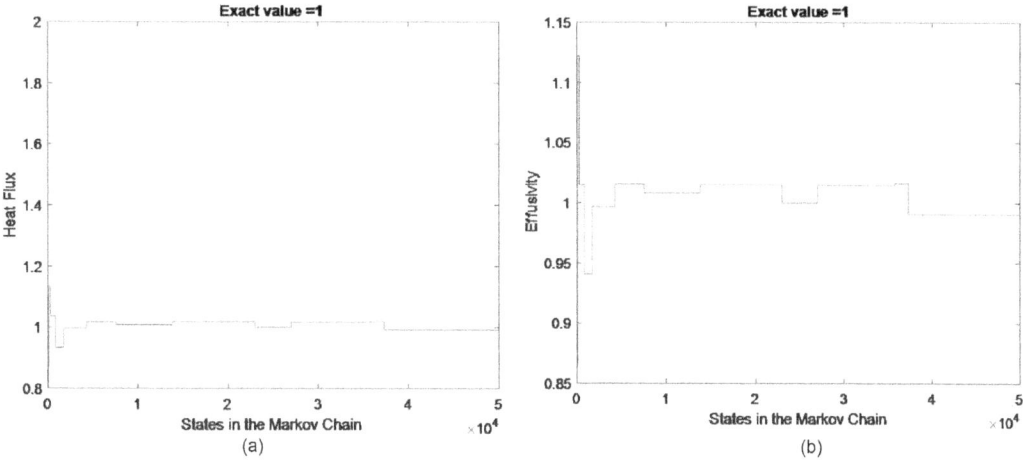

FIGURE 4.23   Markov chains obtained with $\xi_j = 0.5 P_j^{(t)}$ for: (a) ϕ and (b) e^*.

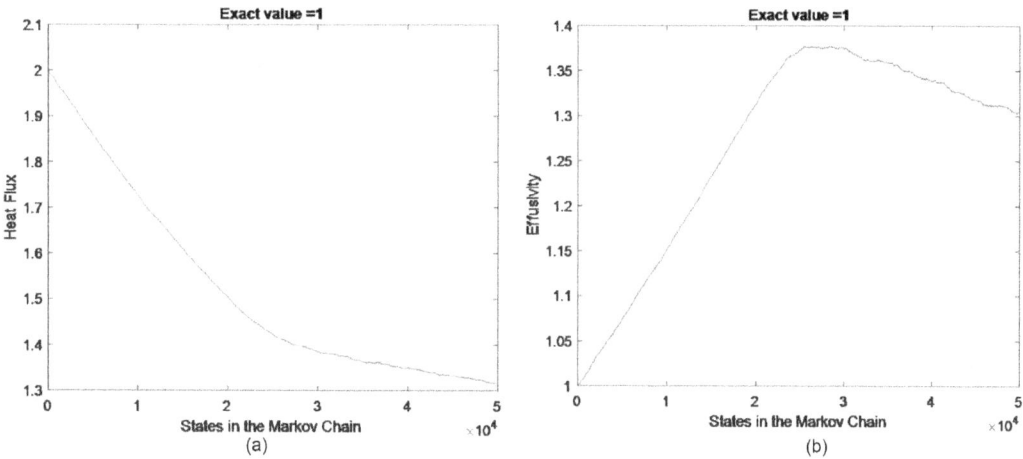

FIGURE 4.24   Markov chains obtained with $\xi_j = 5 \times 10^{-5} P_j^{(t)}$ for: (a) ϕ and (b) e^*.

We now consider a case with standard deviation of the Gaussian proposal distribution given by $\xi_j = 5 \times 10^{-5} P_j^{(t)}$ in equation (4.6.17). With such a small standard deviation, many candidates were accepted for the Markov chain, which moved very slowly and did not exhibit convergence toward an equilibrium distribution even with 50,000 states, as shown in Figures 4.24a and b. The number of candidates accepted was about 38,000, that is, 76% of all states generated. One may think that longer chains should be simulated to reach an equilibrium distribution in this case. On the other hand, the quite high rate of accepted samples suggests that the standard deviation of the Gaussian proposal should be increased, instead of simulating longer chains. In fact, many correlated samples were generated with this small standard deviation of the Gaussian proposal, resulting in IACTs of 11,773 and 8790 for the heat flux and effusivity, respectively.

PROPOSAL DISTRIBUTION

Example 4.2 above shows the importance of the proposal distribution for the success of the Metropolis-Hastings algorithm in the stochastic simulation of the sought posterior distribution. Actually, the values used in this example for the standard deviation ξ_j are only illustrative and may not be applied in general.

Theoretical results available in the literature revealed that the optimal acceptance rate with Gaussian random-walk proposals, like in equation (4.2.6), is 23.4%, for a large number (N) of parameters [102,123,124]. The Gaussian proposal should be centered at the sample of the current state of the Markov chain, $\mathbf{P}^{(t)}$, with covariance matrix $\delta^2 \Sigma$, where Σ is the covariance matrix of the asymptotic posterior distribution and $\delta = 2.38/\sqrt{N}$ [123,124]. On the other hand, Rosenthal [124] shows that the statistical efficiency of the algorithm is high for acceptance rates between 10% and 60%. In practice, numerical experiments are commonly used to obtain acceptance rates around 30% and small IACTs, which characterize a large sampling efficiency.

A Metropolis-Hastings algorithm with an adaptive proposal distribution was proposed by Haario et al. [101]. This algorithm is not Markovian, but results in ergodic distributions. In this adaptive algorithm, a Gaussian proposal is used with center at the sample $\mathbf{P}^{(t)}$ in the form [101,102]:

$$q(\mathbf{P}^* \mid \mathbf{P}^{(t)}) = \begin{cases} N\left(\mathbf{P}^{(t)}, \dfrac{0.1^2}{N}\mathbf{I}\right) & t \leq 2N \\[3mm] (1-\beta)N\left(\mathbf{P}^{(t)}, \dfrac{2.38^2}{N}\Sigma_t\right) + \beta N\left(\mathbf{P}^{(t)}, \dfrac{0.1^2}{N}\mathbf{I}\right) & t > 2N \end{cases} \tag{4.6.19}$$

where $N(\mathbf{a}, \mathbf{B})$ is a Gaussian distribution with mean $\mathbf{a}$ and covariance matrix $\mathbf{B}$, N is the number of parameters, $\mathbf{I}$ is the identity matrix and Σ_t is the covariance matrix of the posterior distribution up to state t. The positive constant β ($0 < \beta < 1$) is used to promote the mixing between $N\left(\mathbf{P}^{(t)}, \dfrac{2.38^2}{N}\Sigma_t\right)$ and $N\left(\mathbf{P}^{(t)}, \dfrac{0.1^2}{N}\mathbf{I}\right)$ in order to avoid that the algorithm halts if Σ_t is not well defined.

4.7 REDUCTION OF THE COMPUTATIONAL TIME FOR SOLVING INVERSE PROBLEMS WITH TECHNIQUE IV

For many practical cases, the computation of the direct problem solution is very time-consuming, like, for example, in nonlinear multi-dimensional problems that involve coupled phenomena. Limitations are then imposed on the number of states of the Markov chain that can be computed within a feasible time, for the solution of the inverse problem. Therefore, the use of reduced models, or even metamodels not related to any of the phenomena accounted for in the direct problem formulation, is attractive for the reduction of the computational time and implementation of MCMC methods.

Radial basis functions were applied in [125] for the interpolation of the likelihood function within a parameter space given by uniform priors. For the inverse problem of estimating the thermal conductivity components of an orthotropic medium, which was presented in Section 4.3, the reduction of computational time was about 25 times when the interpolated likelihood was used in the Metropolis-Hastings algorithm, instead of the actual likelihood. On the other hand, the interpolation had to be carried out with priors of small variances to yield posteriors similar to those obtained without interpolation. The same behavior was also observed for an inverse problem of mass transfer [125].

Modeling errors between the solution of a *complete model*, which is supposed to very accurately represent the phenomena of the problem under analysis, and the solution of an *approximate model*, can be formally taken into account by solving the inverse problem within the Bayesian framework of statistics. We present in this section two approaches to improve the solution of inverse problems obtained with approximate models: (i) The *Delayed Acceptance Metropolis-Hastings (DAMH) algorithm* [126] and (ii) The *Approximation Error Model (AEM)* [17,127–131].

In the DAMH algorithm [126], the Metropolis-Hastings algorithm is applied with the posterior distribution computed with the approximate model. If a proposal candidate is accepted with the approximate model, another test of Metropolis-Hastings is then performed with the complete model to finally decide if such candidate should be accepted or not. In this sense, DAMH can be seen as the combination of two Metropolis-Hastings algorithms, where the first one acts as a filter to pre-evaluate proposal candidates with the approximate model.

In the AEM approach [17,127–131], a statistical representation of the error between the solutions of the complete and approximate models is developed by sampling from the prior distribution for the parameters. Therefore, AEM requires priors with limited variances from which samples of the modeling errors can be generated by an offline Monte Carlo simulation. The modeling errors are then represented as additional noise in the measurement error model, and the Metropolis-Hastings algorithm is applied for the solution of the inverse problem with a modified likelihood function. It should be noted that there is a fundamental difference between the DAMH and the AEM approaches. While AEM uses a posterior distribution modified by the modeling errors, the DAMH generates samples from the correct posterior.

Delayed Acceptance Metropolis-Hastings (DAMH) Algorithm

The DAMH algorithm can be summarized as follows [126]:

1. Let $t = 0$ and start the Markov chain with the sample $\mathbf{P}^{(0)}$ at the initial state.
2. Sample a candidate point $\mathbf{P}^*$ from a proposal distribution $q(\mathbf{P}^* \mid \mathbf{P}^{(t)})$.
3. Calculate the probability $\alpha_{\mathrm{app}}(\mathbf{P}^* \mid \mathbf{P}^{(t)})$ by using the approximate model, where

$$\alpha_{\mathrm{app}}(\mathbf{P}^* \mid \mathbf{P}^{(t)}) = \min\left[1, \frac{\pi_{\mathrm{app}}(\mathbf{P}^* \mid \mathbf{Y})q(\mathbf{P}^{(t-1)} \mid \mathbf{P}^*)}{\pi_{\mathrm{app}}(\mathbf{P}^{(t-1)} \mid \mathbf{Y})q(\mathbf{P}^* \mid \mathbf{P}^{(t-1)})}\right] \qquad (4.7.1a)$$

4. Generate a random value $U_{\mathrm{app}} \sim U(0,1)$.
5. If $U_{\mathrm{app}} \le \alpha_{\mathrm{app}}(\mathbf{P}^* \mid \mathbf{P}^{(t)})$, proceed to step 6. Otherwise, return to step 2.
6. Calculate a new acceptance probability with the complete model, that is,

$$\alpha(\mathbf{P}^* \mid \mathbf{P}^{(t)}) = \min\left[1, \frac{\pi(\mathbf{P}^* \mid \mathbf{Y})q(\mathbf{P}^{(t-1)} \mid \mathbf{P}^*)}{\pi(\mathbf{P}^{(t-1)} \mid \mathbf{Y})q(\mathbf{P}^* \mid \mathbf{P}^{(t-1)})}\right] \qquad (4.7.1b)$$

7. Generate a new random value $U \sim U(0,1)$.
8. If $U \le \alpha(\mathbf{P}^* \mid \mathbf{P}^{(t)})$ set $\mathbf{P}^{(t+1)} = \mathbf{P}^*$. Otherwise, set $\mathbf{P}^{(t+1)} = \mathbf{P}^{(t)}$.
9. Make $t = t + 1$ and return to step 2 in order to generate the sequence $\left\{\mathbf{P}^{(1)}, \mathbf{P}^{(2)}, \ldots, \mathbf{P}^{(n)}\right\}$.

In equations (4.7.1a and b), $\pi_{\mathrm{app}}(\mathbf{P}|\mathbf{Y})$ and $\pi(\mathbf{P}|\mathbf{Y})$ are the posterior distributions with the likelihoods computed by using the approximate model and the complete model, respectively. In the DAMH algorithm, the approximate model is applied in the test of step 5 to possibly generate good candidates to be accepted with the complete model in step 8.

The DAMH algorithm can be quite effective, because the acceptance rates in the regular Metropolis-Hastings algorithm are expected to be around 30%, as discussed above. Hence, depending on the acceptance rate and on how fast and accurate the solution of the approximate model is, the use of the DAMH algorithm might result in significant reductions in computational times, as compared to those of the regular Metropolis-Hastings algorithm applied with the complete model.

Approximation Error Model (AEM) Approach

In this approach, the statistical representation of the approximation error, between the solutions of the approximate and complete models, is constructed and then represented as additional noise in the likelihood [17,127–131]. With the hypotheses that the measurement errors are additive and independent of the parameters **P**, we have:

$$\mathbf{Y} = \mathbf{T}(\mathbf{P}) + \boldsymbol{\varepsilon} \tag{4.7.2}$$

where **T**(**P**) is the accurate solution of the complete model. In this book, the vector of measurement errors, $\boldsymbol{\varepsilon}$, is assumed to be Gaussian, with zero mean and known covariance matrix **W**, so that the likelihood function is given by equation (3.1.6).

If the solution of an approximate model, $\mathbf{T}_{app}(\mathbf{P}_{app})$, is used for the solution of the inverse problem instead of the solution of the complete model, **T**(**P**), equation (4.7.2) becomes:

$$\mathbf{Y} = \mathbf{T}_{app}(\mathbf{P}_{app}) + [\mathbf{T}(\mathbf{P}) - \mathbf{T}_{app}(\mathbf{P}_{app})] + \boldsymbol{\varepsilon} \tag{4.7.3}$$

where $\mathbf{P}_{app}$ is the vector of parameters of the approximate model, which is usually a subset of **P**.

By defining the error between the solutions of the complete and approximate models as:

$$\mathbf{e}(\mathbf{P}) = [\mathbf{T}(\mathbf{P}) - \mathbf{T}_{app}(\mathbf{P}_{app})] \tag{4.7.4}$$

equation (4.7.3) can be written as:

$$\mathbf{Y} = \mathbf{T}_{app}(\mathbf{P}_{app}) + \boldsymbol{\eta}(\mathbf{P}) \tag{4.7.5}$$

where

$$\boldsymbol{\eta}(\mathbf{P}) = \mathbf{e}(\mathbf{P}) + \boldsymbol{\varepsilon} \tag{4.7.6}$$

that is, the total error $\boldsymbol{\eta}(\mathbf{P})$ includes the modeling errors, $\mathbf{e}(\mathbf{P})$, as well as the experimental errors, $\boldsymbol{\varepsilon}$.

For the implementation of the AEM, the statistics of $\mathbf{e}(\mathbf{P})$ is computed before the solution of the inverse problem, as follows:

1. Let $k = 1$. Obtain samples $\mathbf{P}^{(k)}$ and $\mathbf{P}_{app}^{(k)}$ from the prior distribution for the parameters.
2. Solve the complete model with the parameter vector $\mathbf{P}^{(k)}$ and the approximate model with the parameter vector $\mathbf{P}_{app}^{(k)}$.
3. Compute the sample of the approximation error:

$$\mathbf{e}^{(k)}(\mathbf{P}) = [\mathbf{T}(\mathbf{P}^{(k)}) - \mathbf{T}_{app}(\mathbf{P}_{app}^{(k)})] \tag{4.7.7}$$

4. Make $k = k+1$.
5. Repeat the steps above until the statistics of $\mathbf{e}(\mathbf{P})$, obtained with the samples $\mathbf{e}^{(k)}(\mathbf{P})$, converge.

As it will be apparent below, the statistics of $\mathbf{e}^{(k)}(\mathbf{P})$ that are of interest for the AEM are the mean and the covariance matrix. They can be calculated as follows (see Note 1 in Chapter 1):

$$\bar{\mathbf{e}} = \frac{1}{N_s} \sum_{k=1}^{N_s} \mathbf{e}^{(k)}(\mathbf{P}) \tag{4.7.8a}$$

$$\mathbf{W}_{\mathbf{e}} = \frac{1}{N_s - 1} \sum_{k=1}^{N_s} \left[\mathbf{e}^{(k)}(\mathbf{P}) - \bar{\mathbf{e}} \right]\left[\mathbf{e}^{(k)}(\mathbf{P}) - \bar{\mathbf{e}} \right]^T \tag{4.7.8b}$$

where N_s is the number of samples of the approximation error.

We note that the mean $\bar{\mathbf{e}}$ usually converges faster than the variances and covariances in $\mathbf{W}_e$. The convergence of $\mathbf{W}_e$ can be verified in terms of the convergence of its trace or its largest eigenvalues. Also, note that the AEM cannot be applied with improper priors, since they have unbounded variances and the samples $\mathbf{P}^{(k)}$ and $\mathbf{P}_{app}^{(k)}$ cannot be obtained.

After obtaining the samples $\mathbf{e}^{(k)}(\mathbf{P})$ in a sufficient number N_s for which $\bar{\mathbf{e}}$ and $\mathbf{W}_e$ are converged, these random variables are modeled in terms of an analytical distribution. A simple but very effective approach is to consider that they follow a Gaussian distribution [17,127–131], that is,

$$\pi(\mathbf{e}) \propto \exp\left\{-\frac{1}{2}[\mathbf{e}-\bar{\mathbf{e}}]^T \mathbf{W}_e^{-1}[\mathbf{e}-\bar{\mathbf{e}}]\right\} \tag{4.7.9}$$

Consider a Gaussian prior with mean μ and covariance matrix $\mathbf{V}$, and a Gaussian likelihood with zero mean and covariance matrix $\mathbf{W}$, like in equation (3.1.19). Then, the posterior with the error model of equation (4.7.5), with $\eta(\mathbf{P})$ given by equation (4.7.6) and $\mathbf{e}$ modeled as a Gaussian variable such as in equation (4.7.9), is given by [127]:

$$\pi(\mathbf{P}|\mathbf{Y}) \propto \exp\left\{-\frac{1}{2}[\mathbf{Y}-\mathbf{T}_{app}(\mathbf{P}_{app})-\bar{\eta}]^T \tilde{\mathbf{W}}^{-1}[\mathbf{Y}-\mathbf{T}_{app}(\mathbf{P}_{app})-\bar{\eta}]-\frac{1}{2}(\mathbf{P}-\mu)^T \mathbf{V}^{-1}(\mathbf{P}-\mu)\right\} \tag{4.7.10}$$

where

$$\bar{\eta} = \bar{\varepsilon}+\bar{\mathbf{e}}+\Gamma_{\eta\mathbf{P}}\mathbf{V}^{-1}(\mathbf{P}-\mu) \tag{4.7.11a}$$

$$\tilde{\mathbf{W}} = \mathbf{W}_e+\mathbf{W}-\Gamma_{\eta\mathbf{P}}\mathbf{V}^{-1}\Gamma_{\mathbf{P}\eta} \tag{4.7.11b}$$

and $\bar{\varepsilon}$ and $\bar{\mathbf{e}}$ are the means of ε and $\mathbf{e}$, respectively, while $\mathbf{W}_e$ is the covariance of $\mathbf{e}$ and $\Gamma_{\eta\mathbf{P}}$ is the covariance of η and $\mathbf{P}$. Equations (4.7.11a and b) give the *Complete Error Model* [127].

We note that, with the standard hypotheses regarding the measurement errors made above, $\bar{\varepsilon} = 0$. By further neglecting the linear dependence between η and $\mathbf{P}$, that is, $\Gamma_{\eta\mathbf{P}} = 0$, equations (4.7.11a and b) simplify to the so-called *Enhanced Error Model*:

$$\bar{\eta} \approx \bar{\mathbf{e}} \tag{4.7.12a}$$

$$\tilde{\mathbf{W}} \approx \mathbf{W}_e+\mathbf{W} \tag{4.7.12b}$$

Therefore, in the *AEM* approach, the Metropolis-Hastings algorithm can be readily applied with the posterior distribution given by equation (4.7.10), where the original likelihood was modified by the mean and covariance matrix of the total error $\eta(\mathbf{P})$, that is, $\bar{\eta}$ and $\tilde{\mathbf{W}}$, respectively. These two quantities are usually quite well approximated by equations (4.7.12a and b), respectively, and application of the more complicated complete error model given by equations (4.7.11a and b) is not needed.

In the next section, the *AEM* approach is applied to an extension of the problem examined in Section 4.4, involving the measurement of thermophysical properties of liquids with the line heat source probe method.

4.8 APPROXIMATION ERROR MODEL TO ACCOUNT FOR CONVECTIVE EFFECTS IN THE LINE HEAT SOURCE PROBE METHOD

The application of the transient line heat source probe method (described in detail in Section 4.4) to low viscous liquids is more susceptible to erroneous results, due to the early onset of natural convection, particularly for experiments above the ambient temperature. The AEM approach was then used here to compensate for modeling errors, between a natural convection *complete* model and a heat conduction *approximate* model, in the estimation of thermal properties of a liquid with the line heat source probe. Statistics of the modeling errors were estimated over the prior distributions of the different parameters appearing in the mathematical formulation and were accounted for in the inverse analysis by applying the Metropolis-Hastings algorithm with the posterior distribution given by equation (4.7.10). Simulated temperature measurements taken with a sensor located inside the probe were used in the inverse analysis for the case of water [132]. The physical problem under analysis in this work consisted of a thin cylinder (the needle of the probe), which was fully inserted into a liquid with unknown thermal properties, contained in a cylindrical cell. Heat was generated at a constant and uniform rate inside the needle.

For the *complete model*, we assumed axial symmetry around the needle and homogeneous and isotropic media with constant thermal properties. Heat transfer within the probe and the cylindrical cell took place by conduction. Heat transfer within the liquid occurred by natural convection and Boussinesq's approximation was considered valid. At the initial time, all the regions were assumed in thermal equilibrium with the external medium and the liquid was assumed to be a rest. The *reduced model* was formulated by the one-dimensional heat conduction problem given by equations (4.4.1a-h).

The following materials were considered [132]: Pyrex for the cell container ($k_c = 1.3$ W/m K, $c_c = 750$ J/kg K, $\rho_c = 2200$ kg/m^3), water as the fluid whose thermal properties are of interest ($k_f = 0.6$ W/m K, $c_f = 4180$ J/kg m^3, $\rho_f = 1000$ kg/m^3, $v_f = 9 \times 10^{-4}$ Pa s, $\beta_f = 2.569 \times 10^{-4}$ K^{-1}) and a highly conductive material for the probe ($k_p = 20$ W/m K, $C_p = 358,812$ J/kg m^3). In addition, the liquid was assumed to be at atmospheric pressure, and the heat power delivered inside the probe was taken as 0.488 W. The solution of the two-dimensional natural convection complete model was obtained with the finite element method, while the reduced one-dimensional heat conduction model was solved by implicit finite volumes [132].

Figure 4.25a presents the velocity field in water, after 14 s that the heating of the probe was started. This figure clearly shows the hydrodynamic boundary layer formed around the probe. As a result of natural convection, the temperature increase in the probe, which was used for the estimation of the fluid thermophysical properties, was smaller than that for heat conduction, as illustrated in Figure 4.25b. Therefore, larger thermal conductivity values were estimated for the fluid if measurements affected by natural convection were used in the solution of the inverse problem with a heat conduction model.

Gaussian priors were assumed for the parameters, in order to compute the statistics of the approximation error with 1000 samples of a Monte Carlo simulation. The mean and the covariance matrix of the approximation error, given by equations (4.7.8a and b), respectively, were then used in

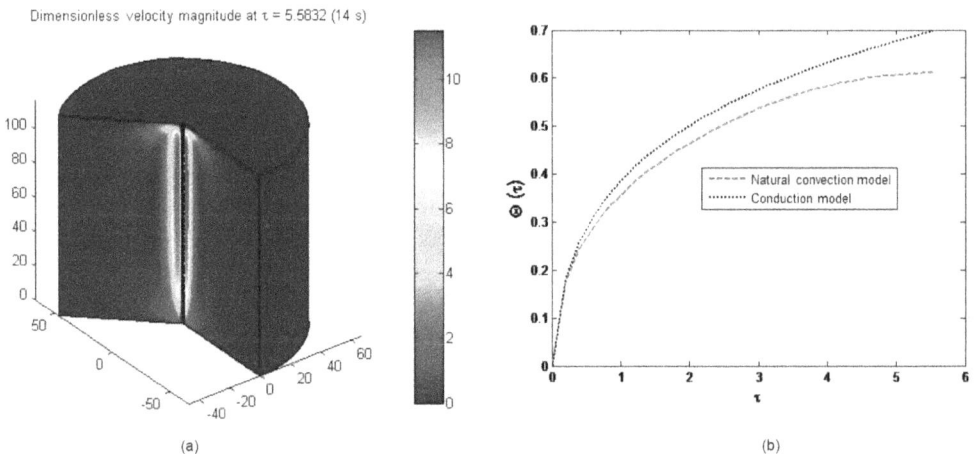

(a)   (b)

FIGURE 4.25 (a) Velocity field in water and (b) temperature variation in the probe with the natural convection model and with the heat conduction model. (From Ref. [132].)

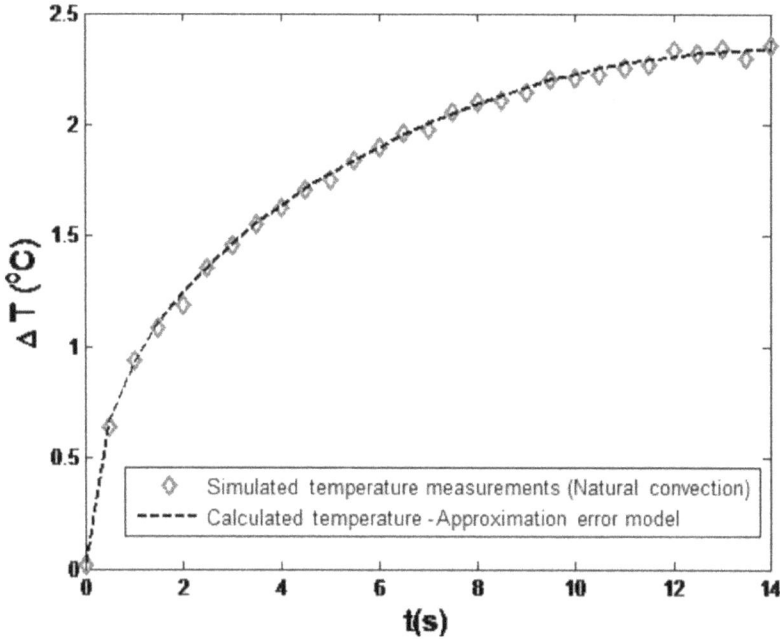

FIGURE 4.26   Comparison of simulated measurements and estimated temperatures. (From Ref. [132].)

the posterior distribution (4.7.10), for the application of the Metropolis-Hastings algorithm. The estimated parameters were in excellent agreement with those used to generate the simulated measurements [132]. Moreover, Figure 4.26 presents a comparison between the simulated measurements that were generated with the complete model, and the estimated temperatures that resulted from the inverse problem solution with the reduced model. Despite the fact that a reduced model was used for the solution of the inverse problem, the AEM approach resulted in accurate estimated parameters and in an excellent agreement between simulated measurements and estimated temperatures.

PROBLEMS

4.1   Consider heat conduction in a semi-infinite medium ($x > 0$), initially at the uniform temperature T_0. For times $t > 0$, the temperature of the surface at $x = 0$ is maintained at $T = 0°C$. The analytical solution for this heat conduction problem is given by:

$$T(x,t) = T_0 \, \text{erf}\left(\frac{x}{\sqrt{4\alpha t}} \right)$$

where α is the thermal diffusivity and erf(.) is the error function [56]. Solve the inverse problem of estimating the model parameters T_0 and α with the Metropolis-Hastings algorithm, considering transient measurements of a sensor located at $x_{\text{meas}} = 0.01$ m. Generate the simulated measurements with $T_0 = 50°C$, $\alpha = 4.9 \times 10^{-7} \text{m}^2/\text{s}$ and additive, uncorrelated, Gaussian errors, with zero mean and standard deviation of 1°C. Use a uniform prior for the parameters, as follows: $40°C < T_0 < 65°C$ and $10^{-7} \text{m}^2/\text{s} < \alpha < 10^{-5} \text{m}^2/\text{s}$. Use a random walk proposal with a uniform distribution and examine the effects of the random walk variation on the Markov chains.

4.2   Repeat Problem 4.1 but use a random walk proposal with a Gaussian distribution.

4.3   Repeat Problem 4.2 but with independent Gaussian prior distributions. Examine the effects of the prior means and standard deviations on the resulting Markov chains.

NOTE 1: METROPOLIS-HASTINGS ALGORITHM WITH SAMPLING BY BLOCKS OF PARAMETERS

Different versions of the Metropolis-Hastings algorithm can be found in the literature (see, for example [60]). In particular, a modified version of the Metropolis-Hastings algorithm has been proposed for cases that involve groups of linearly dependent parameters [59,102]. In this case, the sampling procedure and the acceptance/rejection test are performed separately for each block of parameters, within one iteration of the Metropolis-Hastings algorithm [59,102].

As an example, we consider a case where the vector of parameters $\mathbf{P}$ is split into two groups of parameters: $\mathbf{P}_1$ and $\mathbf{P}_2$. The Metropolis-Hastings algorithm with sampling by block of parameters is summarized by the following steps:

1. Let $t = 0$ and start the Markov chains with the sample $\mathbf{P}^{(0)}$.
2. Sample candidates $\mathbf{P}_1^*$ from the proposal distribution $q_1(\mathbf{P}_1^* \mid \mathbf{P}_1^{(t)})$ for the vector $\mathbf{P}_1$ and make $\mathbf{P}_2^* = \mathbf{P}_2^{(t)}$.
3. Compute the Metropolis-Hastings ratio

$$\alpha_1(\mathbf{P}^* \mid \mathbf{P}^{(t)}) = \min\left[1, \frac{\pi(\mathbf{P}^* \mid \mathbf{Y}) q_1(\mathbf{P}_1^{(t)} \mid \mathbf{P}_1^*)}{\pi(\mathbf{P}^{(t)} \mid \mathbf{Y}) q_1(\mathbf{P}_1^* \mid \mathbf{P}_1^{(t)})}\right]$$

4. Generate a random number with a uniform distribution in $(0,1)$, $U_1 \sim U(0,1)$.
5. If $U_1 \leq \alpha_1(\mathbf{P}^* \mid \mathbf{P}^{(t)})$, make $\mathbf{P}_1^{(t+1)} = \mathbf{P}_1^*$. Otherwise, make $\mathbf{P}_1^{(t+1)} = \mathbf{P}_1^{(t)}$.
6. Sample candidates $\mathbf{P}_2^*$ from the proposal distribution $q_2(\mathbf{P}_2^* \mid \mathbf{P}_2^{(t)})$ for the vector $\mathbf{P}_2$ and make $\mathbf{P}_1^* = \mathbf{P}_1^{(t+1)}$.
7. Compute the Metropolis-Hastings ratio

$$\alpha_2(\mathbf{P}^* \mid \mathbf{P}^{(t)}) = \min\left[1, \frac{\pi(\mathbf{P}^* \mid \mathbf{Y}) q_2(\mathbf{P}_2^{(t)} \mid \mathbf{P}_2^*)}{\pi(\mathbf{P}^{(t)} \mid \mathbf{Y}) q_2(\mathbf{P}_2^* \mid \mathbf{P}_2^{(t)})}\right]$$

8. Generate a random number with a uniform distribution in $(0,1)$, $U_2 \sim U(0,1)$.
9. If $U_2 \leq \alpha_2(\mathbf{P}^* \mid \mathbf{P}^{(t)})$, make $\mathbf{P}_2^{(t+1)} = \mathbf{P}_2^*$. Otherwise, make $\mathbf{P}_2^{(t+1)} = \mathbf{P}_2^{(t)}$.
10. Let $t = t + 1$ and return to step 2 in order to generate the sequence $\{\mathbf{P}^{(1)}, \mathbf{P}^{(2)}, ..., \mathbf{P}^{(n)}\}$.

Part II

Function Estimation

5 Function Estimation
Minimization of an Objective Functional without Prior Information about the Unknown Functions

This book has been devoted so far to the solution of inverse problems where the unknowns were constant parameters appearing in the formulation, and/or parameters in the approximation of a function by a small number of known basis functions. The second part of the book is now aimed at the solution of inverse problems where the unknown functions are not approximated by few basis functions, differently from the procedure used in Section 2.9.

In this chapter, a powerful and robust technique is presented, where the only assumption made for the solution of the inverse problem is that the unknown function belongs to Hilbert space of square-integrable functions in the domain of interest (see Note 1 at the end of this chapter for basic concepts regarding Hilbert spaces). The solution of the inverse problem is obtained here through the minimization of an *objective functional*, which is defined based on statistical hypotheses for the measurement errors. A *functional* can be broadly defined as a *function of functions* and, for the solution of an inverse problem, it maps the space of unknown functions into the real non-negative numbers [133,134].

The solution technique presented in this chapter is not Bayesian, since it does not take into account prior information about the unknown function. The solution of inverse problems of function estimation within the Bayesian framework is addressed in Chapter 6.

5.1 TECHNIQUE V: THE CONJUGATE GRADIENT METHOD WITH ADJOINT PROBLEM FOR FUNCTION ESTIMATION

In this section, we present a powerful iterative minimization scheme called the *conjugate gradient method of minimization with adjoint problem*, for solving inverse heat transfer problems of *function estimation*. In this method, no *a priori* information on the functional form of the unknown function is required, except for the functional space which it belongs to. Regularization of the method is obtained by selecting the stopping criterion for the iterative procedure, so that stable solutions are obtained. This method belongs to the class of *Alifanov's iterative regularization* techniques [2,6,22,29–43,92,135].

The basic steps of Technique V for the solution of function estimation problems are very similar to those of Technique III for parameter estimation problems. They include:

- Direct problem
- Inverse problem
- Sensitivity problem
- Adjoint problem

- Gradient equation
- Iterative procedure
- Stopping criterion
- Computational algorithm

In order to develop such basic steps of Technique V, we use the inverse heat conduction problem of estimating the unknown time-varying strength $g_p(t)$ of a plane energy source examined in Chapter 2 (see Figure 2.1). While in Chapter 2 $g_p(t)$ was approximated in terms of N known basis functions $C_j(t)$, that is,

$$g_p(t) = \sum_{j=1}^{N} P_j C_j(t) \tag{5.1.1}$$

we now only assume that the unknown function belongs to the Hilbert space of square-integrable functions in the time domain, denoted as $L_2(0, t_f)$, where t_f is the duration of the experiment. Functions in such space satisfy the following property [2,6,22,29–43,92,133–135]:

$$\int_{t=0}^{t_f} [g_p(t)]^2 dt < \infty \tag{5.1.2}$$

For the solution of the inverse problem, we consider that transient measurements are available from one single sensor located at the position x_{meas} inside the slab, as illustrated in Figure 2.1. The measured data are assumed to be continuous in time. Besides that, the measurement errors are supposed additive, Gaussian, uncorrelated, with zero mean and known variance given by the function $\sigma^2(t)$. Thus, the inverse problem is solved with the minimization of the following objective functional, which is analogous to the objective function given by equation (2.9.1a):

$$S[g_p(t)] = \int_{t=0}^{t_f} \frac{[Y(t) - T(x_{\text{meas}}, t; g_p(t))]^2}{[\sigma(t)]^2} dt \tag{5.1.3a}$$

where $Y(t)$ is the measured temperature and $T(x_{\text{meas}}, t; g_p(t))$ is the estimated temperature obtained from the solution of the direct problem. If the variance of the measurements is constant, the minimization of (5.1.3a) is equivalent to the minimization of the following objective functional:

$$S[g_p(t)] = \int_{t=0}^{t_f} [Y(t) - T(x_{\text{meas}}, t; g_p(t))]^2 dt \tag{5.1.3b}$$

THE DIRECT PROBLEM

The direct problem involves the determination of the temperature field in the medium when the source term is known. The formulation of the direct problem is given by:

$$\frac{\partial^2 T(x,t)}{\partial x^2} + g_p(t)\delta(x - 0.5) = \frac{\partial T(x,t)}{\partial t} \quad \text{in } 0 < x < 1, \text{ for } t > 0 \tag{5.1.4a}$$

$$\frac{\partial T(0,t)}{\partial x} = 0 \quad \text{at } x = 0, \text{ for } t > 0 \tag{5.1.4b}$$

$$\frac{\partial T(1,t)}{\partial x} = 0 \text{ at } x = 1, \text{ for } t > 0 \tag{5.1.4c}$$

$$T(0,t) = 0 \text{ for } t = 0, \text{ in } 0 < x < 1 \tag{5.1.4d}$$

THE INVERSE PROBLEM

In the inverse problem considered here, the source term $g_p(t)$ is an unknown function of time, while transient measured temperature data, $Y(t)$, taken at the location x_{meas}, are available over the time domain $0 < t \le t_f$, where t_f is the final time. The inverse problem is solved by minimizing the objective functional (5.1.3a) with the conjugate gradient method. Auxiliary problems are required for the implementation of the iterative procedure of the conjugate gradient method, namely the sensitivity problem and the adjoint problem, which are derived below.

THE SENSITIVITY PROBLEM

The derivation of the sensitivity problem for Technique V is very similar to that for Technique III. It is assumed that when $g_p(t)$ undergoes a variation $\Delta g_p(t)$, the temperature $T(x,t)$ varies by an amount $\Delta T(x,t)$. Therefore, we replace $T(x,t)$ by $[T(x,t) + \Delta T(x,t)]$ and $g_p(t)$ by $[g_p(t) + \Delta g_p(t)]$ in the direct Problem (5.1.4) and subtract from it the original Problem (5.1.4), in order to obtain the *sensitivity problem* given by:

$$\frac{\partial^2 \Delta T(x,t)}{\partial x^2} + \Delta g_p(t)\, \delta(x - 0.5) = \frac{\partial \Delta T(x,t)}{\partial t} \text{ in } 0 < x < 1, \text{ for } t > 0 \tag{5.1.5a}$$

$$\frac{\partial \Delta T(0,t)}{\partial x} = 0 \text{ at } x = 0, \text{ for } t > 0 \tag{5.1.5b}$$

$$\frac{\partial \Delta T(1,t)}{\partial x} = 0 \text{ at } x = 1, \text{ for } t > 0 \tag{5.1.5c}$$

$$\Delta T(x,0) = 0 \text{ for } t = 0, \text{ in } 0 < x < 1 \tag{5.1.5d}$$

THE ADJOINT PROBLEM

Such as for the sensitivity problem, the derivation procedure of the adjoint problem for Technique V is very similar to the one for Technique III. To develop the *adjoint problem*, we introduce a *Lagrange multiplier*, $\lambda(x,t)$. We multiply equation (5.1.4a) by $\lambda(x,t)$, integrate the resulting expression over the spatial domain, from $x = 0$ to $x = 1$, and then over the time domain, from $t = 0$ to $t = t_f$. The expression obtained in this manner is added to the functional $S[g_p(t)]$ given by equation (5.1.3a) in order to obtain the following extended functional:

$$S[g_p(t)] = \int_{t=0}^{t_f} \frac{[Y(t) - T(x_{meas}, t; g_p(t))]^2}{[\sigma(t)]^2} \, dt + \int_{x=0}^{1} \int_{t=0}^{t_f} \lambda(x,t) \left[\frac{\partial^2 T}{\partial x^2} + g_p(t)\delta(x - 0.5) - \frac{\partial T}{\partial t} \right] dt\, dx \tag{5.1.6}$$

which is the equivalent form of equation (2.9.5) for parameter estimation.

An expression for the variation $\Delta S[g_p(t)]$ of the functional $S[g_p(t)]$ can be developed by assuming that $T(x, t)$ is perturbed by $\Delta T(x, t)$ when $g_p(t)$ is perturbed by $\Delta g_p(t)$. The variation $\Delta S[g_p(t)]$ gives the directional derivative of $S[g_p(t)]$ in the direction of the perturbation $\Delta g_p(t)$ [6,22,92].

By replacing $T(x,t)$ by $[T(x,t) + \Delta T(x,t)]$, $g_p(t)$ by $[g_p(t)+\Delta g_p(t)]$ and $S[g_p(t)]$ by $\{S[g_p(t)]+\Delta S[g_p(t)]\}$ in equation (5.1.6), subtracting from the resulting expression the original equation (5.1.6), and neglecting second-order terms, we obtain:

$$\Delta S[g_p(t)] = \int_{t=0}^{t_f} \int_{x=0}^{1} 2\frac{[T(x,t;g_p(t))-Y(t)]}{[\sigma(t)]^2}\Delta T(x,t)\delta(x - x_{\text{meas}})dx\,dt$$

$$+ \int_{t=0}^{t_f} \int_{x=0}^{1} \lambda(x,t)\left[\frac{\partial^2 \Delta T}{\partial x^2} + \Delta g_p(t)\delta(x - 0.5) - \frac{\partial \Delta T}{\partial t}\right]dx\,dt \qquad (5.1.7)$$

where $\delta(.)$ is the Dirac delta function. Equation (5.1.7) is analogous to equation (2.9.6) for parameter estimation.

The second integral term on the right-hand side of equation (5.1.7) is simplified with integration by parts and by utilizing the boundary and initial conditions of the sensitivity problem. The integral terms containing $\Delta T(x,t)$ in the resulting expression are then allowed to go to zero, in order to obtain the *adjoint problem* given by:

$$\frac{\partial \lambda(x,t)}{\partial t} + \frac{\partial^2 \lambda(x,t)}{\partial x^2} + 2\frac{[T(x,t;g_p(t))-Y(t)]}{[\sigma(t)]^2}\delta(x - x_{\text{meas}}) = 0 \text{ in } 0 < x < 1, \text{ for } 0 < t < t_f \qquad (5.1.8a)$$

$$\frac{\partial \lambda(0,t)}{\partial x} = 0 \text{ at } x = 0, \text{ for } 0 < t < t_f \qquad (5.1.8b)$$

$$\frac{\partial \lambda(1,t)}{\partial x} = 0 \text{ at } x = 1, \text{ for } 0 < t < t_f \qquad (5.1.8c)$$

$$\lambda(x,t_f) = 0 \text{ for } t = t_f, \text{ in } 0 < x < 1 \qquad (5.1.8d)$$

THE GRADIENT EQUATION

In the limiting process used above to obtain the adjoint problem, the following term is left:

$$\Delta S[g_p(t)] = \int_{t=0}^{t_f} \lambda(0.5,t)\Delta g_p(t)\,dt \qquad (5.1.9)$$

The reader should recall that in Technique III the parameterized form of $\Delta g_p(t)$ (equation 2.9.3b) was substituted into the earlier equation in order to obtain the components of the gradient vector given by equation (2.9.15). Such an approach cannot be used here, since we are now dealing with function estimation, rather than with parameter estimation as in Technique III. However, by invoking the hypothesis that the unknown function $g_p(t)$ belongs to the space of square-integrable functions in the domain $0 < t < t_f$, we can write [6,22,92]:

$$\Delta S[g_p(t)] = \int_{t=0}^{t_f} \nabla S[g_p(t)]\Delta g_p(t)\,dt \qquad (5.1.10)$$

where $\nabla S[g_p(t)]$ is the gradient of the functional $S[g_p(t)]$. Equation (5.1.10) is the definition of the directional derivative of $S[g_p(t)]$ in the direction of the perturbation $\Delta g_p(t)$, that is, the inner product of $\nabla S[g_p(t)]$ and $\Delta g_p(t)$ [6,22,92].

From the comparison of equations (5.1.9) and (5.1.10), we conclude that

$$\nabla S[g_p(t)] = \lambda(0.5, t) \tag{5.1.11}$$

which is the *gradient equation* for the functional.

THE ITERATIVE PROCEDURE FOR TECHNIQUE V

The mathematical development given above provides three distinct problems defined by equations (5.1.4), (5.1.5) and (5.1.8), called the *direct, sensitivity and adjoint problems*, for the computation of the functions $T(x,t)$, $\Delta T(x,t)$ and $\lambda(x,t)$, respectively. The measured data $Y(t)$ are considered available from a sensor located at x_{meas} and the gradient $\nabla S[g_p(t)]$ is given by equation (5.1.11).

The unknown function $g_p(t)$ is estimated through the minimization of the functional $S[g_p(t)]$ given by equation (5.1.3a). This is achieved with an iterative procedure by proper selection of the direction of descent and of the step size in going from iteration k to $k + 1$. Together with the gradient equation (5.1.11), the equations below provide the iterative procedure of the conjugate gradient method for the estimation of $g_p(t)$. A general version of this iterative procedure, for the estimation of functions that can vary spatially as well as in time, is presented in Note 2 at the end of this chapter.

The iterative procedure of the conjugate gradient method [2,6,22,29–43,76–79,92,135] for the estimation of the function $g_p(t)$ is given by:

$$g_p^{k+1}(t) = g_p^k(t) + \beta^k d^k(t) \tag{5.1.12}$$

where β^k is the *search step size* and $d^k(t)$ is the *direction of descent*, defined as

$$d^k(t) = -\nabla S[g_p^k(t)] + \gamma^k d^{k-1}(t) + \psi^k d^q(t) \tag{5.1.13}$$

The superscript q denotes the iteration number where a *restarting strategy* is applied to the iterative procedure of the conjugate gradient method, such as described in Section 2.3.

In the *Fletcher-Reeves* version of the conjugate gradient method, the conjugation coefficients are given by [76]:

$$\gamma^k = \frac{\displaystyle\int_{t=0}^{t_f} \{\nabla S[g_p^k(t)]\}^2 \, dt}{\displaystyle\int_{t=0}^{t_f} \{\nabla S[g_p^{k-1}(t)]\}^2 \, dt} \quad \text{with } \gamma^0 = 0 \text{ for } k = 0 \tag{5.1.14a}$$

$$\psi^k = 0 \text{ for } k = 0,1,2,\dots \tag{5.1.14b}$$

The conjugation coefficients are computed as follows, in *Polak-Ribiere's* version of the conjugate gradient method [77]:

$$\gamma^k = \frac{\displaystyle\int_{t=0}^{t_f} \nabla S[g_p^k(t)]\{\nabla S[g_p^k(t)] - \nabla S[g_p^{k-1}(t)]\} \, dt}{\displaystyle\int_{t=0}^{t_f} \{\nabla S[g_p^{k-1}(t)]\}^2 \, dt} \quad \text{with } \gamma^0 = 0 \text{ for } k = 0 \tag{5.1.15a}$$

$$\psi^k = 0 \text{ for } k = 0,1,2,\dots \tag{5.1.15b}$$

For the *Hestenes-Stiefel* version of the conjugate gradient method we have [78]:

$$\gamma^k = \frac{\displaystyle\int_{t=0}^{t_f} \nabla S[g_p^k(t)]\{\nabla S[g_p^k(t)] - \nabla S[g_p^{k-1}(t)]\}\, dt}{\displaystyle\int_{t=0}^{t_f} d^{k-1}(t)\{\nabla S[g_p^k(t)] - \nabla S[g_p^{k-1}(t)]\}\, dt} \quad \text{with } \gamma^0 = 0 \text{ for } k = 0 \qquad (5.1.16a)$$

$$\psi^k = 0 \text{ for } k = 0,1,2,\ldots \qquad (5.1.16b)$$

In *Powell-Beale's* version of the conjugate gradient method, γ^k and ψ^k are given, respectively, by [79]:

$$\gamma^k = \frac{\displaystyle\int_{t=0}^{t_f} \nabla S[g_p^k(t)]\{\nabla S[g_p^k(t)] - \nabla S[g_p^{k-1}(t)]\}\, dt}{\displaystyle\int_{t=0}^{t_f} d^{k-1}(t)\{\nabla S[g_p^k(t)] - \nabla S[g_p^{k-1}(t)]\}\, dt} \quad \text{with } \gamma^0 = 0 \text{ for } k = 0 \qquad (5.1.17a)$$

$$\psi^k = \frac{\displaystyle\int_{t=0}^{t_f} \nabla S[g_p^k(t)]\{\nabla S[g_p^{q+1}(t)] - \nabla S[g_p^q(t)]\}\, dt}{\displaystyle\int_{t=0}^{t_f} d^q(t)\{\nabla S[g_p^{q+1}(t)] - \nabla S[g_p^q(t)]\}\, dt} \quad \text{with } \psi^0 = 0 \text{ for } k = 0 \qquad (5.1.17b)$$

In *Powell-Beale's* version [79], such as for parameter estimation, restarting is performed by making $\psi^k = 0$ in equation (5.1.17b) when gradients at successive iterations tend to be non-orthogonal or when the direction of descent is not sufficiently downhill. The test of the non-orthogonality of gradients for the present inverse problem of estimating $g_p(t)$ is given by:

$$\text{ABS}\left(\int_{t=0}^{t_f} \nabla S[g_p^k(t)] \nabla S[g_p^{k-1}(t)]\, dt\right) \geq 0.2 \int_{t=0}^{t_f} \{\nabla S[g_p^k(t)]\}^2\, dt \qquad (5.1.18)$$

The tests to verify if the direction of descent is not sufficiently downhill are:

$$\int_{t=0}^{t_f} d^k(t)\nabla S[g_p^k(t)]\, dt \leq -1.2 \int_{t=0}^{t_f} \{\nabla S[g_p^k(t)]\}^2\, dt \qquad (5.1.19a)$$

$$\int_{t=0}^{t_f} d^k(t)\nabla S[g_p^k(t)]\, dt \geq -0.8 \int_{t=0}^{t_f} \{\nabla S[g_p^k(t)]\}^2\, dt \qquad (5.1.19b)$$

The algorithm for the calculation of the direction of descent in Powell-Beale's version of the conjugate gradient method of function estimation is the same as for parameter estimation (see Section 2.3). This algorithm is not repeated here for the sake of brevity, but can also be found in Note 2 of this chapter.

An expression for the step size β^k is now developed, by following a procedure similar to that used in Technique III. The step size β^k is determined by minimizing the functional $S[g_p^{k+1}(t)]$ given by equation (5.1.3a) with respect to β^k, that is,

$$\min_{\beta^k} S[g_p^{k+1}(t)] = \min_{\beta^k} \int_{t=0}^{t_f} \frac{\left\{ Y(t) - T[x_{\text{meas}},t;\ g_p^k(t)+\beta^k d^k(t)] \right\}^2}{[\sigma(t)]^2} dt \qquad (5.1.20)$$

where equation (5.1.12) was used in $T[x_{\text{meas}},t;\ g_p^{k+1}(t)]$.

By linearizing $T[x_{\text{meas}},t;\ g_p^{k+1}(t)] = T[x_{\text{meas}},t;\ g_p^k(t)+\beta^k d^k(t)]$ and making

$$d^k(t) = \Delta g_p^k(t) \qquad (5.1.21)$$

we obtain:

$$T[x_{\text{meas}},t;g_p^k(t)+\beta^k d^k(t)] \approx T[x_{\text{meas}},t;g_p^k(t)] + \beta^k \frac{\partial T}{\partial g_p^k} \Delta g_p^k(t) \qquad (5.1.22)$$

Let

$$\Delta T[d^k(t)] = \frac{\partial T}{\partial g_p^k} \Delta g_p^k(t) \qquad (5.1.23)$$

and then equation (5.1.22) can be written as:

$$T[x_{\text{meas}},t;g_p^k(t)+\beta^k d^k(t)] \approx T[x_{\text{meas}},t;g_p^k(t)] + \beta^k \Delta T[x_{\text{meas}},t;d^k(t)] \qquad (5.1.24)$$

By substituting equation (5.1.24) into equation (5.1.20), we obtain:

$$\min_{\beta^k} S[g_p^{k+1}(t)] = \min_{\beta^k} \int_{t=0}^{t_f} \frac{\left\{ Y(t) - T[x_{\text{meas}},t;g_p^k(t)] - \beta^k \Delta T[x_{\text{meas}},t;d^k(t)] \right\}^2}{[\sigma(t)]^2} dt \qquad (5.1.25)$$

By performing the minimization above, we find the following expression for the search step size of Technique V:

$$\beta^k = \frac{\displaystyle\int_{t=0}^{t_f} \frac{\{Y(t) - T[x_{\text{meas}},t;g_p^k(t)]\}}{[\sigma(t)]^2} \Delta T[x_{\text{meas}},t;d^k(t)]\, dt}{\displaystyle\int_{t=0}^{t_f} \frac{\{\Delta T[x_{\text{meas}},t;d^k(t)]\}^2}{[\sigma(t)]^2} dt} \qquad (5.1.26)$$

where $\Delta T[x_{\text{meas}},t;d^k(t)]$ is the solution of the sensitivity problem given by equation (5.1.5) at the measurement location x_{meas}, obtained by setting $\Delta g_p^k(t) = d^k(t)$.

By examining equations (5.1.8d) and (5.1.11), it can be noticed that the gradient equation is null at the final time t_f. Therefore, the initial guess used for $g_p(t)$ at $t = t_f$ is never changed by the iterative procedure of the conjugate gradient method of function estimation given by equations (5.1.12) and (5.1.13). The estimated function can deviate from the exact solution in a neighborhood of t_f, if the initial guess used is too different from the exact $g_p(t)$ at $t = t_f$. This drawback of the method can be easily overcome by using a final time larger than that of interest, so that the effects of the initial guess are not noticeable in the time interval that the solution is sought. Another approach to overcome this difficulty is to repeat the solution of the inverse problem by using as initial guess a previously estimated value for $g_p(t)$ in the neighborhood of t_f.

THE STOPPING CRITERION FOR TECHNIQUE V

Similarly to Techniques II and III, the stopping criterion based on the *discrepancy principle* gives the conjugate gradient method of function estimation an iterative regularization character. The stopping criterion is given by:

$$S[g_p(t)] < \varepsilon \qquad (5.1.27)$$

where the tolerance ε is selected so that smooth solutions are obtained with measurements containing random errors. By following Morozov's discrepancy principle [2–22], it is assumed that the solution is sufficiently accurate when:

$$\left| Y(t) - T\left[x_{\text{meas}}, t; g_p(t)\right] \right| \approx \sigma(t) \qquad (5.1.28)$$

where $\sigma^2(t)$ is the variance of measurement errors.

If the objective function $S[g_p(t)]$ is given by equation (5.1.3a), the tolerance ε is obtained as:

$$\varepsilon = t_f \qquad (5.1.29)$$

Alternatively, if the objective function is given by equation (5.1.3b) for the case of a constant standard deviation σ, we have:

$$\varepsilon = \sigma^2 t_f \qquad (5.1.30)$$

For those cases involving measurements with unknown standard deviation, an alternative approach based on an additional measurement can be used, as described in Note 3 at the end of this chapter.

THE COMPUTATIONAL ALGORITHM FOR TECHNIQUE V

Suppose an initial guess $g_p^0(t)$ is available for the function $g_p(t)$. Set $k = 0$ and then:

Step 1: Solve the direct Problem (5.1.4) and compute $T(x, t)$ based on $g_p^k(t)$.
Step 2: Check the stopping criterion (5.1.27). Continue if not satisfied.
Step 3: Knowing $T\left[x_{\text{meas}}, t; g_p^k(t)\right]$ and measured temperature $Y(t)$, solve the adjoint problem (5.1.8).
Step 4: Knowing $\lambda(0.5, t)$, compute $\nabla S[g_p^k(t)]$ from equation (5.1.11).
Step 5: Knowing the gradient $\nabla S[g_p^k(t)]$, compute γ^k and ψ^k with one of the pairs of equations (5.1.14)–(5.1.17) and the direction of descent $d^k(t)$ from equation (5.1.13).
Step 6: Set $\Delta g_p^k(t) = d^k(t)$ and solve the sensitivity Problem (5.1.5).
Step 7: Knowing $\Delta T[x_{\text{meas}}, t; d^k(t)]$, compute the search step size β^k from equation (5.1.26).
Step 8: Knowing the search step size β^k and the direction of descent $d^k(t)$, compute the new estimate $g_p^{k+1}(t)$ from equation (5.1.12) and return to step 1.

The extension of the above algorithm to the use of multiple sensors is analogous to that described in Section 2.9 for Technique III. It is a straightforward matter and will not be repeated here.

For the computational implementation of Technique V, the measurements need to be treated as discrete in time. Therefore, the source term in the governing equation of the adjoint problem (equation 5.1.8a) is actually discrete. Similarly, integrals are calculated as summations, and the iterative procedure given by equation (5.1.12) is applied at the discrete times when the measurements are available.

Example 5.1

Apply the conjugate gradient method to solve the inverse problem of estimating the source term $g_p(t)$, as described in this section. Use as initial guess $g_p^0(t) = 0$ in $0 < t \leq t_f$, where $t_f = 2$. Generate 100 simulated measurements at $x_{meas} = 1$ with the function $g_p(t) = 1 + \sin \pi t + \cos \pi t + \sin 2\pi t + \cos 2\pi t$. Use additive, uncorrelated, Gaussian measurement errors with standard deviations $\sigma = 0$ and $\sigma = 0.01 T_{max}$, where T_{max} is the maximum measured temperature.

Solution: Since the gradient equation is null at the final time as described above, the initial guess used for $g_p(t)$ at $t = t_f$ is never changed by the iterative procedure, generating instabilities on the solution in the neighborhood of t_f. One approach to overcome such difficulties is to consider a final time larger than that of interest. We illustrate such an approach by considering $t_f = 2.0$, 2.2 and 2.4. The numbers of measurements were increased accordingly for $t_f > 2.0$.

Figures 5.1a-c show the results obtained with Technique V for final times of 2.0, 2.2 and 2.4, respectively. The deviation of the estimated function from the exact one in the neighborhood of t_f, caused by the null gradient at t_f, is apparent in Figure 5.1a. Note in this figure that the estimated function is zero for $t = t_f$, which is exactly the initial guess used for the iterative procedure of Technique V. As t_f was increased to 2.2, the effects caused by the null gradient at the final time are practically not noticeable in the time domain of interest, $0 < t \leq 2$, as can be seen in Figure 5.1b. In fact, quite accurate estimates were obtained in this case with errorless measurements, as well as with measurements containing random errors. The solution in the neighborhood of $t_f = 2$ can be further improved by increasing the final time from 2.2 to 2.4, as apparent in Figure 5.1c.

The application of the conjugate gradient method with adjoint problem of function estimation is further illustrated in the next sections, with inverse problems involving different heat transfer modes, as well as coupled heat and mass transfer.

5.2 ESTIMATION OF THE SPACEWISE AND TIMEWISE VARIATIONS OF THE WALL HEAT FLUX IN LAMINAR FLOW

In this section, we present the solution of the inverse problem of estimating the wall heat flux in a parallel plate channel, by using Technique V, the conjugate gradient method with adjoint problem of function estimation. We follow Reference [136] and consider the unknown heat flux to vary in time and along the channel flow direction. We examine the accuracy of the present function estimation approach by using transient simulated measurements of several sensors located inside the channel. The inverse problem is solved for different functional forms of the unknown wall heat flux, including those containing sharp corners and discontinuities, which are the most difficult to be recovered by an inverse analysis.

Direct Problem

The physical problem considered here is the laminar hydrodynamically developed flow between parallel plates of a fluid with constant properties. The inlet temperature is maintained at a constant value T_0, which is also assumed to be the initial fluid temperature. For times greater than zero, the plates are subjected to a time- and space-dependent heat flux, as illustrated in Figure 5.2.

By taking into account the symmetry with respect to the x-axis and neglecting conduction along the flow direction, the mathematical formulation of this problem in dimensionless form is given by [136]:

$$\frac{\partial \Theta}{\partial \tau} + U(Y)\frac{\partial \Theta}{\partial X} = \frac{\partial^2 \Theta}{\partial Y^2} \quad \text{in } 0 < Y < 1, 0 < X < L, \text{ for } \tau > 0 \tag{5.2.1a}$$

$$\frac{\partial \Theta}{\partial Y} = 0 \quad \text{at } Y = 0, 0 < X < L, \text{ for } \tau > 0 \tag{5.2.1b}$$

$$\frac{\partial \Theta}{\partial Y} = Q(X,\tau) \quad \text{at } Y = 1, 0 < X < L, \text{ for } \tau > 0 \tag{5.2.1c}$$

$$\Theta = 0 \quad \text{at } X = 0, 0 < X < L, \text{ for } \tau > 0 \tag{5.2.1d}$$

$$\Theta = 0 \quad \text{for } \tau = 0, \text{ in } 0 < Y < 1, 0 < X < L \tag{5.2.1e}$$

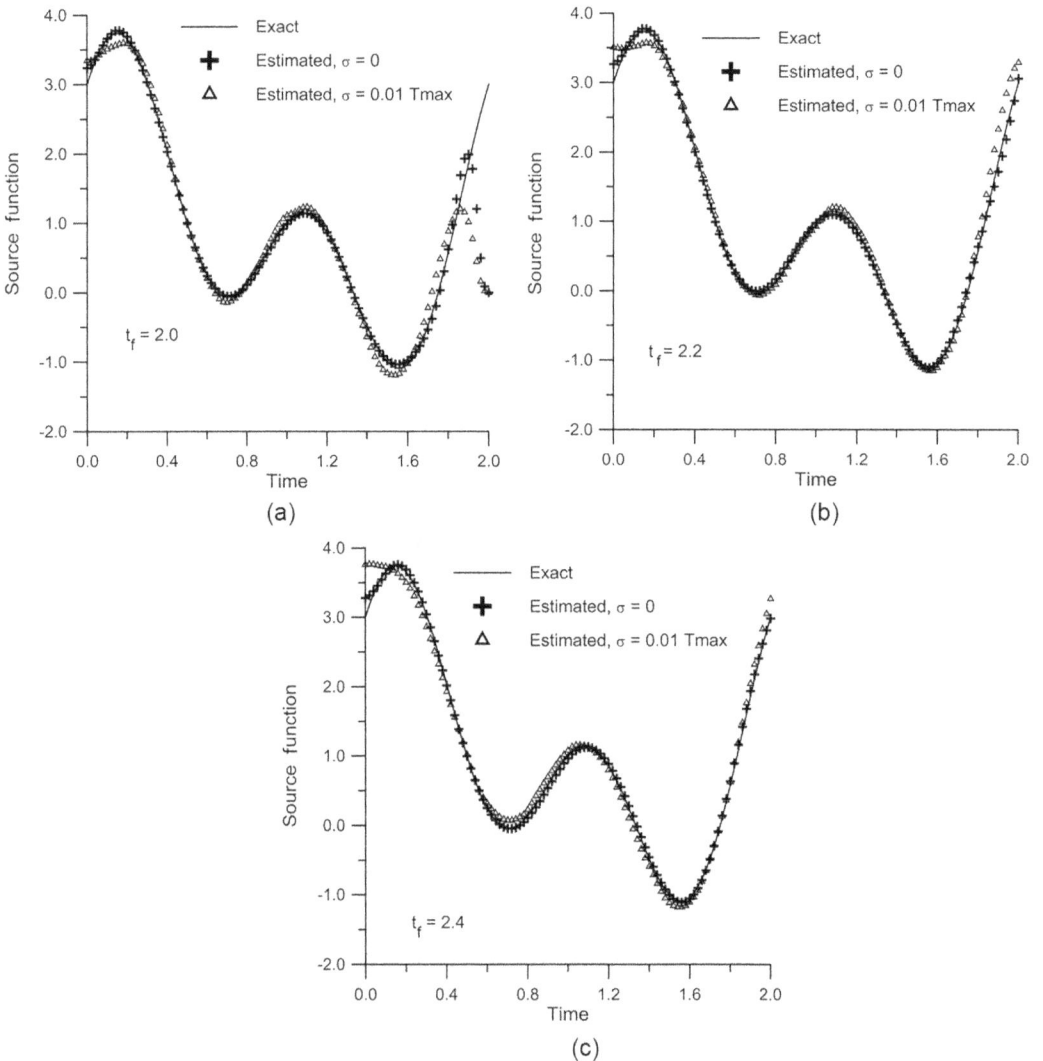

FIGURE 5.1   Estimation of the source term for $t_f = 2.0$ (a); $t_f = 2.2$ (b); $t_f = 2.4$ (c).

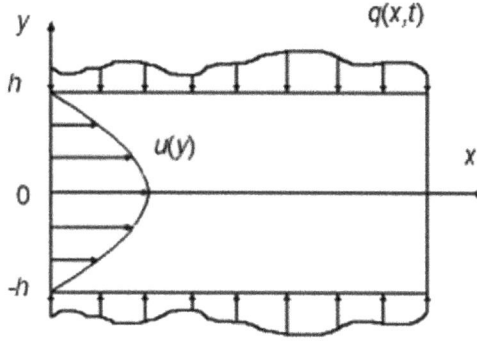

FIGURE 5.2   Physical problem. (From Ref. [136].)

where the following dimensionless groups were introduced:

$$Y = \frac{y}{h}; \quad X = \frac{\alpha x}{u_m h^2}; \quad \Theta = \frac{T - T_0}{\dfrac{q_0 h}{k}}; \quad \tau = \frac{\alpha t}{h^2} \tag{5.2.2a-d}$$

$$U(Y) = \frac{u(y)}{u_m} = \frac{3}{2}\left[1 - \left(\frac{y}{h}\right)^2\right] \tag{5.2.2e}$$

and α and k are the fluid thermal diffusivity and conductivity, respectively, h is the channel half-width and u_m is the mean fluid velocity. The wall heat flux is written as

$$q(x,t) = q_0 Q(X,\tau) \tag{5.2.3}$$

where q_0 is a constant reference value with units of heat flux and $Q(X,\tau)$ is a dimensionless function of X and τ.

The direct problem given by equation (5.2.1) is concerned with the determination of the temperature field of the fluid inside the channel, when the boundary heat flux $Q(X,\tau)$ at $Y = 1$ is known.

INVERSE PROBLEM

For the inverse problem, the heat flux $Q(X,\tau)$ at $Y = 1$ is considered to be unknown and is to be estimated by using the transient readings of M temperature sensors located inside the channel. We assume that no information is available regarding the functional form of the unknown wall heat flux, except that it belongs to the space of square-integrable functions in the domain $0 < \tau < \tau_f$ and $0 < X < L$, where τ_f is the duration of the experiment and L is the length of the channel. The solution of such inverse problem is obtained by minimizing the following functional:

$$S[Q(X,\tau)] = \int_{\tau=0}^{\tau_f} \sum_{m=1}^{M} \left\{ \Theta\left[X_m^*, Y_m^*, \tau; Q(X,\tau)\right] - Z_m(\tau) \right\}^2 d\tau \tag{5.2.4}$$

where $Z_m(\tau)$ is the measured temperature at the sensor location (X_m^*, Y_m^*) inside the channel, which has additive, uncorrelated, Gaussian errors with zero mean and constant standard deviation. The function $\Theta[X_m^*, Y_m^*, \tau; Q(X,\tau)]$ is the estimated temperature at the same location. Such estimated temperature is obtained from the solution of the direct problem given by equation (5.2.1) by using an estimate for the unknown heat flux $Q(X,\tau)$.

The development of the sensitivity and adjoint problems, required for the implementation of the iterative procedure of Technique V, are described next.

SENSITIVITY PROBLEM

The sensitivity problem is obtained by assuming that the heat flux $Q(X,\tau)$ is perturbed by an amount $\Delta Q(X,\tau)$. Such perturbation in the heat flux causes a perturbation $\Delta\Theta(X,Y,\tau)$ in the temperature $\Theta(X,Y,\tau)$. By substituting $\Theta(X,Y,\tau)$ by $[\Theta(X,Y,\tau)+\Delta\Theta(X,Y,\tau)]$ and $Q(X,\tau)$ by $[Q(X,\tau)+\Delta Q(X,\tau)]$ in the direct problem given by equation (5.2.1), and then subtracting from the resulting expressions the original direct problem, we obtain the following sensitivity problem for the determination of the sensitivity function $\Delta\Theta(X,Y,\tau)$:

$$\frac{\partial\Delta\Theta}{\partial\tau}+U(Y)\frac{\partial\Delta\Theta}{\partial X}=\frac{\partial^2\Delta\Theta}{\partial Y^2}\ \text{ in } 0<Y<1,0<X<L,\text{ for }\tau>0 \tag{5.2.5a}$$

$$\frac{\partial\Delta\Theta}{\partial Y}=0\ \text{ at }Y=0,0<X<L,\text{ for }\tau>0 \tag{5.2.5b}$$

$$\frac{\partial\Delta\Theta}{\partial Y}=\Delta Q(X,\tau)\ \text{ at }Y=1,0<X<L,\text{ for }\tau>0 \tag{5.2.5c}$$

$$\Delta\Theta=0\ \text{ at }X=0,0<Y<1,\text{ for }\tau>0 \tag{5.2.5d}$$

$$\Delta\Theta=0\ \text{ for }\tau=0,\text{ in }0<Y<1,0<X<L \tag{5.2.5e}$$

ADJOINT PROBLEM

In order to obtain the adjoint problem, we multiply the differential equation (5.2.1a) of the direct problem by the Lagrange multiplier $\lambda(X,Y,\tau)$ and integrate over the time and space domains. The resulting expression is then added to equation (5.2.4) to obtain the following extended functional:

$$S[Q(X,\tau)]=\int_{\tau=0}^{\tau_f}\int_{X=0}^{L}\int_{Y=0}^{1}\left\{\sum_{m=1}^{M}[\Theta(X,Y,\tau)-Z(\tau)]^2\,\delta\left(X-X_m^*\right)\delta\left(Y-Y_m^*\right)\right.$$

$$\left.+\left[\frac{\partial\Theta}{\partial\tau}+U(Y)\frac{\partial\Theta}{\partial X}-\frac{\partial^2\Theta}{\partial Y^2}\right]\lambda(X,Y,\tau)\right\}dY\,dX\,d\tau \tag{5.2.6}$$

where $\delta(\cdot)$ is the Dirac delta function.

The variation of the extended functional (5.2.6) is obtained and, after manipulations like those described in the previous section, the resulting expression is allowed to go to zero in order to obtain the following adjoint problem for the Lagrange multiplier $\lambda(X, Y, \tau)$:

$$-\frac{\partial\lambda}{\partial\tau}-U(Y)\frac{\partial\lambda}{\partial X}-\frac{\partial^2\lambda}{\partial Y^2}+2\sum_{m=1}^{M}(\Theta-Z)\delta(X-X_m^*)\delta(Y-Y_m^*)=0\ \text{ in }0<Y<1,0<X<L,\text{ for }\tau>0$$

$$\tag{5.2.7a}$$

$$\frac{\partial\lambda}{\partial Y}=0\ \text{ at }Y=0,0<X<L,\text{ for }\tau>0 \tag{5.2.7b}$$

$$\frac{\partial\lambda}{\partial Y}=0\ \text{ at }Y=1,0<X<L,\text{ for }\tau>0 \tag{5.2.7c}$$

$$\lambda=0\ \text{ at }X=L,0<Y<1,\text{ for }\tau>0 \tag{5.2.7d}$$

$$\lambda=0\ \text{ for }\tau=\tau_f,\text{ in }0<Y<1\ 0<X<L \tag{5.2.7e}$$

GRADIENT EQUATION

In the process of obtaining the adjoint problem, the following integral term is left:

$$\Delta S\big[Q(X,\tau)\big] = -\int_{\tau=0}^{\tau_f}\int_{X=0}^{L} \lambda(X,1,\tau)\Delta Q(X,\tau)\,dX\,d\tau \tag{5.2.8a}$$

From the hypothesis that $Q(X,\tau)$ belongs to the space of square-integrable functions in $0<X<L$ and $0<\tau<\tau_f$, we can write

$$\Delta S\big[Q(X,\tau)\big] = \int_{\tau=0}^{\tau_f}\int_{X=0}^{L} \nabla S[Q(X,\tau)]\Delta Q(X,\tau)\,dX\,d\tau \tag{5.2.8b}$$

Therefore, by comparing equations (5.2.8a and b), we obtain the gradient equation for the functional as:

$$\nabla S\big[Q(X,\tau)\big] = -\lambda(X,1,\tau) \tag{5.2.9}$$

ITERATIVE PROCEDURE

The iterative algorithm of Technique V was applied with the Fletcher-Reeves version of the conjugate gradient method. The expression for the search step size β^k was obtained by minimizing the functional given by equation (5.2.4) with respect to β^k (see Note 2 at the end of this chapter). We obtain:

$$\beta^k = \frac{\displaystyle\int_{\tau=0}^{\tau_f}\sum_{m=1}^{M}(\Theta_m - Z_m)\,\Delta\Theta_m(d^k)\,d\tau}{\displaystyle\int_{\tau=0}^{\tau_f}\sum_{m=1}^{M}\big[\Delta\Theta_m(d^k)\big]^2\,d\tau} \tag{5.2.10}$$

where $\Delta\Theta_m(d^k)$ is the solution of the sensitivity Problem (5.2.5) obtained by setting $\Delta Q(X,\tau) = d^k(X,\tau)$.

RESULTS

We used transient simulated measurements in order to assess the accuracy of the present approach of estimating the unknown wall heat flux $Q(X,\tau)$. For the cases considered below, we have taken the total experiment duration τ_f as 0.08 and the channel length L as 0.004, while the heat flux at the boundary $Y=1$ was assumed in the form:

$$Q(X,\tau) = Q_X(X) + Q_\tau(\tau) \tag{5.2.11}$$

By examining equations (5.2.7d) and (5.2.7e), we note that the gradient of the functional given by equation (5.2.9) is null at the final time τ_f and the final longitudinal position L. Therefore, the initial guess used for the iterative process remains unchanged at $\tau = \tau_f$ and at $X = L$. In the examples shown below, we used as an initial guess at the final time and at the final position the exact values for $Q(X,\tau)$, which were assumed available. For other times and axial positions, we take $Q(X,\tau)$ null as the initial guess for the conjugate gradient method. We lose no generality with such an approach, since we can always choose τ_f and L sufficiently larger than the respective experimental time and length of interest, so that the initial guess has no influence on the solution, as illustrated in Section 5.1.

Figures 5.3a-c present the results obtained for a boundary heat flux containing a triangular variation in X and a step variation in time, in the form:

$$Q_X(X) = \begin{cases} 1, & \text{for } X \leq 0.001 \text{ and } X \geq 0.003 \\ 1000X, & \text{for } 0.001 < X \leq 0.002 \\ -1000X + 4, & \text{for } 0.002 < X < 0.003 \end{cases} \qquad (5.2.12)$$

$$Q_\tau(\tau) = \begin{cases} 1, & \text{for } \tau \leq 0.02 \text{ and } \tau \geq 0.06 \\ 2, & \text{for } 0.02 < \tau < 0.06 \end{cases} \qquad (5.2.13)$$

For such case, we have used in the inverse analysis 21 sensors located at $Y^* = 0.95$. The first sensor was located at $X_1^* = 0.00004$ and the last one at $X_{21}^* = 0.00396$. The others were equally spaced, so that $X_m^* = 0.0002(m-1)$, for $m = 2,...,$ 20. Figures 5.3a-c show the results for error-less measurements (dashed lines), as well as for measurements with a constant standard devia-tion $\sigma = 0.01 \, \Theta_{max}$ (symbols), where Θ_{max} is the maximum temperature measured by the sensors. In Figure 5.3a, we have the results for the axial variation for 3 different times, where Q_τ (0.002) $= Q_\tau$ (0.07) = 1 and Q_τ (0.04) = 2 from equation (5.2.13). The unknown heat fluxes for such times were accurately predicted, so that the results for $\tau = 0.002$ and $\tau = 0.07$ fall in the curve at the bot-tom, while those for $\tau = 0.04$ fall in the curve at the top of Figure 5.3a. The predicted heat flux is in good agreement with the exact one for both errorless measurements and measurements with random error. Figures 5.3b and c show the results obtained for the flux variation in time for dif-ferent axial positions. The results for $X = 0.0004$ and $X = 0.0036$ fall on the same curve in Figure 5.3b as expected, since Q_X (0.0004) $= Q_X$ (0.0036) = 1 from equation (5.2.12). The results shown in Figure 5.3c for $X = 0.002$, where $Q(X,\tau)$ has a peak in X, are also in good agreement with the exact functional form assumed for the heat flux.

The RMS error (e_{RMS}) for the results shown in Figure 5.3, obtained with errorless measurements, is 0.014. We define the RMS error here as:

$$e_{RMS} = \sqrt{\frac{1}{I} \sum_{i=1}^{I} \left[Q_{ex}(X_i, \tau_i) - Q_{est}(X_i, \tau_i) \right]^2} \qquad (5.2.14)$$

where I is the total number of measurements used in the inverse analysis, while Q_{exa} and Q_{est} are the exact and estimated heat fluxes, respectively.

Figures 5.4a–c present the results obtained for a heat flux with a step variation in X and with a triangular variation in time in the form:

$$Q_X(X) = \begin{cases} 1, & \text{for } X \leq 0.001 \text{ and } X \geq 0.003 \\ 2, & \text{for } 0.001 < X < 0.003 \end{cases} \qquad (5.2.15)$$

$$Q_\tau(\tau) = \begin{cases} 1, & \text{for } \tau \leq 0.02 \text{ and } \tau \geq 0.06 \\ 50\tau, & \text{for } 0.02 < \tau \leq 0.04 \\ -50\tau + 4, & \text{for } 0.04 < \tau < 0.06 \end{cases} \qquad (5.2.16)$$

where the dashed lines show the results obtained with errorless measurements and the symbols show the results obtained with measurements with a standard deviation of $\sigma = 0.01\,\Theta_{max}$. The 21 sensors used for this case are located at $Y^* = 0.95$ and at the same axial positions as for the case shown in Figure 5.3. Figure 5.4a shows the axial variation of $Q(X,\tau)$ for different times that correspond to $Q_\tau(\tau) = 1$, as given by equation (5.2.16). Similarly, Figure 5.4b shows the axial variation of $Q(X,\tau)$ for $\tau = 0.04$, when $Q_\tau(\tau)$ has a peak, i.e., $Q_\tau(\tau) = 2$ as given by equation (5.2.16). In Figure 5.4c, we have the results for the variation of $Q(X,\tau)$ in time for three different axial positions, so that, in accordance with equation (5.2.15) we have $Q_X(0.0004) = Q_X(0.0036) = 1$ and $Q_X(0.002) = 2$. As for the case presented in Figures 5.3a-c, Figures 5.4a-c show that the present function estimation approach is capable of recovering the unknown heat flux $Q(X,\tau)$ quite accurately for errorless measurements, as well as for measurements containing random errors. The RMS error is 0.045 for the results shown in Figure 5.4, obtained with errorless measurements.

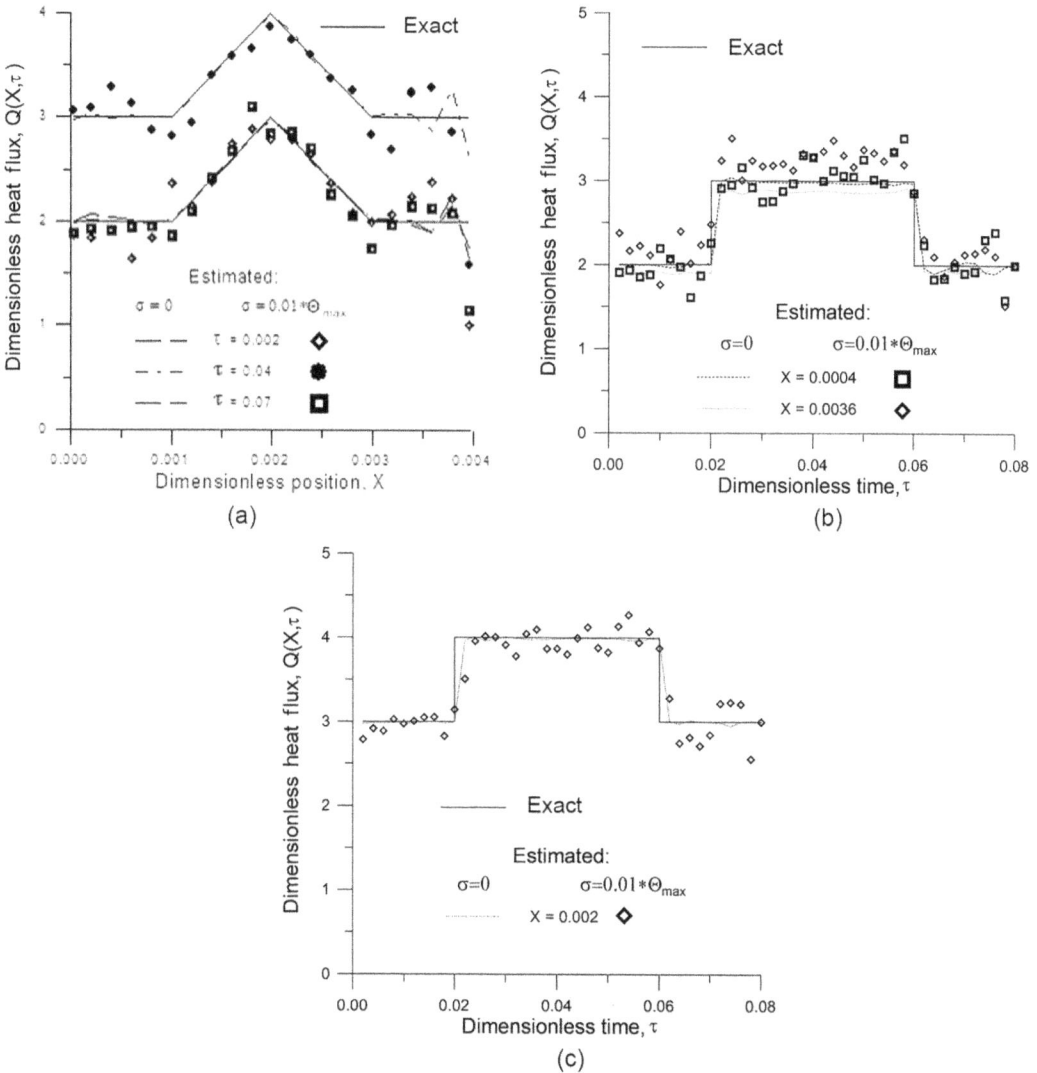

FIGURE 5.3 (a) Inverse problem solution for different times obtained with 21 sensors. Triangular variation in the axial direction given by equation (5.2.12). (From Ref. [136]). (b) Inverse problem solution for different axial positions obtained with 21 sensors. Step variation in time given by equation (5.2.13). (From Ref. [136]). (c) Inverse problem solution for $X = 0.002$ obtained with 21 sensors. Step variation in time given by equation (5.2.13). (From Ref. [136]).

The results shown in Figures 5.3 and 5.4 can be generally improved by using more measurements in the inverse analysis. Let's consider, for example, the estimation of the axial variation of $Q(X,\tau)$ shown in Figure 5.4a. In Figure 5.5, we present the estimation of $Q(X,\tau)$ for the same case studied in Figure 5.4a, but using the errorless measurements of 101 sensors instead of 21. The sensors were equally spaced along the channel length and at $Y^* = 0.95$. The time frequency of measurements was considered to be the same as for Figure 5.4a. By comparing Figures 5.4a and 5.5, we can clearly notice the improvement in the estimation of $Q(X,\tau)$ by using more sensors along the channel. The RMS error obtained with 101 sensors is 0.013 as compared to 0.045 obtained by using 21 sensors.

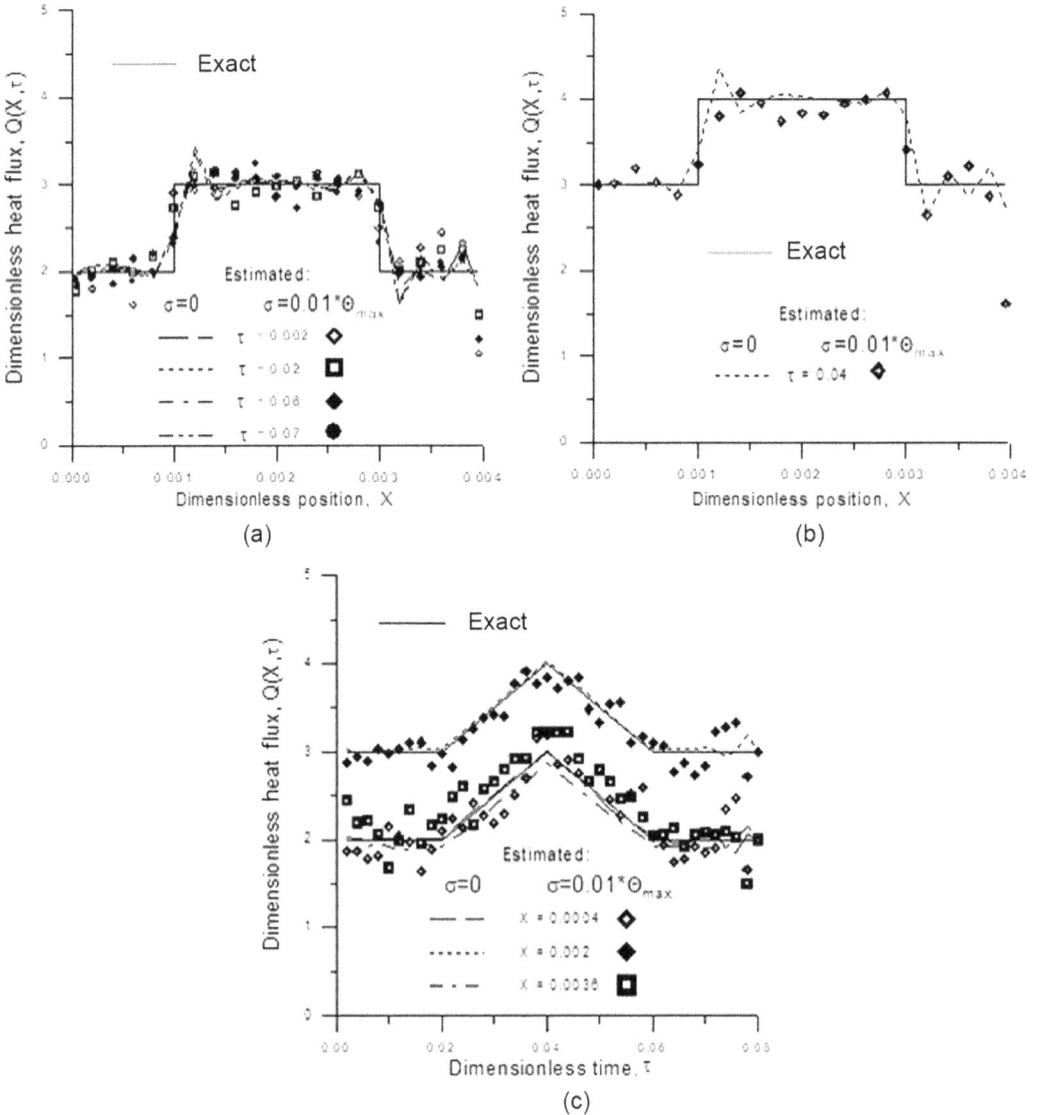

FIGURE 5.4 (a) Inverse problem solution for different times obtained with 21 sensors. Step variation in the axial direction given by equation (5.2.15). (From Ref. [136].) (b) Inverse problem solution for $\tau = 0.04$ obtained with 21 sensors. Step variation in the axial direction given by equation (5.2.15). (From Ref. [136].) (c) Inverse problem solution for different axial positions obtained with 21 sensors. Triangular variation in time given by equation (5.2.16). (From Ref. [136].)

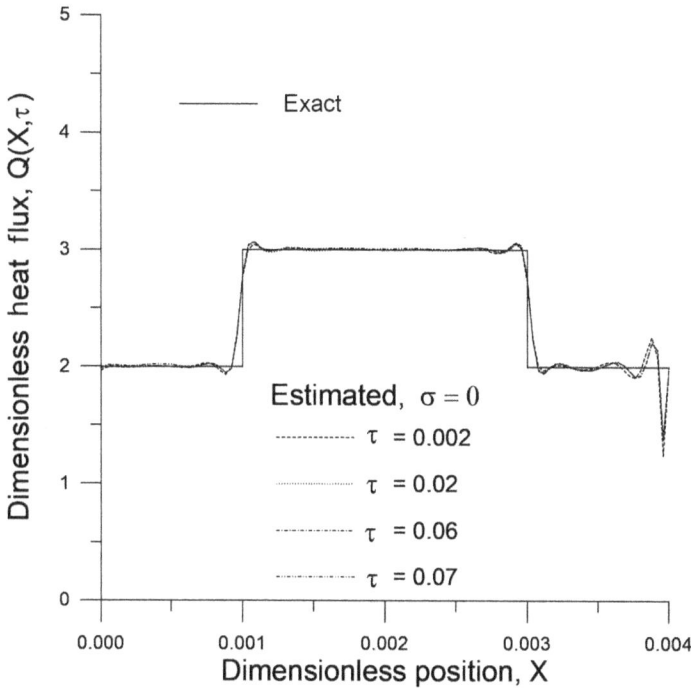

FIGURE 5.5 Inverse problem solution for different times obtained with 101 sensors. Step variation in the axial direction given by equation (5.2.15). (From Ref. [136].)

For inverse heat conduction problems dealing with the estimation of a boundary condition, the sensors should be located as close to the boundary with the unknown condition as possible, in order to improve the estimation. Such is also the case for inverse convection problems. In this case, the sensors must be located within the thermal boundary layer along the channel. Otherwise, the measurements are not sensitive to the boundary heat flux. We have estimated $Q(X,\tau)$ for $Q_X(X)$ and $Q_\tau(\tau)$ given by equations (5.2.15) and (5.2.16), respectively, and by using the errorless measurements of 21 sensors located at the same axial positions as for Figure 5.4, but at $Y^* = 0.9$, instead of at $Y^* = 0.95$. The RMS error has increased to 0.238, as compared to 0.045 obtained with the sensors located at $Y^* = 0.95$.

We note in Figures 5.2–5.5 that generally the agreement between the estimated solutions and the exact functional form assumed for $Q(X, \tau)$ tends to deteriorate near the end of the channel and near the final time. This is due to the very small values of the gradient of the functional, equation (5.2.9), in such regions, as can be noticed by examining equations (5.2.7d and e).

5.3   SIMULTANEOUS ESTIMATION OF SPATIALLY DEPENDENT DIFFUSION COEFFICIENT AND SOURCE TERM IN A DIFFUSION PROBLEM

In this section, the conjugate gradient method with adjoint problem is used for the simultaneous estimation of the spatially varying diffusion coefficient and of the source term distribution in a one-dimensional diffusion problem [137]. This work can be physically associated with the detection of material non-homogeneities, such as inclusions, obstacles or cracks, in heat conduction, groundwater flow and tomography problems. Three versions of the conjugate gradient method presented in Section 5.1 are compared for the solution of the present inverse problem by using simulated measurements containing random errors.

DIRECT PROBLEM

We focus our attention on the following dimensionless diffusion problem:

$$\frac{\partial U}{\partial t} = \frac{\partial}{\partial x}\left(D(x)\frac{\partial U}{\partial x}\right) + \mu(x)U \qquad \text{in } 0 < x < 1, \text{ for } t > 0 \qquad (5.3.1a)$$

$$\frac{\partial U}{\partial x} = 0 \qquad \text{at } x = 0 \text{ for } t > 0 \qquad (5.3.1b)$$

$$D(x)\frac{\partial U}{\partial x} = 1 \qquad \text{at } x = 1 \text{ for } t > 0 \qquad (5.3.1c)$$

$$U = 0 \qquad \text{for } t = 0 \text{ in } 0 < x < 1 \qquad (5.3.1d)$$

In the direct problem, the diffusion coefficient function, $D(x)$, and the source term distribution function, $\mu(x)$, are known quantities. The direct problem is concerned with the computation of $U(x,t)$.

INVERSE PROBLEM

For the inverse problem of interest here, the functions $D(x)$ and $\mu(x)$ are regarded as unknown. Such functions are simultaneously estimated by using measurements of $U(x,t)$ taken at appropriate locations in the medium or on its boundaries. Such measurements may contain random errors that are assumed to be uncorrelated, additive, Gaussian, with zero mean and with a known constant standard deviation.

   For the simultaneous estimation of the functions $D(x)$ and $\mu(x)$, we make use of a minimization procedure involving the following objective functional:

$$S[D(x),\mu(x)] = \frac{1}{2}\int_{t=0}^{t_f} \sum_{m=1}^{M} \left\{U[x_m,t;D(x),\mu(x)] - Y_m(t)\right\}^2 dt \qquad (5.3.2)$$

where $Y_m(t)$ are the transient measurements of $U(x,t)$ taken at the positions x_m, $m = 1,\ldots, M$. The estimated dependent variable $U[x_m,t;D(x),\mu(x)]$ is obtained from the solution of the direct Problem (5.3.1a-d) at the measurement positions with estimates for $D(x)$ and $\mu(x)$.

SENSITIVITY PROBLEMS

The sensitivity function, solution of the sensitivity problem, is defined as the directional derivative of $U(x,t)$ in the direction of the perturbation of the unknown function, as discussed in Section 5.1. Since the present inverse problem involves two unknown functions, two sensitivity problems are required for the estimation procedure, resulting from perturbations in $D(x)$ and $\mu(x)$. The procedure used below to find these sensitivity problems is equivalent to that presented in Section 5.1, but here the second-order terms are naturally discarded from the formulations, as the limits of the derivative definitions are applied.

   The sensitivity problem for the directional derivative of $U(x,t)$ in the direction of the perturbation $\Delta D(x)$, denoted as $U_D(x,t)$, is obtained by assuming that the dependent variable $U(x,t)$ is perturbed by $\varepsilon\Delta U_D(x,t)$ when the diffusion coefficient $D(x)$ is perturbed by $\varepsilon\Delta D(x)$. Here, ε is a real number. The sensitivity problem for $U_D(x,t)$ is then obtained by applying the following limiting process:

$$\lim_{\varepsilon\to 0} \frac{L_\varepsilon(D_\varepsilon) - L(D)}{\varepsilon} = 0 \qquad (5.3.3)$$

where $L_\varepsilon(D_\varepsilon)$ and $L(D)$ are the direct problem formulations written in operator form for perturbed and unperturbed quantities, respectively. The application of the limiting process given by equation (5.3.3) results in the following sensitivity problem:

$$\frac{\partial \Delta U_D}{\partial t} = \frac{\partial}{\partial x}\left(D(x)\frac{\partial \Delta U_D}{\partial x} + \Delta D(x)\frac{\partial U}{\partial x}\right) + \mu(x)\,\Delta U_D \qquad \text{in } 0 < x < 1 \text{ for } t > 0 \qquad (5.3.4a)$$

$$\frac{\partial \Delta U_D}{\partial x} = 0 \qquad \text{at } x = 0 \text{ for } t > 0 \qquad (5.3.4b)$$

$$\Delta D(x)\frac{\partial U}{\partial x} + D(x)\frac{\partial \Delta U_D}{\partial x} = 0 \qquad \text{at } x = 1 \text{ for } t > 0 \qquad (5.3.4c)$$

$$\Delta U_D = 0 \qquad \text{in } 0 < x < 1 \text{ for } t = 0 \qquad (5.3.4d)$$

A limiting process analogous to that given by equation (5.3.3), obtained from the perturbation $\varepsilon \Delta \mu(x)$, results in the following sensitivity problem for the directional derivative of $U(x,t)$ in the direction of the perturbation $\Delta \mu$, $\Delta U_\mu(x,t)$:

$$\frac{\partial \Delta U_\mu}{\partial t} = \frac{\partial}{\partial x}\left(D(x)\frac{\partial \Delta U_\mu}{\partial x}\right) + \mu(x)\,\Delta U_\mu + \Delta \mu(x)U \qquad \text{in } 0 < x < 1 \text{ for } t > 0 \qquad (5.3.5a)$$

$$\frac{\partial \Delta U_\mu}{\partial x} = 0 \qquad \text{at } x = 0 \text{ and } x = 1 \text{ for } t > 0 \quad (5.3.5b,c)$$

$$\Delta U_\mu = 0 \qquad \text{in } 0 < x < 1; \text{ for } t = 0 \qquad (5.3.5d)$$

The sensitivity Problems (5.3.4) and (5.3.5) depend on the unknown functions $D(x)$ and $\mu(x)$. Therefore, the present estimation problem is nonlinear.

Adjoint Problem

A Lagrange multiplier $\lambda(x,t)$ is utilized in the minimization of the functional (5.3.2) because the estimated dependent variable $U[x_m,t;D(x),\mu(x)]$ appearing in such functional needs to satisfy a constraint, which is the solution of the direct problem. Such Lagrange multiplier, needed for the computation of the gradient equations (as will be apparent below), is obtained through the solution of problems *adjoint* to the sensitivity problems, given by equations (5.3.4) and (5.3.5). Despite the fact that the present inverse problem involves the estimation of two unknown functions, thus resulting in two sensitivity problems as discussed above, one single problem, adjoint to Problems (5.3.4) and (5.3.5), is obtained.

In order to derive the adjoint problem, the governing equation of the direct problem, equation (5.3.1a), is multiplied by the Lagrange multiplier $\lambda(x,t)$, integrated in the space and time domains of interest and added to the original functional (5.3.2). The following extended functional is obtained:

$$S[D(x),\mu(x)] = \frac{1}{2}\int_{x=0}^{1}\int_{t=0}^{t_f}\sum_{m=1}^{M}\{U[x,t;D(x),\mu(x)] - Y(t)\}^2\,\delta(x - x_m)\,dt\,dx$$

$$+ \int_{x=0}^{1}\int_{t=0}^{t_f}\left[\frac{\partial U}{\partial t} - \frac{\partial}{\partial x}\left(D(x)\frac{\partial U}{\partial x}\right) - \mu(x)U\right]\lambda(x,t)\,dt\,dx \qquad (5.3.6)$$

where δ is the Dirac delta function.

Directional derivatives of $S[D(x),\mu(x)]$ in the directions of perturbations in $D(x)$ and $\mu(x)$ are, respectively, defined by:

$$\Delta S_D[D(x),\mu(x)] = \lim_{\varepsilon \to 0} \frac{S[D_\varepsilon(x),\mu(x)] - S[D(x),\mu(x)]}{\varepsilon} \tag{5.3.7a}$$

$$\Delta S_\mu[D(x),\mu(x)] = \lim_{\varepsilon \to 0} \frac{S[D(x),\mu_\varepsilon(x)] - S[D(x),\mu(x)]}{\varepsilon} \tag{5.3.7b}$$

where $S[D_\varepsilon(x),\mu(x)]$ and $S[D(x),\mu_\varepsilon(x)]$ denote the extended functional (5.3.6) written for perturbed $D(x)$ and $\mu(x)$, respectively.

After performing lengthy but straightforward manipulations, such as described in Section 5.1, and letting the directional derivatives of $S[D(x),\mu(x)]$ go to zero, which is a necessary condition for the minimization of the extended functional (5.3.6), the following adjoint problem for the Lagrange multiplier $\lambda(x,t)$ is obtained:

$$-\frac{\partial \lambda}{\partial t} - \frac{\partial}{\partial x}\left(D(x)\frac{\partial \lambda}{\partial x}\right) - \mu(x)\lambda + \sum_{m=1}^{M}[U-Y]\delta(x-x_m) = 0 \qquad \text{in } 0 < x < 1, \text{ for } t > 0 \tag{5.3.8a}$$

$$\frac{\partial \lambda}{\partial x} = 0 \qquad \text{at } x = 0 \text{ and } x = 1 \text{ for } t > 0 \tag{5.3.8b,c}$$

$$\lambda = 0 \qquad \text{in } 0 \leq x \leq 1 \text{ at } t = t_f \tag{5.3.8d}$$

GRADIENT EQUATIONS

During the limiting processes used to obtain the adjoint problem, which were applied to the directional derivatives of $S[D(x),\mu(x)]$ in the directions of perturbations in $D(x)$ and $\mu(x)$, the following integral terms, respectively, remained:

$$\Delta S_D[D(x),\mu(x)] = \int_{x=0}^{1}\int_{t=0}^{t_f} \Delta D(x)\frac{\partial U(x,t)}{\partial x}\frac{\partial \lambda(x,t)}{\partial x} dt \ dx \tag{5.3.9a}$$

$$\Delta S_\mu[D(x),\mu(x)] = -\int_{x=0}^{1}\int_{t=0}^{t_f} \Delta \mu(x)\lambda(x,t)U(x,t)dt \ dx \tag{5.3.9b}$$

By invoking the hypotheses that $D(x)$ and $\mu(x)$ belong to the Hilbert space of square-integrable functions in the domain $0 < x < 1$, it is possible to write:

$$\Delta S_D[D(x),\mu(x)] = \int_{x=0}^{1} \nabla S[D(x)]\Delta D(x) \ dx \tag{5.3.10a}$$

$$\Delta S_\mu[D(x),\mu(x)] = \int_{x=0}^{1} \nabla S[\mu(x)]\Delta \mu(x) \ dx \tag{5.3.10b}$$

Hence, by comparing equations (5.3.9a and b) and (5.3.10a and b), we obtain the gradient components of $S[D(x), \mu(x)]$ with respect to $D(x)$ and $\mu(x)$, respectively, as:

$$\nabla S[D(x)] = \int_{t=0}^{t_f} \frac{\partial U(x,t)}{\partial x} \frac{\partial \lambda(x,t)}{\partial x} dt \tag{5.3.11}$$

$$\nabla S[\mu(x)] = -\int_{t=0}^{t_f} \lambda(x,t) U(x,t) dt \tag{5.3.12}$$

An analysis of equations (5.3.11) and (5.3.8b and c) reveals that the gradient component with respect to $D(x)$ is null at $x = 0$ and $x = 1$. As a result, the initial guess used for $D(x)$ is never changed by the iterative procedure of the conjugate gradient method at such points, which can create instabilities in the inverse problem solution in their neighborhoods.

ITERATIVE PROCEDURE

For the simultaneous estimation of $D(x)$ and $\mu(x)$, the iterative procedure of the conjugate gradient method is written, respectively, as (see equation N2.5.2):

$$D^{k+1}(x) = D^k(x) + \beta_D^k d_D^k(x) \tag{5.3.13a}$$

$$\mu^{k+1}(x) = \mu^k(x) + \beta_\mu^k d_\mu^k(x) \tag{5.3.13b}$$

where $d_D^k(x)$ and $d_\mu^k(x)$ are the directions of descent for $D(x)$ and $\mu(x)$, respectively, β_D^k and β_μ^k are the search step sizes for $D(x)$ and $\mu(x)$, respectively, and k is the number of iterations.

The directions of descent for the conjugate gradient method for $D(x)$ and $\mu(x)$ can be written, respectively, as (see equation N2.5.3):

$$d_D^k(x) = -\nabla S[D^k(x)] + \gamma_D^k d_D^{k-1}(x) + \psi_D^k d_D^{qD}(x) \tag{5.3.14a}$$

$$d_\mu^k(x) = -\nabla S[\mu^k(x)] + \gamma_\mu^k d_\mu^{k-1}(x) + \psi_\mu^k d_\mu^{q\mu}(x) \tag{5.3.14b}$$

where γ_D^k, γ_μ^k, ψ_D^k and ψ_μ^k are the conjugation coefficients.

The conjugation coefficients for Powell-Beale's version of the conjugate gradient method are given by (see equations N2.5.7a and b):

$$\gamma_D^k = \frac{\displaystyle\int_{x=0}^{1} \left\{ \nabla S[D^k(x)] - \nabla S[D^{k-1}(x)] \right\} \nabla S[D^k(x)] \; dx}{\displaystyle\int_{x=0}^{1} \left\{ \nabla S[D^k(x)] - \nabla S[D^{k-1}(x)] \right\} d_D^{k-1}(x) \; dx} \tag{5.3.15a}$$

$$\gamma_\mu^k = \frac{\displaystyle\int_{x=0}^{1} \left\{ \nabla S[\mu^k(x)] - \nabla S[\mu^{k-1}(x)] \right\} \nabla S[\mu^k(x)] \; dx}{\displaystyle\int_{x=0}^{1} \left\{ \nabla S[\mu^k(x)] - \nabla S[\mu^{k-1}(x)] \right\} d_\mu^{k-1}(x) \; dx} \tag{5.3.15b}$$

$$\psi_D^k = \frac{\int_{x=0}^{1} \left\{\nabla S[D^{qD+1}(x)] - \nabla S[D^{qD}(x)]\right\} \nabla S[D^k(x)] \, dx}{\int_{x=0}^{1} \left\{\nabla S[D^{qD+1}(x)] - \nabla S[D^{qD}(x)]\right\} d_D^{qD}(x) \, dx} \qquad (5.3.15c)$$

$$\psi_\mu^k = \frac{\int_{x=0}^{1} \left\{\nabla S[\mu^{q\mu+1}(x)] - \nabla S[\mu^{q\mu}(x)]\right\} \nabla S[\mu^k(x)] \, dx}{\int_{x=0}^{1} \left\{\nabla S[\mu^{q\mu+1}(x)] - \nabla S[\mu^{q\mu}(x)]\right\} d_\mu^{q\mu}(x) \, dx} \qquad (5.3.15d)$$

where $\gamma_D^k = \gamma_\mu^k = \psi_D^k = \psi_\mu^k = 0$, for $k = 0$. The procedure for the calculation of the directions of descent $d_D^k(x)$ and $d_\mu^k(x)$ can be found in Note 2 at the end of this chapter.

For Fletcher-Reeves' version of the conjugate gradient method, the conjugation coefficients are computed as (see equations N2.5.4a):

$$\gamma_D^k = \frac{\int_{x=0}^{1} \{\nabla S[D^k(x)]\}^2 \, dx}{\int_{x=0}^{1} \{\nabla S[D^{k-1}(x)]\}^2 \, dx} \qquad \gamma_\mu^k = \frac{\int_{x=0}^{1} \{\nabla S[\mu^k(x)]\}^2 \, dx}{\int_{x=0}^{1} \{\nabla S[\mu^{k-1}(x)]\}^2 \, dx} \qquad (5.3.16a,b)$$

For Polak-Ribiere's version of the conjugate gradient method, the conjugation coefficients are computed as (see equations N2.5.5):

$$\gamma_D^k = \frac{\int_{x=0}^{1} \nabla S[D^k(x)]\{\nabla S[D^k(x)] - \nabla S[D^{k-1}(x)]\} dx}{\int_{x=0}^{1} \{\nabla S[D^{k-1}(x)]\}^2 \, dx} \qquad (5.3.17a)$$

$$\gamma_\mu^k = \frac{\int_{x=0}^{1} \nabla S[\mu^k(x)]\{\nabla S[\mu^k(x)] - \nabla S[\mu^{k-1}(x)]\} \, dx}{\int_{x=0}^{1} \{\nabla S[\mu^{k-1}(x)]\}^2 \, dx} \qquad (5.3.17b)$$

In Fletcher-Reeves' and Polak-Ribiere's versions of the conjugate gradient method, $\gamma_D^k = \gamma_\mu^k = 0$ for $k = 0$. Furthermore, in these two versions of the conjugate gradient method $\psi_D^k = \psi_\mu^k = 0$ for any k, so that a restarting strategy is not performed.

The search step sizes β_D^k and β_μ^k appearing in the expressions of the iterative procedures for the estimation of $D(x)$ and $\mu(x)$, equations (5.3.13a and b), respectively, are obtained by minimizing the objective functional at each iteration along the specified directions of descent. If the objective functional given by equation (5.3.2) is linearized with respect to β_D^k and β_μ^k, closed form expressions can be obtained for such quantities as follows [137]:

$$\beta_d^k = \frac{F_1 A_{22} - F_2 A_{12}}{A_{11} A_{22} - A_{12}^2}; \quad \beta_\mu^k = \frac{F_2 A_{11} - F_1 A_{12}}{A_{11} A_{22} - A_{12}^2} \tag{5.3.18a,b}$$

where

$$A_{11} = \int_{t=0}^{t_f} \sum_{m=1}^{M} [\Delta U_D^k(x_m, t)]^2 \, dt \tag{5.3.19a}$$

$$A_{22} = \int_{t=0}^{t_f} \sum_{m=1}^{M} [\Delta U_\mu^k(x_m, t)]^2 \, dt \tag{5.3.19b}$$

$$A_{12} = \int_{t=0}^{t_f} \sum_{m=1}^{M} \Delta U_D^k(x_m, t) \Delta U_\mu^k(x_m, t) \, dt \tag{5.3.19c}$$

$$F_1 = \int_{t=0}^{t_f} \sum_{m=1}^{M} [Y_m^k - U^k(x_m, t)][\Delta U_D^k(x_m, t)] dt \tag{5.3.19d}$$

$$F_2 = \int_{t=0}^{t_f} \sum_{m=1}^{M} [Y_m^k - U^k(x_m, t)][\Delta U_\mu^k(x_m, t)] dt \tag{5.3.19e}$$

In equations (5.3.19a-e), $\Delta U_D^k(x,t)$ and $\Delta U_\mu^k(x,t)$ are the solutions of the sensitivity problems given by equations (5.3.4a-d) and (5.3.5a-d), respectively, obtained by setting $\Delta D^k(x) = d_D^k(x)$ and $\Delta \mu^k(x) = d_\mu^k(x)$.

RESULTS

The accuracy of the present solution approach was examined by using simulated transient measurements containing random errors in the inverse analysis. Different functional forms, including those containing sharp corners and discontinuities that are the most difficult to be recovered by the inverse analysis, were used to generate the simulated measurements.

The test cases examined below in dimensionless form were associated with a heat conduction problem in a homogeneous steel bar of length 0.050 m. The diffusion coefficient and the spatial distribution of the source term were supposed to vary from base values 54 W/mK and 10^5 W/m³K, which result in dimensionless values of 1 and 5, respectively. The final time was assumed to be 60 s, resulting in a dimensionless value $t_f = 0.36$, and 50 transient measurements were supposed available per sensor.

Figure 5.6 shows the results obtained with the measurements of two non-intrusive sensors for a step variation of $D(x)$ and for constant $\mu(x)$. The results presented in this figure were obtained with Powell-Beale's version of the conjugate gradient method. The simulated measurements in this case contained random errors with standard deviation $\sigma = 0.01 Y_{max}$, where Y_{max} is the maximum absolute value of the measured variable. The initial guesses used for the iterative procedure of the conjugate gradient method were $D(x) = 0.9$ and $\mu(x) = 4.5$. We note in Figure 5.6 that quite accurate results were obtained for such strict test case, involving a discontinuous variation for $D(x)$ and only non-intrusive measurements. Although some blurring is observed near the discontinuity of $D(x)$ at $x = 0.25$, the locations of the discontinuities and the maximum value of the function were quite accurately estimated. Furthermore, the estimated function for $\mu(x)$ oscillated about its constant exact value with an amplitude smaller than the original distance of the initial guess to the exact function.

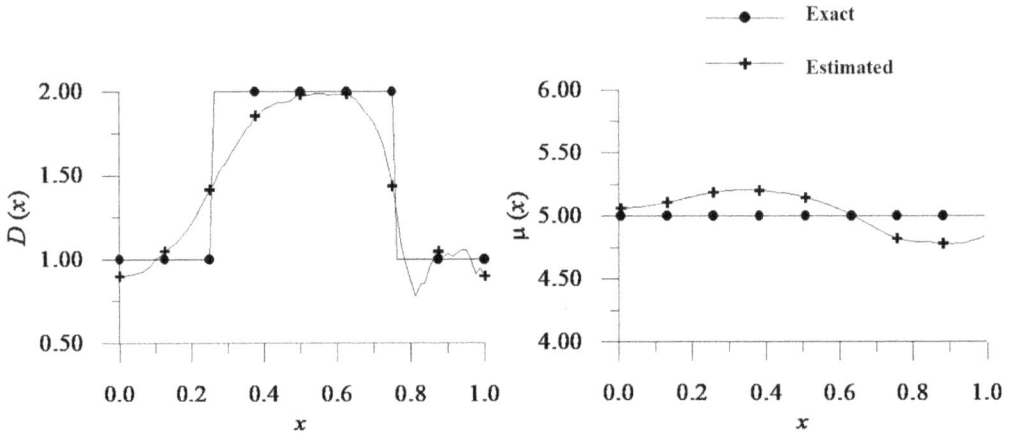

FIGURE 5.6 Simultaneous estimation of $\mu(x)$ and $D(x)$ obtained with measurements of two non-intrusive sensors. (From Ref. [137].)

The accuracy of the estimated functions improved when measurements of more sensors were used in the inverse analysis, as illustrated in Figure 5.7, which was obtained with measurements containing random errors ($\sigma = 0.01Y_{max}$) of ten sensors evenly located inside the medium.

Figure 5.8 presents the results obtained for a second-degree polynomial variation for $D(x)$ and for constant $\mu(x)$. The results presented in Figure 5.8 were obtained with measurements containing random errors ($\sigma = 0.01Y_{max}$) of ten sensors evenly located inside the medium by using Powell-Beale's version of the conjugate gradient method. A comparison of Figures 5.7 and 5.8 shows that the accuracy of estimated functions generally improves for continuous and smooth variations of the unknown diffusion coefficient.

Figure 5.9 illustrates the results obtained for a constant exact functional form for $D(x)$ and a triangular variation for $\mu(x)$. The results presented in Figure 5.9 were obtained with measurements containing random errors ($\sigma = 0.01Y_{max}$) of ten sensors evenly located inside the medium by using Powell-Beale's version of the conjugate gradient method. Differently from the results shown in Figures 5.7 and 5.8, we note in Figure 5.9 that the present solution approach failed to estimate the peak value of the exact triangular function for $\mu(x)$. The locations of the discontinuities in the first derivative of the exact function, which characterize the change of $\mu(x)$ from its base value, could not

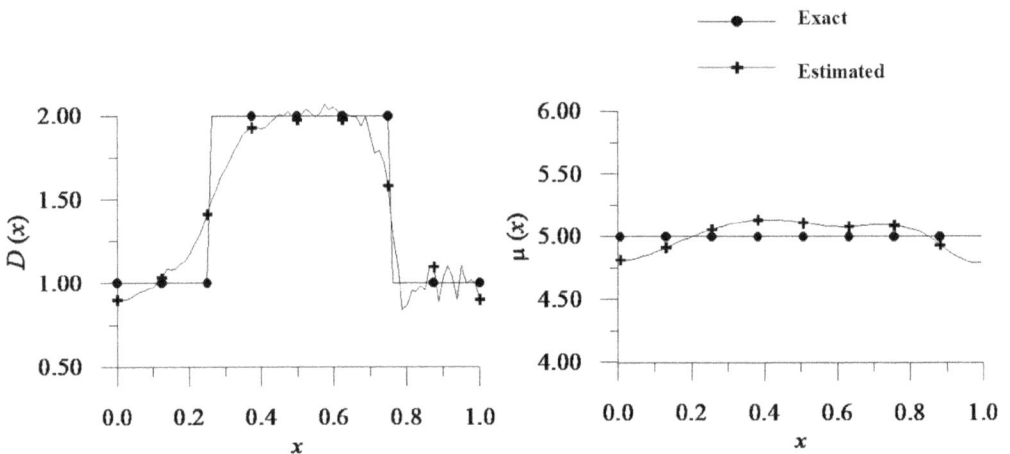

FIGURE 5.7 Simultaneous estimation of $\mu(x)$ and $D(x)$ obtained with measurements of ten sensors equally spaced in the medium. (From Ref. [137].)

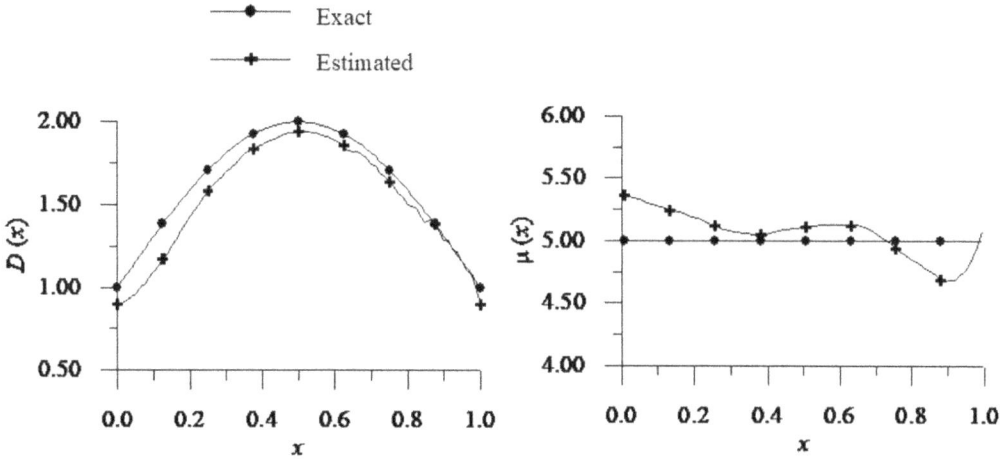

FIGURE 5.8 Simultaneous estimation of $\mu(x)$ and $D(x)$ obtained with measurements of ten sensors equally spaced in the medium for a second degree polynomial variation for $D(x)$ and for constant $\mu(x)$. (From Ref. [137].)

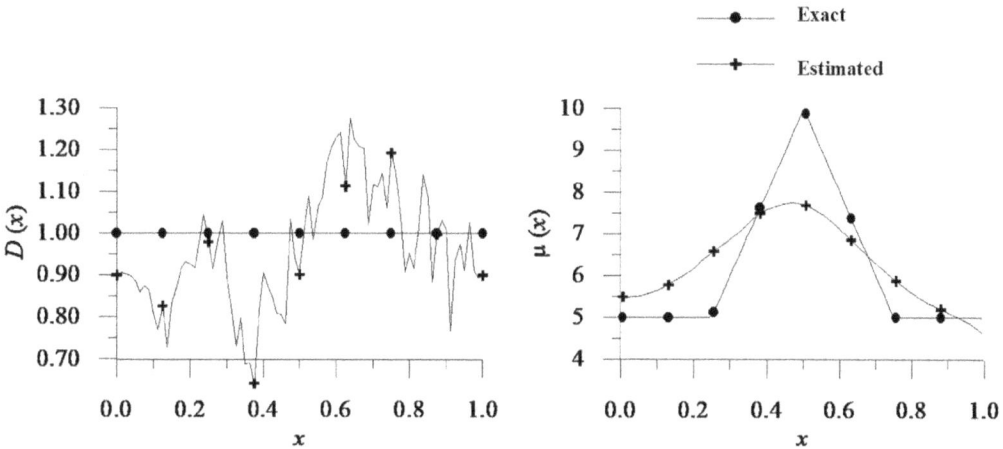

FIGURE 5.9 Simultaneous estimation of $\mu(x)$ and $D(x)$ obtained with measurements of ten sensors equally spaced in the medium for constant $D(x)$ and a triangular variation for $\mu(x)$. (From Ref. [137].)

be accurately estimated. Furthermore, the function estimated for $D(x)$ was characterized by large oscillations. This is due to the lower sensitivity of the measured variable with respect to $\mu(x)$ as compared to the sensitivity with respect to $D(x)$.

Figure 5.10 presents the reduction of the objective functional with respect to the number of iterations obtained with Powell-Beale's, Polak-Ribiere's and Fletcher-Reeves' versions of the conjugate gradient. The results presented in Figure 5.10 correspond to the test case shown in Figure 5.9 but with errorless measurements. Figure 5.10 shows that the prescribed tolerance for the iterative procedure of the conjugate gradient method was only reached with Powell-Beale's version; the other two versions did not effectively reduce the objective functional and the iterative procedure was stopped when the specified maximum number of iterations (100) was reached. In fact, Powell-Beale's version of the conjugate gradient method resulted in the largest rate of reduction of the objective functional, so that the tolerance prescribed for the stopping criterion was reached in the smallest number of iterations. However, for some test cases the use of Polak-Ribiere's version of the conjugate gradient method resulted in reduction rates for the functional comparable to those obtained with Powell-Beale's version, but unexpected oscillations were observed on the values of the functional [137].

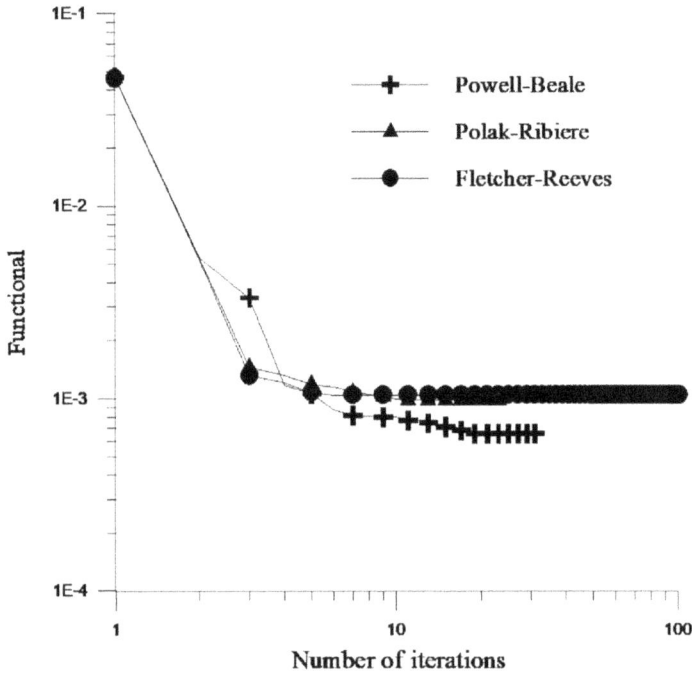

FIGURE 5.10 Comparison of different versions of the conjugate gradient method for the simultaneous estimation of $\mu(x)$ and $D(x)$. (From Ref. [137].)

5.4 SIMULTANEOUS ESTIMATION OF THE SPACEWISE AND TIMEWISE VARIATIONS OF MASS AND HEAT TRANSFER COEFFICIENTS IN DRYING

In this section, the conjugate gradient method with adjoint problem is applied for the identification of the heat and mass transfer coefficients at the surface of drying capillary-porous bodies [138]. The unknown functions are supposed to vary in time and along the surface open to the surrounding environment. The effects of temperature and moisture content measurements on the inverse analysis are examined. A comparison of different versions of the conjugate gradient method is also presented, as applied to this inverse problem.

DIRECT PROBLEM

The physical problem involves a two-dimensional capillary porous medium in Cartesian coordinates initially at uniform temperature and uniform moisture content. The lateral surfaces of the body are impervious to moisture transfer and thermally insulated. The bottom boundary, which is impervious to moisture transfer, is in direct contact with a heater. The top boundary is in contact with the dry surrounding air, thus resulting in a convective boundary condition for both the energy and the mass conservation equations. The mass and heat transfer coefficients at this boundary may vary in time and along the surface open to the surrounding environment. The linear system of equations proposed by Luikov [139], with associated initial and boundary conditions, for the modeling of such physical problem involving heat and mass transfer in capillary porous media, can be written in dimensionless form as:

$$\frac{\partial \theta}{\partial \tau} = \alpha \left(\frac{\partial^2 \theta}{\partial X^2} + \frac{\partial^2 \theta}{\partial Y^2} \right) - \beta \left(\frac{\partial^2 \phi}{\partial X^2} + \frac{\partial^2 \phi}{\partial Y^2} \right) \text{ in } 0 < X < r_a, 0 < Y < 1 \text{ and } \tau > 0 \quad (5.4.1a)$$

$$\frac{\partial \phi}{\partial \tau} = Lu\left(\frac{\partial^2 \phi}{\partial X^2} + \frac{\partial^2 \phi}{\partial Y^2}\right) - Lu\,Pn\left(\frac{\partial^2 \theta}{\partial X^2} + \frac{\partial^2 \theta}{\partial Y^2}\right) \text{ in } 0 < X < r_a, 0 < Y < 1 \text{ and } \tau > 0, \quad (5.4.1b)$$

$$\frac{\partial \theta}{\partial Y} = -Q \text{ at } Y = 0, \text{ for } \tau > 0 \qquad (5.4.1c)$$

$$\frac{\partial \phi}{\partial Y} = -Pn\,Q \text{ at } Y = 0, \text{ for } \tau > 0 \qquad (5.4.1d)$$

$$\frac{\partial \theta}{\partial Y} = Bi_q(X,\tau)(1-\theta) - (1-\varepsilon)Ko\,Lu\,Bi_m(X,\tau)(1-\phi) \text{ at } Y = 1, \text{ for } \tau > 0 \qquad (5.4.1e)$$

$$\frac{\partial \phi}{\partial Y} = Pn\frac{\partial \theta}{\partial Y} + Bi_m(X,\tau)(1-\phi) \text{ at } Y = 1, \text{ for } t > 0 \qquad (5.4.1f)$$

$$\frac{\partial \theta}{\partial X} = \frac{\partial \phi}{\partial X} = 0 \text{ at } X = 0 \text{ and } X = r_a \text{ , for } \tau > 0 \qquad (5.4.1g,h)$$

$$\theta(X,Y,0) = \phi(X,Y,0) = 0 \text{ for } \tau = 0, \text{ in } 0 < X < r_a, 0 < Y < 1 \qquad (5.4.1i, j)$$

where the following dimensionless variables were defined:

$$\theta = \frac{T - T_0}{T_s - T_0} \qquad \phi = \frac{u_0 - u}{u_0 - u^*} \qquad Q = \frac{qh}{k(T_s - T_0)}$$

$$\tau = \frac{at}{h^2} \qquad Lu = \frac{a_m}{a} \qquad Pn = \delta\frac{T_s - T_0}{u_0 - u^*} \qquad Ko = \frac{r(u_0 - u^*)}{c(T_s - T_0)}$$

$$X = \frac{x}{h} \qquad Y = \frac{y}{h} \qquad r_a = \frac{L}{h}$$

$$Bi_q(X,\tau) = \frac{h_q(X,\tau)\,h}{k} \qquad Bi_m(X,\tau) = \frac{h_m(X,\tau)\,h}{k_m}$$

$$\alpha = 1 + \varepsilon\,Ko\,Lu\,Pn \qquad \beta = \varepsilon\,Ko\,Lu \qquad \gamma = 1 - (1-\varepsilon)Ko\,Lu\,Pn \qquad (5.4.2a\text{-}o)$$

The properties of the porous medium appearing above include the thermal diffusivity (a), the moisture diffusivity (a_m), the thermal conductivity (k), the moisture conductivity (k_m) and the specific heat (c). Other physical quantities appearing in the dimensionless groups of equations (5.4.2) are the heat transfer coefficient (h_q), the mass transfer coefficient (h_m), the thickness of porous medium (h), the width of the porous medium (L), the prescribed heat flux (q), the latent heat of evaporation of water (r), the temperature of the surrounding air (T_s), the uniform initial temperature in the medium (T_0), the moisture content in equilibrium with the surrounding air (u^*), the uniform initial moisture content in the medium (u_0), the thermogradient coefficient (δ) and the phase conversion factor (ε). *Lu, Pn* and *Ko* denote the Luikov, Posnov and Kossovitch numbers, respectively [139].

Problem (5.4.1) is referred to as a direct problem when initial and boundary conditions, as well as all parameters appearing in the formulation, are known. The objective of the direct problem is to determine the dimensionless temperature and moisture content fields, $\theta(X,Y,\tau)$ and $\phi(X,Y,\tau)$, respectively, in the capillary porous media.

INVERSE PROBLEM

For the inverse problem of interest here, the functions $Bi_q(X,\tau)$ and $Bi_m(X,\tau)$ are regarded as unknown quantities. Despite the fact that such functions are physically correlated, they are treated as independent quantities in order to examine the strictest case involving the simultaneous estimation of $Bi_q(X,\tau)$ and $Bi_m(X,\tau)$.

For the estimation of such functions, we consider available the transient temperature measurements $M_i(\tau)$ taken at the locations (X_i,Y_i), $i = 1,...,I$, as well as the transient moisture content measurements $C_n(\tau)$ taken at the locations (X_n^*,Y_n^*), $n = 1,...,N$. We note that the measurements may contain random errors, but all the other quantities appearing in the formulation of the direct problem are supposed to be exactly known.

The objective functional is given by:

$$S[Bi_m(X,\tau),Bi_q(X,\tau)] = \int_{\tau=0}^{\tau_f} \left\{ \sum_{i=1}^{I} [\theta(X_i,Y_i,\tau;Bi_m,Bi_q) - M_i(\tau)]^2 w_\theta \right\} d\tau$$

$$+ \int_{\tau=0}^{\tau_f} \left\{ \sum_{n=1}^{N} [\phi(X_n^*,Y_n^*,\tau;Bi_m,Bi_q) - C_n(\tau)]^2 w_\phi \right\} d\tau \qquad (5.4.3)$$

where $\theta(X_i,Y_i,\tau;Bi_m,Bi_q)$ and $\phi(X_n^*,Y_n^*,\tau;Bi_m,Bi_q)$ are the estimated temperature and moisture content, respectively, which are obtained from the solution of the direct problem with estimates for the unknown functions. In equation (5.4.3), w_θ and w_ϕ are weights for the temperature and moisture content measurements, respectively, given by the inverse of the corresponding variances.

For the minimization of such objective functional, we use here the conjugate gradient method with adjoint problem, as described next.

SENSITIVITY PROBLEMS

Like in the example presented in Section 5.3, the present inverse problem deals with the estimation of two unknown functions. Thus, two sensitivity problems are required in the analysis. They are derived by considering perturbations in the heat and mass transfer coefficients, each at a time, as described next.

Let us consider that the temperature $\theta(X,Y,\tau)$ and the moisture content $\phi(X,Y,\tau)$ undergo variations $\Delta\theta_1(X,Y,\tau)$ and $\Delta\phi_1(X,Y,\tau)$, respectively, when the mass transfer coefficient $Bi_m(X,\tau)$ is perturbed by $\Delta Bi_m(X,\tau)$. By substituting in the direct Problem (5.4.1) $\theta(X,Y,\tau)$ by $[\theta(X,Y,\tau)+\Delta\theta_1(X,Y,\tau)]$, $\phi(X,Y,\tau)$ by $[\phi(X,Y,\tau)+\Delta\phi_1(X,Y,\tau)]$ and $Bi_m(X,\tau)$ by $[Bi_m(X,\tau)+\Delta Bi_m(X,\tau)]$, and then subtracting from the resulting problem the original direct problem, we obtain the following sensitivity problem for the sensitivity functions $\Delta\theta_1(X,Y,\tau)$ and $\Delta\phi_1(X,Y,\tau)$ [138]:

$$\frac{\partial \Delta\theta_1}{\partial \tau} = \alpha\left(\frac{\partial^2 \Delta\theta_1}{\partial X^2} + \frac{\partial^2 \Delta\theta_1}{\partial Y^2}\right) - \beta\left(\frac{\partial^2 \Delta\phi_1}{\partial X^2} + \frac{\partial^2 \Delta\phi_1}{\partial Y^2}\right) \text{ in } 0<X<r_a, 0<Y<1 \text{ and } \tau>0 \quad (5.4.4a)$$

$$\frac{\partial \Delta\phi_1}{\partial \tau} = Lu\left(\frac{\partial^2 \Delta\phi_1}{\partial X^2} + \frac{\partial^2 \Delta\phi_1}{\partial Y^2}\right) - Lu\,Pn\left(\frac{\partial^2 \Delta\theta_1}{\partial X^2} + \frac{\partial^2 \Delta\theta_1}{\partial Y^2}\right) \text{ in } 0<X<r_a, 0<Y<1 \text{ and } \tau>0 \quad (5.4.4b)$$

$$\frac{\partial \Delta\theta_1}{\partial Y} = \frac{\partial \Delta\phi_1}{\partial Y} = 0 \text{ at } Y=0, \text{ for } \tau>0 \qquad (5.4.4c,d)$$

$$\frac{\partial \Delta\theta_1}{\partial Y} = -Bi_q(X,\tau)\Delta\theta_1 + (1-\varepsilon)Ko\,Lu\,\Delta Bi_m(X,\tau)\big[\phi-1\big]$$

$$+ (1-\varepsilon)Ko\,Lu\,Bi_m(X,\tau)\,\Delta\phi_1 \qquad \text{at } Y = 1, \text{ for } \tau > 0 \qquad (5.4.4e)$$

$$\frac{\partial \Delta\phi_1}{\partial Y} = \big\{\Delta Bi_m(X,\tau) - Bi_m(X,\tau)\Delta\phi_1 - \phi Bi_m(X,\tau)\big\}\gamma - Bi_q(X,\tau)\,Pn\,\Delta\theta_1 \text{ at } Y = 1, \text{ for } \tau > 0 \quad (5.4.4f)$$

$$\frac{\partial \Delta\theta_1}{\partial X} = \frac{\partial \Delta\phi_1}{\partial X} = 0 \text{ at } X = 0 \text{ and } X = r_a, \text{ for } \tau = 0 \qquad (5.4.4g,h)$$

$$\Delta\theta_1(X,Y,0) = \Delta\phi_1(X,Y,0) = 0 \text{ for } \tau = 0, \text{ in } 0 < X < r_a, 0 < Y < 1 \qquad (5.4.4i,j)$$

Similarly, the sensitivity problem for the sensitivity functions $\Delta\theta_2(X,Y,\tau)$ and $\Delta\phi_2(X,Y,\tau)$, which result from a perturbation $\Delta Bi_q(X,\tau)$ in $Bi_q(X,\tau)$, can be obtained as [138]:

$$\frac{\partial \Delta\theta_2}{\partial \tau} = \alpha\left(\frac{\partial^2 \Delta\theta_2}{\partial X^2} + \frac{\partial^2 \Delta\theta_2}{\partial Y^2}\right) - \beta\left(\frac{\partial^2 \Delta\phi_2}{\partial X^2} + \frac{\partial^2 \Delta\phi_2}{\partial Y^2}\right) \text{ in } 0 < X < r_a, 0 < Y < 1 \text{ and } \tau > 0 \qquad (5.4.5a)$$

$$\frac{\partial \Delta\phi_2}{\partial \tau} = Lu\left(\frac{\partial^2 \Delta\phi_2}{\partial X^2} + \frac{\partial^2 \Delta\phi_2}{\partial Y^2}\right) - Lu\,Pn\left(\frac{\partial^2 \Delta\theta_2}{\partial X^2} + \frac{\partial^2 \Delta\theta_2}{\partial Y^2}\right) \text{ in } 0 < X < r_a, 0 < Y < 1 \text{ and } \tau > 0 \quad (5.4.5b)$$

$$\frac{\partial \Delta\theta_2}{\partial Y} = \frac{\partial \Delta\phi_2}{\partial Y} = 0 \text{ at } Y = 0, \text{ for } \tau > 0 \qquad (5.4.5c,d)$$

$$\frac{\partial \Delta\theta_2}{\partial Y} = \Delta Bi_q(X,\tau) - \big[Bi_q(X,\tau)\Delta\theta_2 + \theta\,\Delta Bi_q(X,\tau)\big] + (1-\varepsilon)Ko\,Lu\,Bi_m(X,\tau)\Delta\phi_2 \text{ at } Y = 1, \text{ for } \tau > 0$$

$$(5.4.5e)$$

$$\frac{\partial \Delta\phi_2}{\partial Y} = -Bi_m(X,\tau)\gamma\,\Delta\phi_2 + \Delta Bi_q(X,\tau)\,Pn - \big[Bi_q(X,\tau)\Delta\theta_2 + \theta\,\Delta Bi_q(X,\tau)\big]Pn \text{ at } Y = 1, \text{ for } \tau > 0$$

$$(5.4.5f)$$

$$\frac{\partial \Delta\theta_2}{\partial X} = \frac{\partial \Delta\phi_2}{\partial X} = 0 \text{ at } X = 0 \text{ and } X = r_a, \text{ for } \tau > 0 \qquad (5.4.5g,h)$$

$$\Delta\theta_2(X,Y,0) = \Delta\phi_2(X,Y,0) = 0 \text{ for } \tau = 0, \text{ in } 0 < X < r_a, 0 < Y < 1 \qquad (5.4.5i,j)$$

ADJOINT PROBLEM

The adjoint problem is derived by multiplying the governing equations of the direct problem by Lagrange multipliers, integrating in the spatial and time domains that they are valid and then adding the resultant equation to the original functional (5.4.3). The directional derivative of the functional in the direction of the perturbation of each of the unknown functions is then obtained and the resultant expression is allowed to go to zero, similarly to what was performed in Section 5.1. The same adjoint problem is obtained for perturbations in $Bi_q(X,\tau)$ and $Bi_m(X,\tau)$. The adjoint problem, for the computation of the Lagrange multipliers $\lambda_1(X,Y,\tau)$ and $\lambda_2(X,Y,\tau)$, is given by:

$$\frac{\partial \lambda_1}{\partial \tau} = \alpha\left(\frac{\partial^2 \lambda_1}{\partial X^2} + \frac{\partial^2 \lambda_1}{\partial Y^2}\right) + Lu\,Pn\left(\frac{\partial^2 \lambda_2}{\partial X^2} + \frac{\partial^2 \lambda_2}{\partial Y^2}\right)$$

$$- \sum_{i=1}^{I} 2(\theta_i - M_i)\delta(X - X_i)\delta(Y - Y_i)w_\theta \text{ in } 0 < X < r_a, 0 < Y < 1 \text{ and } \tau > 0 \qquad (5.4.6a)$$

$$\frac{\partial \lambda_2}{\partial \tau} = Lu\left(\frac{\partial^2 \lambda_1}{\partial X^2} + \frac{\partial^2 \lambda_1}{\partial Y^2}\right) + \beta\left(\frac{\partial^2 \lambda_2}{\partial X^2} + \frac{\partial^2 \lambda_2}{\partial Y^2}\right)$$

$$-\sum_{n=1}^{N} 2\left(\phi_n - C_n\right)\delta(X - X_n^*)\delta(Y - Y_n^*)w_\phi \text{ in } 0 < X < r_a, 0 < Y < 1 \text{ and } \tau > 0 \quad (5.4.6b)$$

$$\frac{\partial \lambda_1}{\partial Y} = \frac{\partial \lambda_2}{\partial Y} = 0 \text{ at } Y = 0, \text{ for } \tau > 0 \quad (5.4.6c,d)$$

$$\frac{\partial \lambda_1}{\partial Y} = -\lambda_1 Bi_q(X,\tau) - \frac{Lu\,\lambda_2\, Bi_q(X,\tau)\,Pn}{\alpha} \text{ at } Y = 1, \text{ for } \tau > 0 \quad (5.4.6e)$$

$$\frac{\partial \lambda_2}{\partial Y} = -\alpha\,\lambda_1\,(1 - \varepsilon)\,Ko\,Bi_m(X,\tau) - \lambda_2\,Bi_m(X,\tau)\gamma \text{ at } Y = 1, \text{ for } \tau > 0 \quad (5.4.6f)$$

$$\frac{\partial \lambda_1}{\partial Y} = \frac{\partial \lambda_2}{\partial Y} = 0 \text{ at } X = 0 \text{ and } X = r_a, \text{ for } \tau > 0 \quad (5.4.6g,h)$$

$$\lambda_1(X,Y,\tau_f) = \lambda_2(X,Y,\tau_f) = 0 \text{ for } \tau = \tau_f \text{ in } 0 < X < r_a, 0 < Y < 1 \quad (5.4.6i,j)$$

GRADIENT EQUATIONS

With the limiting process used to obtain the adjoint Problem (5.4.6), we can also identify the following expressions for the gradient directions, where it was taken into account the hypotheses that $Bi_q(X,\tau)$ and $Bi_m(X,\tau)$ belong to the Hilbert space of square-integrable functions in the domain $0 < X < r_a$ and $0 < \tau < \tau_f$:

$$\nabla S[Bi_m(X,\tau)] = \alpha\,\lambda_1(X,1,\tau)(1 - \varepsilon)\,Ko\,Lu\,[\phi(X,1,\tau) - 1] + Lu\,\lambda_2(X,1,\tau)[1 - \phi(X,1,\tau)\gamma] \quad (5.4.7a)$$

$$\nabla S[Bi_q(X,\tau)] = \alpha\,\lambda_1(X,1,\tau)[1 - \theta(X,1,\tau)] + Lu\,\lambda_2(X,1,\tau)\,Pn\,[1 - \theta(X,1,\tau)] \quad (5.4.7b)$$

ITERATIVE PROCEDURE

The iterative procedure of the conjugate gradient method of function estimation is given by:

$$Bi^{k+1}(X,\tau) = Bi^k(X,\tau) + \beta^k d^k(X,\tau) \quad (5.4.8)$$

where the superscript k denotes the number of iterations, β^k is the search step size, $d^k(X,\tau)$ is the direction of descent and $Bi(X,\tau)$ represents either $Bi_q(X,\tau)$ or $Bi_m(X,\tau)$.

The direction of descent $d^k(X,\tau)$ is given by equation (N2.5.3) in Note 2. For the present inverse problem, the conjugation coefficients for *Powell-Beale's* version of the conjugate gradient method are given by (see equations N2.5.7a,b):

$$\gamma^k = \frac{\displaystyle\int_{\tau=0}^{\tau_f}\int_{X=0}^{r_a} \{\nabla S[Bi^k] - \nabla S[Bi^{k-1}]\}\nabla S[Bi^k]dX\,d\tau}{\displaystyle\int_{\tau=0}^{\tau_f}\int_{X=0}^{r_a} \{\nabla S[Bi^k] - \nabla S[Bi^{k-1}]\}\,d^{k-1}(X,\tau)dX\,d\tau} \quad \text{with } \gamma^k = 0 \text{ for } k = 0 \quad (5.4.9a)$$

$$\psi^k = \frac{\displaystyle\int_{\tau=0}^{\tau_f}\int_{X=0}^{r_a} \{\nabla S[Bi^{q+1}] - \nabla S[Bi^q]\} \, \nabla S[Bi^k] dX \, d\tau}{\displaystyle\int_{\tau=0}^{\tau_f}\int_{X=0}^{r_a} \{\nabla S[Bi^{q+1}] - \nabla S[Bi^q]\} \, d^q(X,\tau) dX \, d\tau} \quad \text{with } \psi^k = 0 \text{ for } k = 0 \qquad (5.4.9b)$$

In Polak-Ribiere's and Fletcher-Reeves' versions of the conjugate gradient method, the conjugation coefficients are given, respectively, by (see equations N2.5.4a and N2.5.5a):

$$\gamma^k = \frac{\displaystyle\int_{\tau=0}^{\tau_f}\int_{X=0}^{r_a} \{\nabla S[Bi^k] - \nabla S[Bi^{k-1}]\}\nabla S[Bi^k] dX \, d\tau}{\displaystyle\int_{\tau=0}^{\tau_f}\int_{X=0}^{r_a} \{\nabla S[Bi^{k-1}]\}^2 \, dX \, d\tau} \qquad (5.4.9c)$$

and

$$\gamma^k = \frac{\displaystyle\int_{\tau=0}^{\tau_f}\int_{X=0}^{r_a} \{\nabla S[Bi^k]\}^2 \, dX \, d\tau}{\displaystyle\int_{\tau=0}^{\tau_f}\int_{X=0}^{r_a} \{\nabla S[Bi^{k-1}]\}^2 \, dX \, d\tau} \qquad (5.4.9d)$$

with $\gamma^k = 0$ for $k = 0$ and $\psi^k = 0$ for any k.

Expressions are obtained for the search step sizes of the iterative procedures for the estimation of $Bi_q(X,\tau)$ and $Bi_m(X,\tau)$ by minimizing the objective functional at each iteration. These expressions are omitted here for the sake of brevity, but they follow the same steps outlined in Note 2 at the end of this chapter.

RESULTS

For the results presented below, we examined test cases involving the drying of a capillary-porous body with dimensions $h = 0.05$ m and $L = 0.5$ m, made of ceramics, with properties [138]: $k = 0.34$ W/mK, $k_m = 2.4\times10^{-7}$ kg/ms°M, $c = 607$ J/kg K, $r = 2.5\times10^6$ J/kg, $T_0 = 24°C$, $u_0 = 80°M$, $\delta = 0.56$ °M/K and $\varepsilon = 0.8$. The air conditions were taken as $T_s = 30°C$ and $u^* = 40°M$ and the applied heat flux as $q = 40$ W/m². Therefore, the dimensionless numbers appearing in the formulation were $Lu = 0.2$, $Pn = 0.084$, $Ko = 49$ and $Q = 0.9$. The final time was taken as 1785 s, so that the dimensionless final time was $\tau_f = 0.2$.

For the estimation of the unknown heat and mass transfer coefficients, we made use of simulated temperature and moisture content measurements. The simulated measurements contained additive, uncorrelated, Gaussian errors with constant standard deviation of 1% of the maximum value of the measured quantity. For each sensor, one measurement was considered to be available every $\Delta\tau = 0.001$, which corresponds to a frequency of 0.11 Hz.

Let us consider initially in the analysis the estimation of the heat transfer coefficient $Bi_q(X,\tau)$, by assuming that the mass transfer coefficient $Bi_m(X,\tau)$ is exactly known. For this case, $Bi_m(X,\tau)$ was given by the same function of $Bi_q(X,\tau)$. Figure 5.11 presents the spatial variation for $Bi_q(X,\tau)$ at selected times, obtained with temperature measurements ($w_\phi = 0$ in equation 5.4.3) of 13 sensors evenly spaced along the body length, at $Y = 0.85$ (which corresponds to 7.5 mm below the top surface). We note in Figure 5.11 that the estimated function deviated from the exact one near $\tau = 0.2$, because of the null gradient at the final time (see equations 5.4.6i and j and 5.4.7a and b). On the other hand, quite accurate estimates could be obtained for $Bi_q(X, \tau)$ at other times by using temperature measurements.

We now consider the estimation of $Bi_m(X, \tau)$ by using temperature measurements ($w_\phi = 0$ in equation 5.4.3) and assuming $Bi_q(X, \tau)$ as exactly known for the inverse analysis. For this case, $Bi_q(X, \tau)$ was given by the same function of $Bi_m(X, \tau)$. Figure 5.12 presents the estimated function obtained with measurements of 13 sensors equally spaced along the body length at the position $Y = 0.85$. Figure 5.12 shows the interesting fact that temperature measurements provided useful information for the estimation of $Bi_m(X, \tau)$. This is an important result, because quite involved and inaccurate techniques for the measurement of moisture content can be avoided, in favor of inexpensive and accurate temperature measurements, for the estimation of $Bi_m(X, \tau)$ if $Bi_q(X, \tau)$ is known. The use of moisture content measurements resulted in accurate estimations for $Bi_m(X, \tau)$, as illustrated in Figure 5.13, but not for the estimation of $Bi_q(X, \tau)$. The estimated function shown in Figure 5.13 was obtained with moisture content measured data ($w_\theta = 0$ in equation 5.4.3), of 13 sensors equally spaced along the body length, at the position $Y = 0.85$.

The simultaneous estimation of $Bi_q(X, \tau)$ and $Bi_m(X, \tau)$ is now examined. For this case, the use of only temperature measurements or only moisture content measurements did not result on accurate estimated functions. Therefore, temperature, as well as moisture content measurements, were required for the simultaneous estimation of $Bi_q(X, \tau)$ and $Bi_m(X, \tau)$. Figures 5.14a and b present the results obtained for $Bi_q(X, \tau)$ and $Bi_m(X, \tau)$, respectively, by using simulated measurements of 13 temperature sensors and 13 moisture content sensors. In this case, we used $w_\theta = 1/(0.01M_{max})^2$ and $w_\phi = 1/(0.01C_{max})^2$ in equation (5.4.3), where M_{max} and C_{max} are the maximum measured values of temperature and moisture content, respectively. The temperature sensors and the moisture content sensors were located at $Y = 0.85$ evenly spaced along the length of the body. Figures 5.15a and b show that quite accurate results could be obtained for the simultaneous estimation of $Bi_q(X, \tau)$ and $Bi_m(X, \tau)$, if both temperature and moisture content measurements were used in the inverse analysis. We note however, when we compare Figures 5.14a and b with Figures 5.11–5.13, that more accurate estimations could be obtained in cases involving one single unknown function.

Finally, we present in Table 5.1 a comparison of Powell-Beale's, Polak-Ribiere's and Fletcher-Reeves' versions of the conjugate gradient method, as applied to the simultaneous estimation of $Bi_q(X, \tau)$ and $Bi_m(X, \tau)$. Table 5.1 presents the number of iterations, final value of the functional and RMS errors, obtained with these three versions, for the same test case examined in Figures 5.14a and b. However, in order to avoid any bias resulting from the simulated noise, errorless measurements were used for the results presented in Table 5.1. The RMS error is given by:

$$e_{RMS}(Bi) = \sqrt{\frac{1}{I\,J} \sum_{j=1}^{J} \sum_{i=1}^{I} \left[Bi_{estimated}(X_j, \tau_i) - Bi_{exact}(X_j, \tau_i) \right]^2} \qquad (5.4.10)$$

where $Bi(X, \tau)$ represents either $Bi_q(X, \tau)$ or $Bi_m(X, \tau)$, while I and J refer to the number of samples in time and space used to compute the RMS error.

We note in Table 5.1 that, for the test case examined, the use of Powell-Beale's version of the conjugate gradient method resulted in the smallest final value of the functional and the smallest RMS errors. In addition, the smallest number of iterations, which was required to reach the final functional value, was obtained with the version of Powell-Beale.

The reduction of the objective functional, obtained with each of the three versions of the conjugate gradient method, is presented in Figure 5.15. This figure shows that Fletcher-Reeves' version did not significantly reduce the functional after 30 iterations. Although Powell-Beale's version also stalled for few iterations around iteration 20, the reduction rate increased afterward, so that the

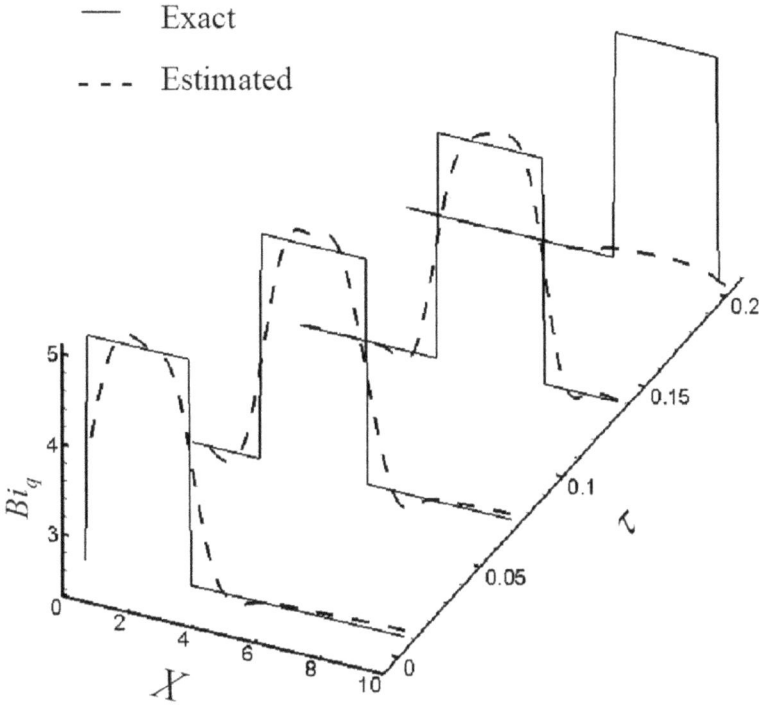

FIGURE 5.11 Results obtained for $Bi_q(X,\tau)$ with known $Bi_m(X,\tau)$ by using temperature measurements. (From Ref. [138].)

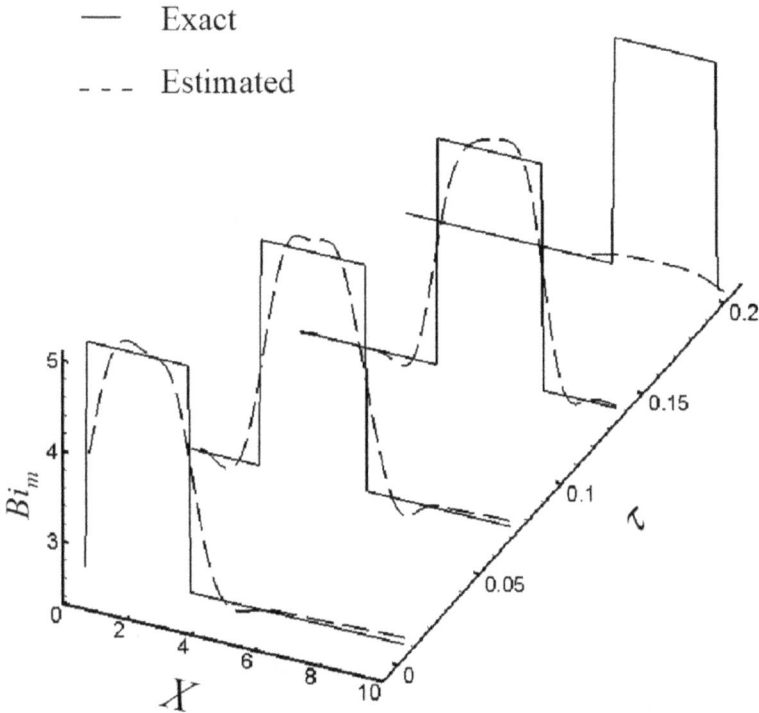

FIGURE 5.12 Results obtained for $Bi_m(X,\tau)$ with known $Bi_q(X,\tau)$ by using temperature measurements. (From Ref. [138].)

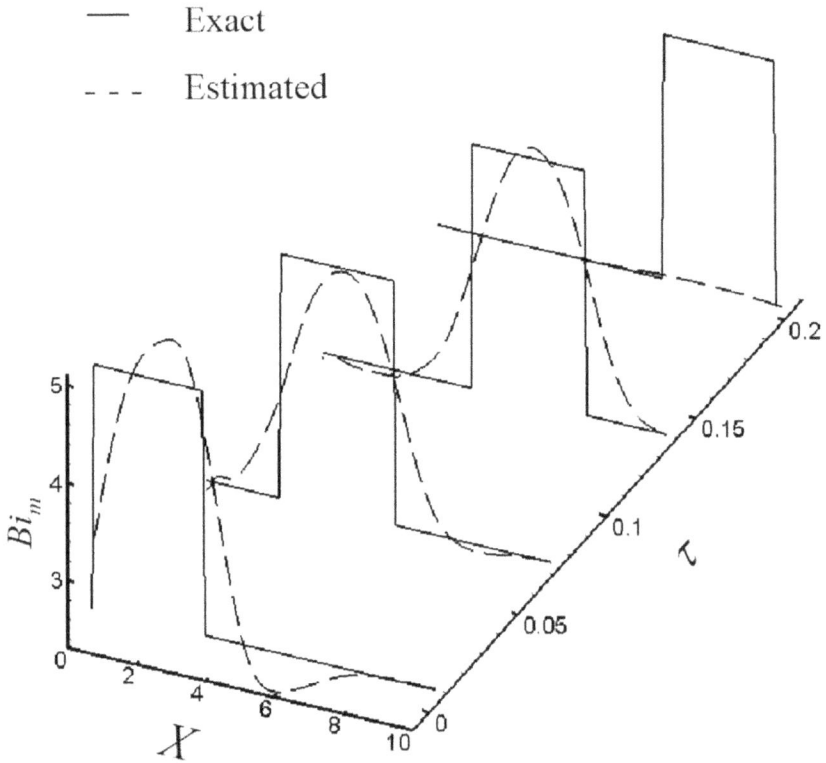

FIGURE 5.13 Results obtained for $Bi_m(X,\tau)$ with known $Bi_q(X,\tau)$ by using moisture content measurements. (From Ref. [138].)

TABLE 5.1

Comparison of Different Versions of the Conjugate Gradient Method

Method	Number of Iterations	Final Functional	RMS Error $Bi_m(X,\tau)$	RMS Error $Bi_q(X,\tau)$
Powell-Beale	63	151.9	0.43	0.68
Polak-Ribiere	75	197.7	0.51	0.70
Fletcher-Reeves	95	430.3	0.57	0.84

Source: From Ref. [138].

---

minimum functional value was reached at iteration 63 (see Table 5.1). We notice in Figure 5.15 that the reduction rate of the functional decreased significantly after 60 iterations for Polak-Ribiere's version and it was not capable of reducing the functional to the same level reached with Powell-Beale's version.

PROBLEMS

5.1   Consider the following heat conduction problem in dimensionless form:

$$\frac{\partial T}{\partial t} = \frac{\partial^2 T}{\partial x^2} \text{ in } 0 < x < 1 \text{ for } t > 0$$

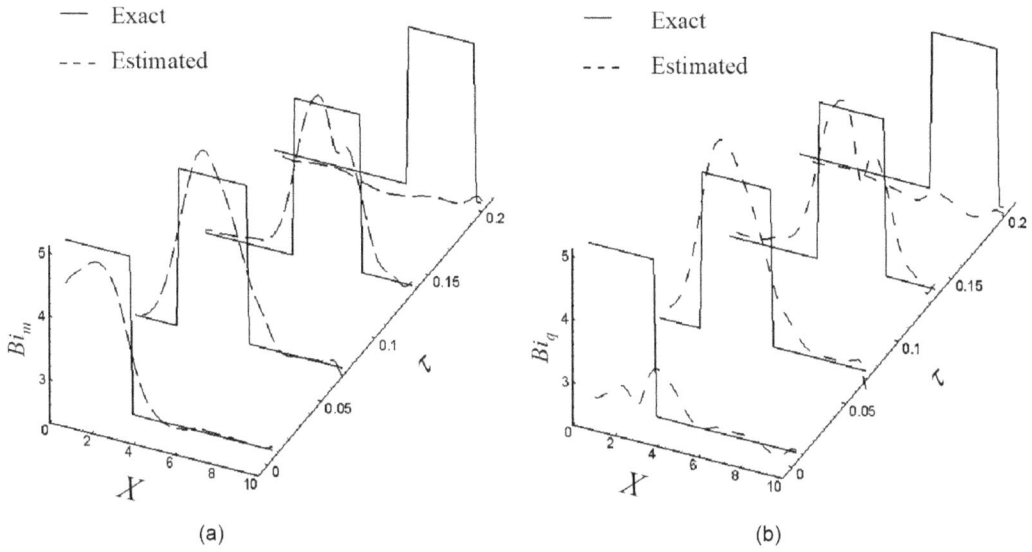

FIGURE 5.14   (a) Results obtained for $Bi_m(X,\tau)$ by using temperature and moisture content measurements—simultaneous estimation of $Bi_q(X,\tau)$ and $Bi_m(X,\tau)$. (From Ref. [138].) (b) Results obtained for $Bi_q(X,\tau)$ by using temperature and moisture content measurements—simultaneous estimation of $Bi_q(X,\tau)$ and $Bi_m(X,\tau)$. (From Ref. [138].)

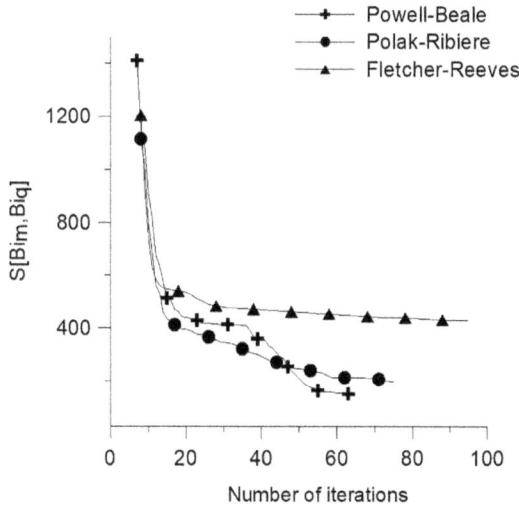

FIGURE 5.15   Reduction of the objective functional. (From Ref. [138].)

$$\frac{\partial T}{\partial x} = 0 \text{ at } x = 0 \text{ for } t > 0$$

$$\frac{\partial T}{\partial x} = q(t) \text{ at } x = 1 \text{ for } t > 0$$

$$T = 0 \text{ for } t = 0 \text{ in } 0 < x < 1$$

Formulate all the steps for the solution of the inverse problem of estimating the unknown heat flux $q(t)$ by using Technique V.

5.2 Solve the inverse problem of estimating the boundary heat flux $q(t)$ described in Problem 5.1 by using Technique V of function estimation. Assume that no information is available on the functional form of $q(t)$, except that it belongs to the space of square-integrable functions in the domain $0 < t < 1$. Use for the inverse analysis 100 equally spaced transient measurements in $0 < t \leq 1$ of a sensor located at $x_{meas} = 0$.

In order to generate the simulated measurements, utilize the following functional forms:

I. $q(t) = 1 + t$

II. $q(t) = 1 + t + t^2$

III. $q(t) = \begin{cases} 1, & t \leq 0.3 \text{ and } t \geq 0.7 \\ 2, & 0.3 < t < 0.7 \end{cases}$

IV. $q(t) = \begin{cases} 1, & t \leq 0.3 \text{ and } t \geq 0.7 \\ 5t - 0.5, & 0.3 < t \leq 0.5 \\ -5t + 4.5, & 0.5 < t < 0.7 \end{cases}$

Use as initial guess a constant function $q(t) = 0$ and examine the effects of random measurement errors on the solution.

5.3 Try to improve the estimated functions of Problem 5.2 in the time interval $0 < t \leq 1$ by using the following approaches:

   i. Use $q(t) = 0$ as initial guess, but consider a time interval larger than that of interest. For example, use for the final time $t_f = 1.1, 1.25, 1.5$, etc. Does the quality of the estimated functions in the time domain $0 < t \leq 1$ improve? Remember to increase the number of measurements accordingly, so that 100 measurements appear in the interval $0 < t \leq 1$.

   ii. Repeat the calculations with $t_f = 1$ and with an initial guess for $q(t)$ equal to the value estimated in Problem 5.2 for a time in the neighborhood of t_f. Let's say, use now as initial guess the value estimated in Problem 5.2 for $q(0.9)$. Repeat this procedure until sufficiently accurate estimates are obtained in the interval $0 < t \leq 1$.

5.4 Repeat Problems 5.2 and 5.3 by using fewer transient measurements in the inverse analysis. Take, as an example, 20 measurements of the sensor located at $x_{meas} = 0$ in the time interval $0 < t \leq 1$. Are the final solutions sensitive to the number of measurements?

5.5 Repeat Problems 5.2 and 5.3 by using in the inverse analysis the transient readings of two sensors located at $x = 0$ and $x = 0.5$. Compare the results obtained with two sensors to the results obtained in Problems 5.2 and 5.3 with a single sensor.

5.6 Repeat Problem 5.2 by using the *steepest descent method* (the conjugation coefficients $\gamma^k = \psi^k = 0$ for all iterations), instead of the conjugate gradient method. How do the two methods compare with respect to the number of iterations required for convergence?

5.7 Repeat Problem 5.2 by using a very small number for the tolerance ε in the stopping criterion, instead of using the discrepancy principle, for cases involving measurements with random errors. What happens to the stability of the estimated functions? Why?

NOTE 1: HILBERT SPACES

We present in this note some definitions and properties regarding Hilbert spaces. For further details on the subject, the reader should consult references [84,133,134].

A *Hilbert space* is a Banach space in which the norm is given by an inner (or scalar) product $\langle .,. \rangle$, that is,

$$\|u\| = \langle u, u \rangle^{1/2} \tag{N1.5.1}$$

where $\|\cdot\|$ designates the norm in the space.

For u belonging to a linear space V, a *norm* in this space is a mapping from V into the non-negative real axis, $[0, \infty)$, satisfying the following properties:

  i. $\|u\| = 0$ if and only if $u = 0$;
 ii. $\|\lambda u\| = |\lambda| \|u\|$, where λ is a scalar;
iii. $\|u + v\| \le \|u\| + \|v\|$, for any u and v in V.

Property (iii) is the so-called *triangle inequality*.

In a Hilbert space V, the *inner product* is given by the following symmetric bilinear form:

$$\langle u, v \rangle \equiv \frac{1}{4} (\|u + v\|^2 - \|u - v\|^2), \quad \forall u, v \in V \tag{N1.5.2}$$

The *vector space* $\mathbb{R}^N$ with the Euclidean norm

$$\|\mathbf{P}\| = \left(\sum_{j=1}^{N} P_j^2 \right)^{1/2} \tag{N1.5.3a}$$

is a Hilbert space, with inner product given by:

$$\langle \mathbf{P}, \mathbf{R} \rangle = \sum_{j=1}^{N} P_j R_j = \mathbf{P}^T \mathbf{R} \tag{N1.5.3b}$$

where $\mathbf{P}^T = [P_1, P_2, ..., P_N]$, $\mathbf{R}^T = [R_1, R_2, ..., R_N]$ and the superscript T denotes transpose.

Similarly, the *space of square-integrable real-valued functions in a domain* Ω, $L_2(\Omega)$, satisfying

$$\int_{\Omega} [f(w)]^2 \, dw < \infty \quad \text{for} \quad w \text{ in } \Omega$$

is a Hilbert space with norm

$$\|f(w)\| = \left\{ \int_{\Omega} [f(w)]^2 \, dw \right\}^{1/2} \tag{N1.5.4a}$$

and inner product

$$\langle f(w), g(w) \rangle = \int_{\Omega} f(w) g(w) \, dw, \quad \text{for} \ f(w) \text{ and } g(w) \in L_2(\Omega) \tag{N1.5.4b}$$

If Ω refers to the time domain, $0 < t < t_f$, equations (N1.5.4a,b), respectively, become:

$$\|f(t)\| = \left\{ \int_{t=0}^{t_f} [f(t)]^2 \, dt \right\}^{1/2} \tag{N1.5.5a}$$

and

$$\langle f(t), g(t) \rangle = \int\limits_{t=0}^{t_f} f(t)\, g(t)\, dt, \quad \text{for } f(t) \text{ and } g(t) \text{ in } L_2(0, t_f) \qquad \text{(N1.5.5b)}$$

Similarly, if Ω refers to the joint spatial and time domains, $\Omega = V \times (0, t_f)$, where $\mathbf{r}$ denotes spatial coordinates, equations (N1.5.4a,b) can be written, respectively, as:

$$\|f(\mathbf{r}, t)\| = \left\{ \int\limits_{V} \int\limits_{t=0}^{t_f} [f(\mathbf{r}, t)]^2\, dt\, d\mathbf{r} \right\}^{1/2} \qquad \text{(N1.5.6a)}$$

and

$$\langle f(\mathbf{r}, t), g(\mathbf{r}, t) \rangle = \int\limits_{V} \int\limits_{t=0}^{t_f} f(\mathbf{r}, t)\, g(\mathbf{r}, t)\, dt\, d\mathbf{r}, \quad \text{for } f(\mathbf{r}, t) \text{ and } g(\mathbf{r}, t) \text{ in } L_2[V \times (0, t_f)] \qquad \text{(N1.5.6b)}$$

Other expressions for the norm and inner product can be developed from equations (N1.5.4a,b) for various domains Ω of interest.

The reader should note that the expressions for the conjugation coefficients (2.3.7–2.3.10) and (5.1.14–5.1.17) are, respectively, analogous. They are given by inner products in the $\mathbb{R}^N$ and $L_2(0, t_f)$ spaces, equations (N1.5.3b) and (N1.5.5b), respectively. Also, expressions (2.9.13) and (5.1.9) are inner products of the gradient direction with the direction of perturbed parameters $\Delta \mathbf{P}$ in $\mathbb{R}^N$, and with the direction of the perturbed function $\Delta g_p(t)$ in $L_2(0, t_f)$, respectively. Therefore, they give the directional derivatives of $S_{ML}(\mathbf{P})$ and $S[g_p(t)]$, respectively, in the direction of the perturbed unknown quantities.

NOTE 2: CONJUGATE GRADIENT METHOD OF FUNCTION ESTIMATION

Let us consider that one single unknown function $f(\mathbf{r}, t)$ is to be estimated in the domain $\Omega = V \times (0, t_f)$, where $\mathbf{r}$ denotes spatial coordinates. Transient measurements of M sensors are available for the solution of the inverse problem. Measurement errors are additive, uncorrelated, Gaussian, with zero mean and known variance functions $\sigma_m^2(t)$, $m = 1, \ldots, M$. The following objective functional is thus defined:

$$S[f(\mathbf{r}, t)] = \sum_{m=1}^{M} \int\limits_{t=0}^{t_f} \frac{[Y_m(t) - T(\mathbf{r}_m, t; f(\mathbf{r}, t))]^2}{[\sigma_m(t)]^2}\, dt \qquad \text{(N2.5.1)}$$

In equation (N2.5.1), $Y_m(t)$ and $T[\mathbf{r}_m, t; f(\mathbf{r}, t)]$ are the measured and estimated variables at the measurement position $\mathbf{r}_m$, respectively, and t_f is the duration of the experiment.

For the minimization of the objective functional given by equation (N2.5.1), we present below the iterative procedure of the conjugate gradient method [2,6,22,29–43, 76–79,92,135]:

$$f^{k+1}(\mathbf{r}, t) = f^k(\mathbf{r}, t) + \beta^k d^k(\mathbf{r}, t) \qquad \text{(N2.5.2)}$$

where the superscript k denotes the number of iterations, β^k is the search step size and $d^k(\mathbf{r}, t)$ is the direction of descent.

The direction of descent $d^k(\mathbf{r}, t)$ is a conjugation of the gradient direction, $\nabla S[f^k(\mathbf{r}, t)]$, with previous directions of descent. It is given in the following general form [2,6,22,29–43,76–79,92,135]:

$$d^k(\mathbf{r},t) = -\nabla S[f^k(\mathbf{r},t)] + \gamma^k d^{k-1}(\mathbf{r},t) + \psi^k d^q(\mathbf{r},t) \qquad (N2.5.3)$$

where γ^k and ψ^k are conjugation coefficients. The superscript q in equation (N2.5.3) denotes the iteration number where a restarting strategy is applied to the iterative procedure of the conjugate gradient method.

Different versions of the conjugate gradient method can be found in the literature depending on the form used for the computation of the conjugation coefficients [2,6,22,29–43,76–79,92,135].

In the Fletcher-Reeves version, the conjugation coefficients are given by [76]:

$$\gamma^k = \frac{\left\|\nabla S[f^k(\mathbf{r},t)]\right\|_\Omega^2}{\left\|\nabla S[f^{k-1}(\mathbf{r},t)]\right\|_\Omega^2} \quad \text{with } \gamma^k = 0 \text{ for } k = 0 \qquad (N2.5.4a)$$

$$\psi^k = 0 \quad \text{for } k = 0,1,2,\dots \qquad (N2.5.4b)$$

where $\left\|\cdot\right\|_\Omega$ designates the L_2 norm in the domain Ω given by equation (N1.5.4a).

In the *Polak-Ribiere* version of the conjugate gradient method, the conjugation coefficients are given by [77]:

$$\gamma^k = \frac{\left\langle \left\{\nabla S[f^k(\mathbf{r},t)] - \nabla S[f^{k-1}(\mathbf{r},t)]\right\}, \nabla S[f^k(\mathbf{r},t)] \right\rangle_\Omega}{\left\|\nabla S[f^{k-1}(\mathbf{r},t)]\right\|_\Omega^2} \quad \text{with } \gamma^k = 0 \text{ for } k = 0 \quad (N2.5.5a)$$

$$\psi^k = 0 \quad \text{for } k = 0,1,2,\dots \qquad (N2.5.5b)$$

where $\left\langle .,. \right\rangle_\Omega$ is the L_2 inner product in the domain Ω given by equation (N1.5.4b).

For the *Hestenes-Stiefel* version of the conjugate gradient method, we have [78]:

$$\gamma^k = \frac{\left\langle \left\{\nabla S[f^k(\mathbf{r},t)] - \nabla S[f^{k-1}(\mathbf{r},t)]\right\}, \nabla S[f^k(\mathbf{r},t)] \right\rangle_\Omega}{\left\langle \left\{\nabla S[f^k(\mathbf{r},t)] - \nabla S[f^{k-1}(\mathbf{r},t)]\right\}, d^{k-1}(\mathbf{r},t) \right\rangle_\Omega} \quad \text{with } \gamma^k = 0 \text{ for } k = 0 \quad (N2.5.6a)$$

$$\psi^k = 0 \quad \text{for } k = 0,1,2,\dots \qquad (N2.5.6b)$$

Restarting strategies were suggested for the conjugate gradient method in order to improve its convergence rate. In accordance with Powell [79], the application of the conjugate gradient method with the conjugation coefficients given by:

$$\gamma^k = \frac{\left\langle \left\{\nabla S[f^k(\mathbf{r},t)] - \nabla S[f^{k-1}(\mathbf{r},t)]\right\}, \nabla S[f^k(\mathbf{r},t)] \right\rangle_\Omega}{\left\langle \left\{\nabla S[f^k(\mathbf{r},t)] - \nabla S[f^{k-1}(\mathbf{r},t)]\right\}, d^{k-1}(\mathbf{r},t) \right\rangle_\Omega} \quad \text{with } \gamma^k = 0 \text{ for } k = 0 \quad (N2.5.7a)$$

$$\psi^k = \frac{\left\langle \left\{\nabla S[f^{q+1}(\mathbf{r},t)] - \nabla S[f^q(\mathbf{r},t)]\right\}, \nabla S[f^k(\mathbf{r},t)] \right\rangle_\Omega}{\left\langle \left\{\nabla S[f^{q+1}(\mathbf{r},t)] - \nabla S[f^q(\mathbf{r},t)]\right\}, d^q(\mathbf{r},t) \right\rangle_\Omega} \quad \text{with } \psi^k = 0 \text{ for } k = 0 \quad (N2.5.7b)$$

requires restarting when gradients at successive iterations tend to be non-orthogonal (which is a measure of the local non-linearity of the problem) or when the direction of descent is not sufficiently downhill. Restarting is performed by making $\psi^k = 0$ in equation (N2.5.7b).

The non-orthogonality of gradients at successive iterations is tested by using:

$$\text{ABS}\left(\left\langle \nabla S[f^{k-1}(\mathbf{r},t)], \nabla S[f^k(\mathbf{r},t)] \right\rangle_\Omega \right) \geq 0.2 \left\|\nabla S[f^k(\mathbf{r},t)]\right\|_\Omega^2 \qquad (N2.5.8a)$$

where ABS (.) denotes the absolute value.

A non-sufficiently downhill direction of descent (i.e., the angle between the direction of descent and the negative gradient direction is too large) is identified if either of the following inequalities are satisfied:

$$\left\langle \nabla S[f^k(\mathbf{r},t)], d^k(\mathbf{r},t) \right\rangle_\Omega \le -1.2 \left\| \nabla S[f^k(\mathbf{r},t)] \right\|_\Omega^2 \qquad (\text{N2.5.8b})$$

or

$$\left\langle \nabla S[f^k(\mathbf{r},t)], d^k(\mathbf{r},t) \right\rangle_\Omega \ge -0.8 \left\| \nabla S[f^k(\mathbf{r},t)] \right\|_\Omega^2 \qquad (\text{N2.5.8c})$$

We note that the coefficients 0.2, 1.2 and 0.8 appearing in equations (N2.5.8a-c) are empirical and are the same used by Powell [79]. In the so-called Powell-Beale's version of the conjugate gradient method, the direction of descent is computed in accordance with the following steps for $k \ge 1$:

  i. Test the inequality (N2.5.8a). If it is satisfied, set $q = k - 1$.
  ii. Compute γ^k with equation (N2.5.7a).
  iii. If $k = q + 1$, set $\psi^k = 0$. If $k \ne q + 1$, compute ψ^k with equation (N2.5.7b).
  iv. Compute the search direction $d^k(\mathbf{r},t)$ with equation (N2.5.3).
  v. If $k \ne q + 1$, test the inequalities (N2.5.8b,c). If either one of them is satisfied, set $q = k - 1$ and $\psi^k = 0$. Then recompute the search direction with equation (N2.5.3).

Although the above versions of the conjugate gradient method are identical for linear inverse problems, numerical experiments revealed the fastest convergence rate and improved robustness of Powell-Beale's version [82,137,138].

Similarly to Techniques II and III, the search step size for Technique V is obtained as the one that minimizes the objective functional given by equation (N2.5.1) at each iteration, that is,

$$\min_{\beta^k} S[f^{k+1}(\mathbf{r},t)] = \min_{\beta^k} \sum_{m=1}^M \int_{t=0}^{t_f} \frac{\left\{ Y_m(t) - T[\mathbf{r}_m,t; f^{k+1}(\mathbf{r},t)] \right\}^2}{[\sigma_m(t)]^2} \, dt \qquad (\text{N2.5.9})$$

By substituting the iterative procedure of the conjugate gradient method of function estimation given by equation (N2.5.2) into $T[\mathbf{r}_m,t; f^{k+1}(\mathbf{r},t)]$, the above equation becomes:

$$\min_{\beta^k} S[f^{k+1}(\mathbf{r},t)] = \min_{\beta^k} \sum_{m=1}^M \int_{t=0}^{t_f} \frac{\left\{ Y_m(t) - T[\mathbf{r}_m,t; f^k(\mathbf{r},t) + \beta^k d^k(\mathbf{r},t)] \right\}^2}{[\sigma_m(t)]^2} \, dt \qquad (\text{N2.5.10})$$

By linearizing $T[\mathbf{r}_m,t; f^k(\mathbf{r},t) + \beta^k d^k(\mathbf{r},t)]$ and making

$$d^k(\mathbf{r},t) = \Delta f^k(\mathbf{r},t) \qquad (\text{N2.5.11})$$

we obtain:

$$T[\mathbf{r}_m,t; f^k(\mathbf{r},t) + \beta^k d^k(\mathbf{r},t)] \approx T[\mathbf{r}_m,t; f^k(\mathbf{r},t)] + \beta^k \frac{\partial T}{\partial f^k} \Delta f^k(\mathbf{r},t) \qquad (\text{N2.5.12})$$

Let

$$\Delta T[\mathbf{r}_m,t; f^k(\mathbf{r},t)] = \frac{\partial T}{\partial f^k} \Delta f^k(\mathbf{r},t) \qquad (\text{N2.5.13})$$

and then equation (N2.5.12) can be written as:

$$T[\mathbf{r}_m,t; f^k(\mathbf{r},t)+\beta^k d^k(\mathbf{r},t)] \approx T[\mathbf{r}_m,t; f^k(\mathbf{r},t)]+\beta^k \Delta T[\mathbf{r}_m,t; d^k(\mathbf{r},t)] \qquad (\text{N2.5.14})$$

where $\Delta T[\mathbf{r}_m,t; d^k(\mathbf{r},t)]$ is the solution of the sensitivity problem obtained by setting $\Delta f^k(\mathbf{r},t) = d^k(\mathbf{r},t)$.

By substituting equation (N2.5.14) into equation (N2.5.10), we obtain:

$$\min_{\beta^k} S[f^{k+1}(\mathbf{r},t)] = \min_{\beta^k} \sum_{m=1}^{M} \int_{t=0}^{t_f} \frac{\left\{ Y_m(t) - T[\mathbf{r}_m,t; f^k(\mathbf{r},t)] - \beta^k \Delta T[\mathbf{r}_m,t; d^k(\mathbf{r},t)] \right\}^2}{[\sigma_m(t)]^2} dt \qquad (\text{N2.5.15})$$

By performing the minimization above, we find the following expression of the search step size for the estimation of $f^k(\mathbf{r},t)$ with Technique V:

$$\beta^k = \frac{\displaystyle\sum_{m=1}^{M} \int_{t=0}^{t_f} \frac{\left\{ Y_m(t) - T[\mathbf{r}_m,t; f^k(\mathbf{r},t)] \right\}}{[\sigma_m(t)]^2} \Delta T[\mathbf{r}_m,t; d^k(\mathbf{r},t)] dt}{\displaystyle\sum_{m=1}^{M} \int_{t=0}^{t_f} \frac{\left\{ \Delta T[\mathbf{r}_m,t; d^k(\mathbf{r},t)] \right\}^2}{[\sigma_m(t)]^2} dt} \qquad (\text{N2.5.16})$$

NOTE 3: ADDITIONAL MEASUREMENT FOR SELECTING THE STOPPING CRITERION OF THE CONJUGATE GRADIENT METHOD

The stopping criterion based on Morozov's discrepancy principle requires the *a priori* knowledge of the standard deviation of the measurement errors. However, there are several practical situations in which scarce information is available regarding this quantity. For such cases, an alternative stopping criterion approach based on an additional measurement can be used, which also provides the conjugate gradient method with an iterative regularization character [6].

In order to illustrate the additional measurement approach for the stopping criterion, we take as an example the estimation of the boundary heat flux $q(t)$ in a slab of unitary thickness by using Technique V. The formulation of the dimensionless heat conduction problem considered here is given by:

$$\frac{\partial T}{\partial t} = \frac{\partial^2 T}{\partial x^2} \text{ in } 0 < x < 1, \text{ for } t > 0 \qquad (\text{N3.5.1a})$$

$$\frac{\partial T}{\partial x} = 0 \text{ at } x = 0, \text{ for } t > 0 \qquad (\text{N3.5.1b})$$

$$\frac{\partial T}{\partial x} = q(t) \text{ at } x = 1, \text{ for } t > 0 \qquad (\text{N3.5.1c})$$

$$T = 0 \text{ at } t = 0, \text{ in } 0 < x < 1 \qquad (\text{N3.5.1d})$$

The unknown function $q(t)$ is estimated with Technique V by minimizing the following functional:

$$S[q(t)] = \int_{t=0}^{t_f} \left\{ Y(t) - T[x_{\text{meas}}, t; q(t)] \right\}^2 dt \qquad (\text{N3.5.2})$$

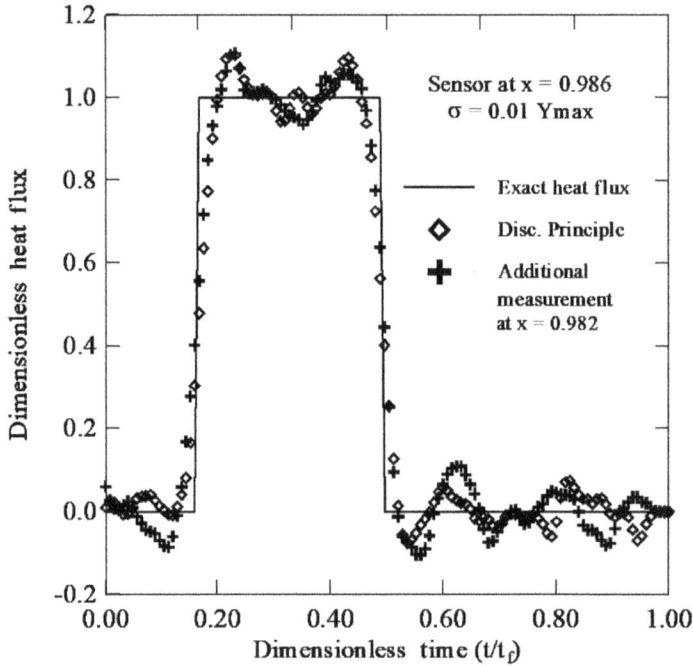

FIGURE 5.16 A comparison of the discrepancy principle and additional measurement approaches for the stopping criterion.

where $Y(t)$ are the measured temperatures at the location x_{meas}, while $T[x_{meas}, t; q(t)]$ are the estimated temperatures at the same location.

Consider now that the additional measured data $Y_c(t)$ of a sensor located at x_c are also available for the analysis. The functional $S_c[q(t)]$ based on such data is given by:

$$S_c[q(t)] = \int_{t=0}^{t_f} \{Y_c(t) - T[x_c, t; q(t)]\}^2 \, dt \qquad \text{(N3.5.3)}$$

The examination of the behavior of the functional $S_c[q(t)]$, as the minimization of $S[q(t)]$ is performed, can be used to detect the point where the errors in the measured data $Y(t)$ start to cause instabilities on the estimated function $q(t)$. If both measurements are sensitive to variations in the heat flux $q(t)$, the value of $S_c[q(t)]$ passes through a minimum and then increases, as a result of ill-posed character of the inverse problem. The iterative procedure is then stopped at the iteration corresponding to the minimum value of $S_c[q(t)]$, so that sufficiently stable solutions can be obtained for the inverse problem.

Results for the estimation of a step variation of the boundary heat flux $q(t)$, obtained by using the discrepancy principle and the additional measurement approach, are illustrated in Figure 5.16. The simulated measured data $Y(t)$ and $Y_c(t)$ were generated with a constant standard deviation $\sigma = 0.01Y_{max}$, where Y_{max} is the maximum value of $Y(t)$. The measurements $Y(t)$ used in the minimization of the functional $S[q(t)]$ were considered taken at the position $x_{meas} = 0.986$. For the case involving the additional measurements for the stopping criterion, the additional sensor was supposed to be located at $x_c = 0.982$. Figure 5.16 shows that the two approaches for the stopping criterion are equivalent. Both provide quite accurate and stable estimates for the step variation of the heat flux, which represents a very strict test function.

6 Function Estimation

Solution within the Bayesian Framework of Statistics with Prior Information about the Unknown Functions

In the previous chapter, the conjugate gradient method with adjoint problem was applied for the solution of inverse problems of function estimation, without any prior information about the unknown function, except for the functional space that it belonged to. The solution of the inverse problem then involved two auxiliary problems, which were required to develop expressions for the gradient direction and for the search step size, in order to apply the iterative procedure of the conjugate gradient method. All mathematical derivations were analytically performed by considering that the transient measurements were a function of time. On the other hand, for the computational implementation of the method, the measurements needed to be treated as discrete in time. Therefore, in the previous chapter, derivations were first analytically performed within the Hilbert space of square integrable functions, but the implementation of the estimation approach later required the discretization of the unknown function. This is a quite unique characteristic of Technique V, which was advanced by the group of the Moscow Aviation Institute led by Dean Oleg M. Alifanov and, for this reason, it is one of the so-called *Alifanov's iterative regularization* techniques [2,6,29–43].

In this chapter, we present a different approach where the unknown function is discretized before the solution of the inverse problem, as follows. Consider the estimation of the heat source term $g_p(t)$ (see Figure 2.1) parameterized as:

$$g_p(t) = \sum_{j=1}^{N} P_j C_j(t)$$

where $C_j(t), j = 1,\ldots, N$, are the following functions with local support:

$$C_j(t) = \begin{cases} 1 & \text{for } \left(t_j - \dfrac{\Delta t}{2}\right) < t < \left(t_j + \dfrac{\Delta t}{2}\right) \\ 0 & \text{elsewhere} \end{cases}$$

Each parameter then represents the local value of the function in the time interval Δt around consecutive measurements. Therefore, $g_p(t_j) = P_j$ (see Figure 1.7) and the number of parameters to be estimated is equal to the number of transient measurements.

The minimization of the *maximum likelihood* objective function described in Chapter 2 is not applicable in cases where the unknown function is parameterized like above, because the ill-posed character of the inverse problem becomes apparent. One classical technique to cope with the ill-posed character of the inverse problem is Tikhonov's regularization, which was briefly described in Section 1.4. This is a technique alternative to that described in Chapter 5, which involves the minimization of objective functions with parameters that represent local values of the unknown function. On the other hand, Alifanov's and Tikhonov's regularization approaches do not belong to the class of Bayesian techniques for the solution of inverse problems, because they are not based on the modeling of prior information and related uncertainties about the unknown functions.

In this chapter, the minimization of the maximum a posteriori (MAP) objective function and the Markov chain Monte Carlo (MCMC) method, which were presented in Chapters 3 and 4, respectively, are applied to the solution of inverse problems of function estimation. These methods are applied to cases where the unknown functions are represented in terms of local constant values.

The solution of inverse problems within the Bayesian framework of statistics is regularized by the prior distribution for the unknown function. Hence, in addition to the prior distributions already discussed in Chapter 3, other priors suitable for function estimation are presented in the first section of this chapter. The concepts of hyperpriors and hyperparameters are also presented in Section 6.1 (although they could have been introduced in Chapters 3 and 4). The other sections of this chapter illustrate the application of the minimization of the MAP objective function and the MCMC method of stochastic simulation to inverse heat transfer problems of practical interest.

6.1 PRIOR DISTRIBUTIONS

A multivariate prior is required for the solution of inverse problems in situations where the parameters represent constant values of a function, either within a time interval for time-varying functions, or within a control volume for spatially varying functions. The use of Gaussian priors for function estimation is of great interest, because they tend to smooth out the oscillations in the solution caused by the ill-posed character of the inverse problem. The reader should recall that the Gaussian prior is given by:

$$\pi(\mathbf{P}) = (2\pi)^{-N/2} |\mathbf{V}|^{-1/2} \exp\left\{-\frac{1}{2}[\mathbf{P} - \boldsymbol{\mu}]^T \mathbf{V}^{-1}[\mathbf{P} - \boldsymbol{\mu}]\right\} \qquad (6.1.1)$$

where $\boldsymbol{\mu}$ and $\mathbf{V}$ are the known mean and covariance matrix, respectively.

If the parameters are independent, the prior covariance matrix is diagonal, with elements given by the variances, σ_i^2, that is,

$$[\mathbf{V}]_{i,j} = \begin{cases} \sigma_i^2, \ i = j \\ 0, \ i \neq j \end{cases} \qquad (6.1.2)$$

However, rarely there is such independence in practice when the parameters are local function values, especially among neighboring parameters. Consider, for example, a function that varies spatially. In this case, the parameters correspond to the mean values of the function inside finite volumes used for the discretization of the spatial domain. The correlation between the parameters of different finite volumes must be taken into account in the covariance matrix of the prior information.

Works related to imaging [140–142] have demonstrated that the *Matérn class* [143] of covariance functions may be appropriate for taking into account the correlation between spatially distributed parameters. The elements of the Matérn covariance matrix for the Gaussian prior distribution can be written as [143]:

$$[\mathbf{V}]_{i,j} = \begin{cases} \sigma_i^2 & i = j \\ \sigma_i^2 \dfrac{2^{1-\alpha}}{\Gamma(\alpha)} \left(\dfrac{\sqrt{2\alpha}\,|\mathbf{r}_i - \mathbf{r}_j|}{l}\right)^\alpha K_\alpha \left(\dfrac{\sqrt{2\alpha}\,|\mathbf{r}_i - \mathbf{r}_j|}{l}\right) & i \neq j \end{cases} \qquad (6.1.3)$$

where $\mathbf{r}_i$ is the position vector of finite volume i, $|\mathbf{r}_i - \mathbf{r}_j|$ is the distance between finite volumes i and j, $\alpha > 0$ is a parameter that controls the smoothness of the random field, l is the characteristic length scale that controls the spatial range of correlation, Γ is the gamma function and K_α is the modified Bessel function of the second kind of order α. With the Matérn covariance matrix, the correlation is more significant for neighboring finite volumes and decreases for large distances between them. The correlation decay rate is controlled by the characteristic length scale l and the smoothness parameter α. Other kinds of covariance matrices for Gaussian priors can be found in [143].

Markov random fields are also popular for priors in inverse problems of estimating spatially distributed functions or time-varying functions [17]. A set $\{P_1, P_2, \ldots, P_N\}$ is a Markov random field if the conditional distribution of P_j depends only on the set of its neighbors [59].

A common use of a Markov random field is with priors that resemble Tikhonov's regularization [17], written in the following general form:

$$\pi(\mathbf{P}) \propto \exp\left[-\frac{1}{2}\gamma \left\| \mathbf{D}(\mathbf{P} - \tilde{\mathbf{P}}) \right\|^2 \right] \qquad (6.1.4)$$

where $\|.\|$ denotes the L_2 norm, the constant γ is a parameter associated with uncertainties in the prior and $\tilde{\mathbf{P}}$ is a reference value for $\mathbf{P}$.

The vector $\tilde{\mathbf{P}}$ is commonly taken as zero. As for the matrix $\mathbf{D}$, it should be such that $\mathbf{D}(\mathbf{P} - \tilde{\mathbf{P}})$ involves a parameter P_j and its neighbors in order to characterize a Markov random field. For cases that $\mathbf{P}$ represents local values of a one-dimensional function (such as a function varying in time or in one single spatial coordinate), the following matrices can be used:

$$\mathbf{D} = \begin{bmatrix} -1 & 1 & 0 & \cdots & 0 \\ 0 & -1 & 1 & \cdots & 0 \\ \vdots & & \ddots & \ddots & \vdots \\ 0 & \cdots & 0 & -1 & 1 \end{bmatrix} \text{ with size } (N-1) \times N \qquad (6.1.5a)$$

or

$$\mathbf{D} = \begin{bmatrix} 1 & -2 & 1 & 0 & \cdots & 0 \\ 0 & 1 & -2 & 1 & \cdots & 0 \\ \vdots & & \ddots & \ddots & \ddots & \vdots \\ 0 & \cdots & 0 & 1 & -2 & 1 \end{bmatrix} \text{ with size } (N-2) \times N \qquad (6.1.5b)$$

which are analogous to the matrices used in first-order and second-order Tikhonov's regularization, respectively [3–22].

Equation (6.1.4) can be rewritten as:

$$\pi(\mathbf{P}) \propto \exp\left[-\frac{1}{2}\gamma (\mathbf{P} - \tilde{\mathbf{P}})^T \mathbf{Z} (\mathbf{P} - \tilde{\mathbf{P}}) \right] \qquad (6.1.6)$$

where

$$\mathbf{Z} = \mathbf{D}^T \mathbf{D} \qquad (6.1.7)$$

Equation (6.1.6) is in a form similar to that of the Gaussian distribution (see equation 6.1.1). For this reason, this prior is also called *Gaussian Markov random field* [59] or *Gaussian smoothness prior* [17]. By comparing equation (6.1.6) with the canonical Gaussian multivariate distribution given by equation (6.1.1), one can notice that the mean and the covariance matrix of this prior are given by $\tilde{\mathbf{P}}$ and $\gamma^{-1}\mathbf{Z}^{-1}$, respectively. Therefore, we could write the Gaussian smoothness prior as:

$$\pi(\mathbf{P}) = (2\pi)^{-N/2}\gamma^{N/2}\left|\mathbf{Z}^{-1}\right|^{-1/2}\exp\left[-\frac{1}{2}\gamma(\mathbf{P}-\tilde{\mathbf{P}})^T\mathbf{Z}(\mathbf{P}-\tilde{\mathbf{P}})\right] \qquad (6.1.8)$$

The prior given by equation (6.1.8) is improper with the matrix $\mathbf{D}$ given by either one of the equations (6.1.5a,b), that is, this prior variance is unbounded since the matrix $\mathbf{Z}$ is singular and $\mathbf{Z}^{-1}$ does not exist.

We now present another Markov random field prior, which gives high probabilities for piecewise constant functions: the *total variation (TV) prior*. This prior is appropriate, for example, for spatially varying functions that exhibit large variations at few surfaces inside the domain and with small variations within the regions limited by these surfaces. The TV prior is given by [17]:

$$\pi(\mathbf{P}) \propto \exp\left[-\gamma\,TV(\mathbf{P})\right] \qquad (6.1.9)$$

where

$$TV(\mathbf{P}) = \sum_{j=1}^{N}V_j(\mathbf{P}) \quad V_j(\mathbf{P}) = \frac{1}{2}\sum_{i\in N_j}l_{ij}\left|P_i - P_j\right| \qquad (6.1.10a,b)$$

In the above equation, N_j is the set of neighbors to P_j and l_{ij} is the length of the edge between neighbors.

The TV prior is improper, such as the Gaussian smoothness prior. The representation of equation (6.1.9) in terms of a canonical probability density would require the derivation of an expression for the normalizing constant $\int_{\mathbb{R}^N}\pi(\mathbf{P})d\mathbf{P}$, or, at least, practical means for its numerical computation.

Improper priors do not pose difficulties for the application of the Metropolis-Hastings algorithm, since the normalizing constants of such densities are cancelled when $\alpha(\mathbf{P}^* \mid \mathbf{P}^{(t)})$ is computed with equation (4.2.3). On the other hand, the above priors include additional parameters that need to be specified for the application of MCMC methods, like γ for the Gaussian smoothness prior and the TV prior, or the parameters α and l in Matérn's covariance matrix (6.1.3). The specification of values for such parameters can be made by numerical experiments by using simulated experimental data that serve as a reference for the inverse problem under analysis. However, within the Bayesian framework, if a parameter is not known it can be regarded as part of the inference problem leading to the use of *hierarchical* or *hyperprior* models, as described below.

HIERARCHICAL MODELS

The parameter γ appearing in the Gaussian smoothness prior given by equation (6.1.6) can be treated as a *hyperparameter*, that is, an unknown parameter of the model of the posterior distribution. As a hyperparameter, it must be provided with a prior distribution (*hyperprior*) and be estimated as part of the solution of the inverse problem [17].

Consider, for example, the *hyperprior density* for γ in the form of the Rayleigh distribution (see Note 1 in Chapter 1):

$$\pi(\gamma) = \frac{\gamma}{\gamma_0^2}\exp\left[-\frac{1}{2}\left(\frac{\gamma}{\gamma_0}\right)^2\right] \qquad (6.1.11)$$

where the scale parameter γ_0 needs to be fixed in advance.

Therefore, the posterior distribution with the Gaussian likelihood (3.1.6), Gaussian smoothness prior (6.1.8) and the Rayleigh hyperprior (6.1.11) is given by:

$$\pi(\gamma,\mathbf{P}|\mathbf{Y}) \propto \gamma^{(N+2)/2} \exp\left\{-\frac{1}{2}[\mathbf{Y}-\mathbf{T}(\mathbf{P})]^T \mathbf{W}^{-1}[\mathbf{Y}-\mathbf{T}(\mathbf{P})] - \frac{1}{2}\gamma(\mathbf{P}-\tilde{\mathbf{P}})^T \mathbf{Z}(\mathbf{P}-\tilde{\mathbf{P}}) - \frac{1}{2}\left(\frac{\gamma}{\gamma_0}\right)^2\right\}$$

(6.1.12)

The parameters α and l in Matérn's covariance matrix (6.1.3), as well as the parameter γ appearing in the TV prior given by equation (6.1.9), can also be treated as hyperparameters, but with extreme caution. For example, the normalizing constant of the TV prior also depends on γ, but differently from the Gaussian smoothness prior, an analytical expression for such normalizing constant is not available. Therefore, without the numerical computation of the normalizing constant for this case, the effects of γ as a hyperparameter are not correctly accounted for in the posterior distribution.

Examples of the solution of inverse function estimation problems within the Bayesian framework of statistics are presented in the next sections of this chapter, where the application of the above priors will be more clear. The algorithms used for the solution of such inverse problems are those presented in Chapters 3 and 4, and are not repeated here. Similarly, techniques for the analysis of Markov chains can be readily found in Chapter 4.

6.2 ESTIMATION OF THE KIDNEY METABOLIC HEAT GENERATION RATE

Thermogenesis results from the cellular metabolism and has a fundamental role for body thermoregulation in endothermic species. The motivation for the inverse problem presented in this section is the analysis of the kidneys' contribution for thermoregulation [144]. The MCMC method is applied for the solution of the inverse problem, which presents inherent difficulties associated with low sensitivity of the parameters of main interest that represent the transient heat source term, and strong correlation of the remaining model parameters. Such difficulties were dealt with by using the version of the Metropolis-Hastings algorithm that samples the parameters in blocks, presented in Note 1 of Chapter 4. Simulated temperature measurements were used for the inverse problem solution and the convergence of the Markov chains was verified with the techniques presented in Section 4.6.

DIRECT PROBLEM

The main simplifications for the mathematical model were to consider the kidney as a homogeneous medium with a uniform temperature. Mass and energy balances were written for a control volume surrounding the kidney. The model included mass transfer processes to/from this control volume, which involved the flow of arterial blood, venous blood and urine. Blood and urine were considered as homogeneous fluids. Figure 6.1 illustrates the kidney model, where m' and h denote mass flow rate and specific enthalpy, respectively. The subscripts a, v and u represent the arterial blood, the venous blood and urine, respectively, while $Q_m(t)$ is the metabolic heat generation rate in the kidney and $Q_l(t)$ is the heat transfer rate lost from the kidney to the surroundings. We further assumed that the kidney volume was constant, as well as that variations in kinetic and potential energies inside the kidney were negligible.

Heat transfer from the kidney to the surrounding organs and tissues was modeled in terms of a global heat transfer coefficient, U, that is,

$$Q_l(t) = UA_k[T_k(t) - T_s]$$

(6.2.1)

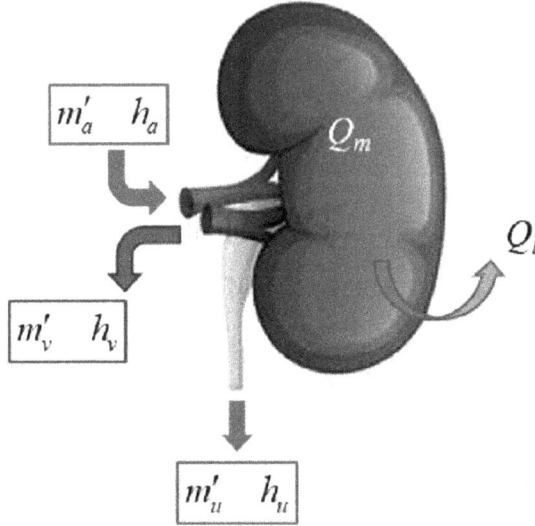

FIGURE 6.1   Sketch of the kidney model. (From Ref. [144].)

where A_k is the surface area of the kidney and T_s is the temperature of the surroundings, which was assumed as constant, so that we can write our model as [144]:

$$\frac{dT_k(t)}{dt} + (\alpha + \gamma + \psi)T_k(t) = \beta T_a(t) + \psi T_s + \phi Q_m(t) \text{ in } V_k \text{ for } t > 0 \qquad (6.2.2a)$$

$$T_k = T_{k,0} \text{ in } V_k \text{ at } t = 0 \qquad (6.2.2b)$$

where

$$\alpha = \frac{m_u' c_u}{m_k c_k}; \quad \beta = \frac{m_a' c_b}{m_k c_k}; \quad \gamma = \frac{(m_a' - m_u')c_b}{m_k c_k} \qquad (6.2.3a\text{-}c)$$

$$\phi = \frac{1}{m_k c_k}; \quad \psi = \frac{UA_k}{m_k c_k} \qquad (6.2.3d,e)$$

The mathematical formulation given by equations (6.2.2a,b) is a *direct problem* when the parameters $(\alpha, \beta, \gamma, \phi, \psi)$, the initial condition $(T_{k,0})$ and the functions $T_a(t)$ and $Q_m(t)$ are known.

INVERSE PROBLEM

This work aims at the solution of the inverse problem of estimating the kidney metabolic heat generation rate, $Q_m(t)$, with measurements of the urine temperature. The urine was considered in thermal equilibrium with the kidney. Moreover, the urine temperature was assumed to be measured sufficiently close to the kidney, so that heat transfer between the urine and the surroundings was not taken into account in the model. The MCMC method presented in Chapter 4 was used for the solution of the inverse problem.

The measurement errors were assumed additive, uncorrelated, Gaussian, with zero mean and constant variance. The parameters of our mathematical model are given by:

$$\mathbf{P}^T = \left[\alpha, \beta, \gamma, \phi, \psi, T_a, T_s, Q_{m0}, Q_{m1}, ..., Q_{ml} \right] \qquad (6.2.4)$$

where the arterial blood temperature, T_a, was assumed constant. The transient metabolic heat generation rate was parameterized with piecewise constant basis functions so that $Q_{mj} \equiv Q_m(t_j)$ in $t_j - \dfrac{\Delta t}{2} \le t \le t_j + \dfrac{\Delta t}{2}, j = 0,\ldots, I$, where Δt is the time interval between consecutive measurements.

The vector $\mathbf{P}$ given by equation (6.2.4) includes a set of constant parameters in equation (6.2.2a) and a set of parameters that represent the transient variation of $Q_m(t)$ given, respectively, by:

$$\mathbf{P}_1^T = \left[\alpha, \beta, \gamma, \phi, \psi, T_a, T_s \right] \tag{6.2.5a}$$

$$\mathbf{P}_2^T = \left[Q_{m0}, Q_{m1}, \ldots, Q_{mI} \right] \tag{6.2.5b}$$

so that we can write:

$$\mathbf{P}^T = \left[\mathbf{P}_1^T, \mathbf{P}_2^T \right] \tag{6.2.6}$$

Therefore, the present inverse problem provides a natural selection of two blocks of model parameters. Moreover, several parameters in $\mathbf{P}_1$ are linearly dependent and with sensitivity coefficients much larger than those for the parameters of major interest, which are given by $\mathbf{P}_2$ [144]. Linearly dependent parameters generally result in correlated and periodic Markov chains, with small statistical efficiencies. For this reason, we used the modified version of the Metropolis-Hastings algorithm presented in Note 1 of Chapter 4, where the sampling procedure and the acceptance/rejection test were performed separately for each block of parameters within one iteration of the Metropolis-Hastings algorithm.

RESULTS

The parameters $\alpha, \beta, \gamma, \phi$ and ψ are composed of other input parameters (see equations 6.2.3a-e). Therefore, the priors for $\alpha, \beta, \gamma, \phi$ and ψ are based on the priors for $\left[m_k, c_k, \rho_u, F_u, \rho_a, F_a, c_u, c_b, U, A_k \right]$, where $m_u' = \rho_u F_u$ and $m_a' = \rho_a F_a$. The priors for the input parameters were prescribed by using reference values available in the literature [144]. Most of the parameters were assumed Gaussian, with means given by the values presented by Table 6.1 and with standard deviations of 1% of these means (see Table 6.2). However, the kidney-specific heat and the urine density were assumed with uniform distributions, with their minimum and maximum values given by (3653, 3891) J/kg K and (1012, 1035) kg/m^3, respectively.

By using a Monte Carlo simulation with 10^5 samples from the priors for $\left[m_k, c_k, \rho_u, F_u, \rho_a, F_a, c_u, c_b, U, A_k \right]$, the histograms and the statistics of the distributions for $\alpha, \beta, \gamma, \phi$ and ψ were obtained. The histograms resembled Gaussian distributions. Therefore, the priors for $\alpha, \beta, \gamma, \phi$ and ψ were assumed as Gaussian with means and standard deviations obtained from the Monte Carlo simulation. These priors, which were truncated at the 2.5% and 97.5% percentiles, are presented in Table 6.3. The arterial blood temperature, as well as the temperature of the organs/tissues surrounding the kidney, was assumed as truncated Gaussian, with mean of 309.4 K, standard deviation of 0.05 K and limited to the interval between 307 K and 313 K.

A non-informative prior was used for the parameters that represent the metabolic heat generation rate, $\mathbf{P}_2^T = \left[Q_{m0}, Q_{m1}, \ldots, Q_{mI} \right]$, in the form of the Gaussian Markov random field, like in equation (6.1.4). The hyperparameter of this prior was modeled by a Rayleigh distribution, so that the posterior distribution can be written as:

$$\pi\left(\mathbf{P}_1, \mathbf{P}_2, \varphi \mid \mathbf{Y}\right) \propto \varphi^{(I+2)/2} \exp\Bigg\{ -\frac{1}{2}[\mathbf{Y} - \mathbf{T}(\mathbf{P})]^T \mathbf{W}^{-1}[\mathbf{Y} - \mathbf{T}(\mathbf{P})]$$

$$-\frac{1}{2}[\mathbf{P}_1 - \boldsymbol{\mu}_1]^T \mathbf{V}_1^{-1}[\mathbf{P}_1 - \boldsymbol{\mu}_1] - \frac{1}{2}\varphi\|\mathbf{D}\mathbf{P}_2\|^2 - \frac{1}{2}\left(\frac{\varphi}{\varphi_0}\right)^2 \Bigg\} \tag{6.2.7}$$

TABLE 6.1
Reference Values Used for the Local Analysis of the Sensitivity Coefficients

Kidney mass $m_k = 0.150\,\text{kg}$
Kidney specific heat $c_k = 3763\,\text{J/kg K}$
Blood specific heat $c_b = 3617\,\text{J/kg K}$
Urine specific heat $c_u = 4178\,\text{J/kg K}$
Arterial blood temperature $T_a = 309.4\,\text{K}$
Temperature of the surrounding tissues and organs $T_s = 309.4\,\text{K}$
Kidney initial temperature $T_{k,0} = 309.4\,\text{K}$
Arterial blood flow rate $F_a = 8.33 \times 10^{-6}\,\text{m}^3/\text{s}$
Urine flow rate $F_u = 8.33 \times 10^{-9}\,\text{m}^3/\text{s}$
Kidney density $\rho_k = 1066\,\text{kg/m}^3$
Blood density $\rho_b = 1050\,\text{kg/m}^3$
Urine density $\rho_u = 1024\,\text{kg/m}^3$
Global heat transfer coefficient $U = 40\,\text{W/m}^2\,\text{K}$
Kidney surface area $A_k = 0.02\,\text{m}^2$
Kidney metabolic heat generation rate $Q_m = 10\,\text{W}$

Source: From Ref. [144].

TABLE 6.2
Gaussian Prior Distributions for Some Input Parameters

Parameter	Mean	Standard Deviation
m_k (kg)	0.150	0.0015
c_b (J/kg K)	3617	36.17
c_u (J/kg K)	4178	41.78
F_a (m³/s)	8.33×10^{-6}	8.33×10^{-8}
F_u (m³/s)	8.33×10^{-9}	8.33×10^{-11}
ρ_b (kg/m³)	1050	10.5
U (W/m² K)	40	0.4
A_k (m²)	0.02	0.0002

Source: From Ref. [144].

TABLE 6.3
Truncated Gaussian Priors for $\alpha, \beta, \gamma, \phi$ and ψ

Parameter	Mean	Standard Deviation	Minimum	Maximum
α (s⁻¹)	6.2986×10^{-5}	1.6343×10^{-6}	0.5991×10^{-4}	0.6618×10^{-4}
β (s⁻¹)	0.0559	0.0015	0.0531	0.0589
γ (s⁻¹)	0.0559	0.0015	0.0530	0.0588
ϕ (K/J)	0.001768	3.6783×10^{-5}	0.0017	0.0018
ψ (s⁻¹)	0.0014	3.5635×10^{-5}	0.0013	0.0015

Source: From Ref. [144].

where **D** is given by equation (6.1.5a).

The following proposal distributions were applied for the estimation of $\mathbf{P}_1^T = \left[\alpha, \beta, \gamma, \phi, \psi, T_a, T_s\right]$ and $\mathbf{P}_2^T = \left[Q_{m0}, Q_{m1}, \ldots, Q_{mI}\right]$, respectively, by using the Metropolis-Hastings algorithm with sampling by block of parameters:

$$q_1(\mathbf{P}_1^* \mid \mathbf{P}_1^{(t)}) = \pi_{\text{prior}}(\mathbf{P}_1^*) \propto \exp\left\{-\frac{1}{2}[\mathbf{P}_1^* - \boldsymbol{\mu}_1]^T \mathbf{V}_1^{-1}[\mathbf{P}_1^* - \boldsymbol{\mu}_1]\right\} \tag{6.2.8a}$$

$$q_2(\mathbf{P}_2^* \mid \mathbf{P}_2^{(t)}) = N\left(\mathbf{P}_2^{(t)}, \frac{0.1^2}{2I}\mathbf{I}\right) \tag{6.2.8b}$$

where N(**a**, **B**) is a Gaussian distribution with mean **a** and covariance matrix **B**. The candidates $\mathbf{P}_1^*$ were generated from the prior for $\mathbf{P}_1$ due to the strong linear dependence of the parameters in this vector.

The convergence of the Markov chains to equilibrium posterior distributions for $\mathbf{P}_2^T = \left[Q_{m0}, Q_{m1}, \ldots, Q_{mI}\right]$ was assessed by using the techniques proposed by Geweke [120] and by Gelman and Rubin [121], which were described in Section 4.6. For the results presented below, one chain was initially generated. Then, ten other multiple chains were generated by using samples of the equilibrium distribution obtained with the initial chain. The number of states in each multiple chain was taken as the number of states in the initial chain divided by 10. The number of states was fixed as 7×10^5 in the initial chain and 7×10^4 in each multiple chain.

The integrated autocorrelation time given by equation (4.6.5), calculated for the posterior distributions involving different functional forms for $Q_m(t)$, ranged from 1800 to 2800 in the initial Markov chain. Therefore, the sampling statistical efficiency was low due to the strong linear dependence among the parameters $\mathbf{P}_1^T = \left[\alpha, \beta, \gamma, \phi, \psi, T_a, T_s\right]$. The acceptance rate of the Metropolis-Hastings algorithm used in this work was around 30% and practically constant throughout the initial Markov chain,

A comparison of the means for a step variation of the metabolic heat generation rate, at the beginning and at the end of the initial chains, $\bar{P}_j^a$ and $\bar{P}_j^b$ given, respectively, by equations (4.6.9a,b), is presented in Figure 6.2. Figures 6.3 and 6.4 present the multiple chains for $Q_m(t)$ at $t = 60\,\text{s}$ and the scale reduction coefficient given by equation (4.6.13), respectively. Figure 6.2 shows that the step variation was correctly identified and there is an excellent agreement between the mean values calculated with samples in different ranges of the Markov chains, corresponding to the first 10% and last 50% of states (see equations 4.6.9a,b). The convergence of the multiple chains for $Q_m(t)$ at $t = 60\,\text{s}$

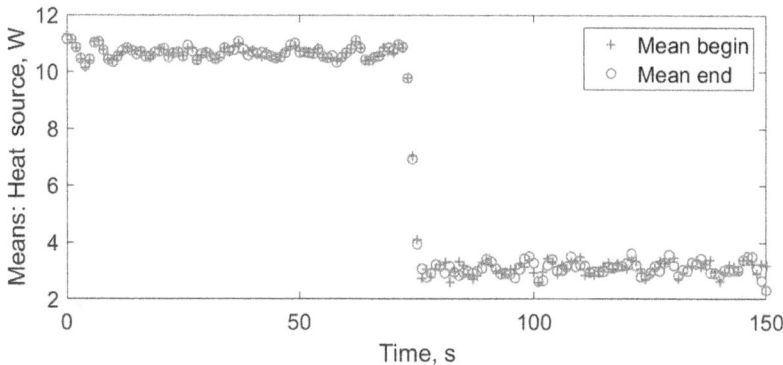

FIGURE 6.2 Comparison of the means in the initial Markov chains. (From Ref. [144].)

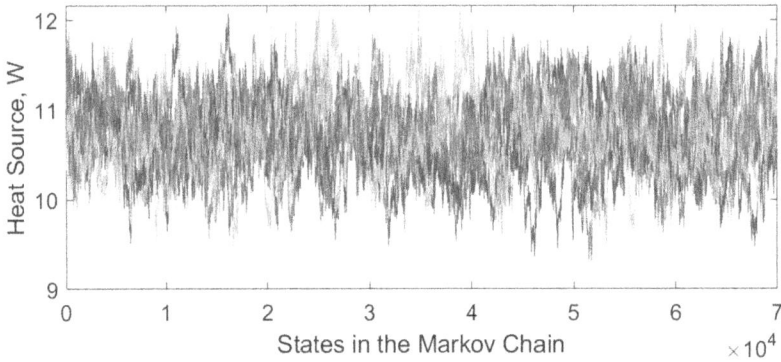

FIGURE 6.3 Multiple chains for $Q_m(t)$ at $t = 60$ s. (From Ref. [144].)

can be verified by the analysis of Figure 6.3. Multiple chains for other times behave similarly to those presented by this figure. The convergence of the multiple chains results in scale reduction coefficients close to 1, except for the region with small values of $Q_m(t)$, particularly, near the final time (see Figure 6.4). As can be noticed in Figures 6.2 and 6.4, the convergences of the initial chain and of the multiple chains are better for large values of $Q_m(t)$, because the correspondent sensitivity coefficients are larger. Lack of convergence near the final time is attributed to the physics of the problem, since the heat generation rate affects the temperatures at later times, when measurements are not available for the inverse analysis.

The results obtained with the initial chain are presented by Figure 6.5. This figure compares the exact function with the means and the 0.5% and 99.5% percentiles of the samples of the Markov chains. The estimated heat generation rate was stable and very accurately recuperated the function discontinuity, thus demonstrating that the non-informative Gaussian Markov random field prior, with the Rayleigh hyperprior for the parameter φ, was appropriate for the inverse problem under analysis. The estimated means were larger than the exact function values. However, the estimated 99% credible intervals enclosed the exact function values. Such facts can also be observed with the Markov chains and the histograms of the samples (with the burn-in period discarded) of the solution at times $t = 60$ s and $t = 90$ s, which are presented by Figures 6.6a and b and 6.7a and b, respectively. As expected, the Markov chain for the solution at $t = 90$ s was more correlated than that for $t = 60$ s due to the smaller sensitivity coefficients for smaller values of the heat generation rate (see Figures 6.6a and b). The histograms of the marginal posterior distributions for the estimated heat source terms resembled Gaussian distributions.

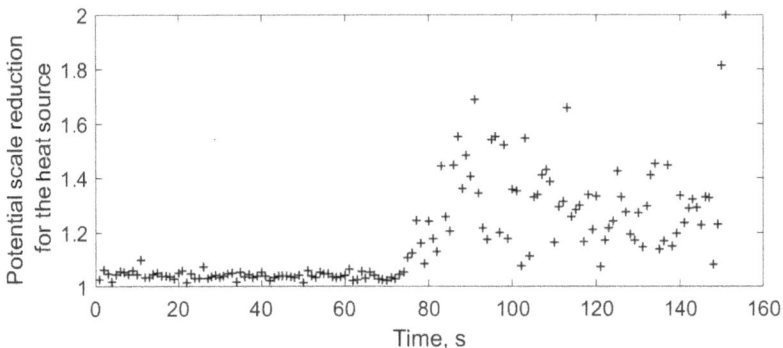

FIGURE 6.4 Scale reduction coefficient given by equation (4.6.13). (From Ref. [144].)

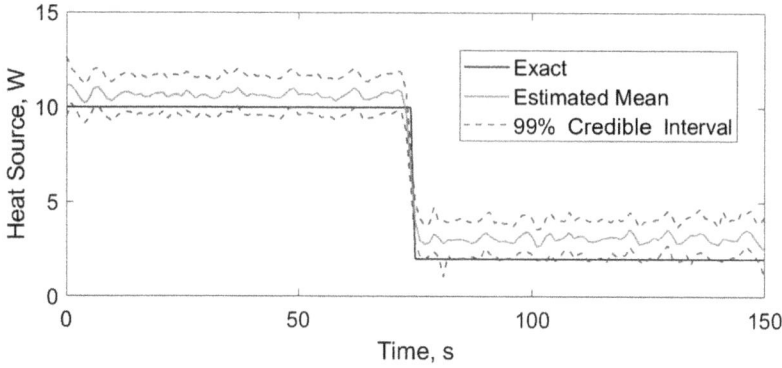

FIGURE 6.5 Estimation of the metabolic heat generation rate. (From Ref. [144].)

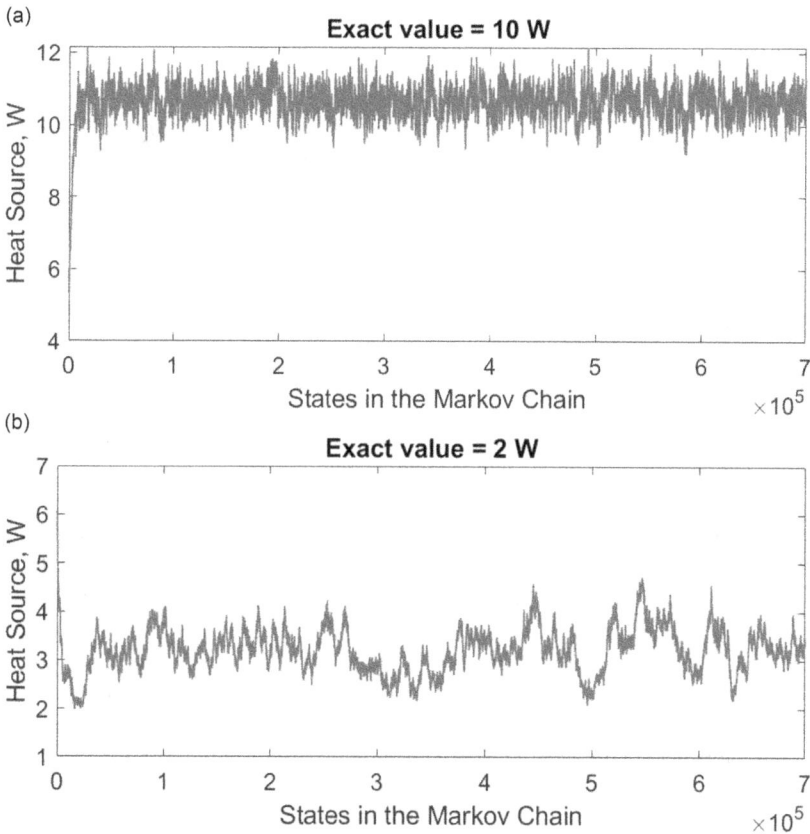

FIGURE 6.6 Markov Chains for the metabolic heat generation rate: (a) $t = 60$ s, (b) $t = 90$ s. (From Ref. [144].)

The means of all multiple chains, as well as the 99% confidence interval calculated with the variance given by equation (4.6.12) is shown by Figure 6.8. Similar to the results obtained with the initial chain, the means of the multiple chains overestimated the exact function due to the underestimation of the artery blood temperature. However, the exact function was within the estimated 99% confidence interval.

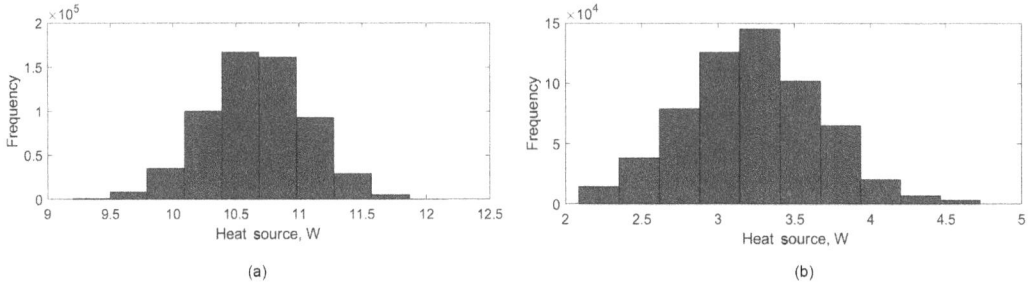

(a)   (b)

FIGURE 6.7 Histograms of the converged states of the Markov chains for the metabolic heat generation rate: (a) $t = 60\,\text{s}$, (b) $t = 90\,\text{s}$. (From Ref. [144].)

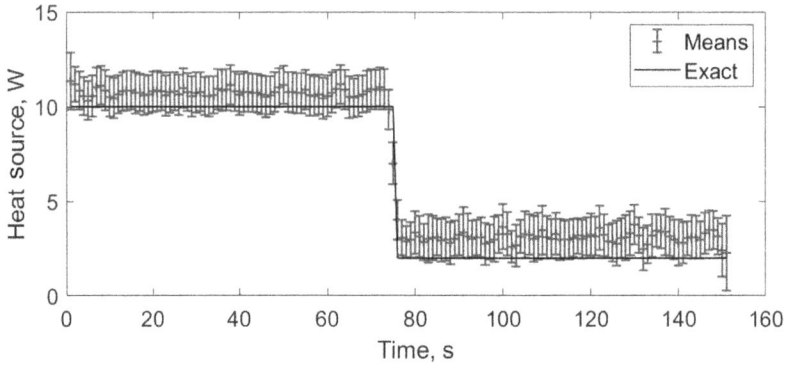

FIGURE 6.8 Means and 99% confidence intervals obtained with the multiple chains. (From Ref. [144].)

6.3 TEMPERATURE ESTIMATION OF INFLAMED BOWEL

Local temperature increase is one of the five classical signs of regions with inflammations. This section is focused on the application of the photoacoustic technique for the estimation of the temperature field in the colon, as the solution of an inverse problem, for the detection of inflamed regions [145]. Photoacoustic techniques have been applied for non-invasive medical diagnosis and imaging [146]. These techniques rely on the generation of pressure waves from the absorption of the energy of a laser pulse. The amplitude of the resulting acoustic wave is proportional to the Grüneisen parameter that includes the local sound velocity, thermal expansion coefficient and specific heat [145–156]. As a consequence that the Grüneisen parameter is temperature dependent, the amplitude of the photoacoustic wave is proportional to the local temperature. The inverse problem was solved here for a rotating laser inside the intestine lumen, which imposed pulses for the generation of the acoustic waves. One single ultrasound detector, also located at the laser rotating shaft, provided the simulated measurements for the inverse analysis [145]. The inverse problem was solved with the minimization of the MAP objective function.

DIRECT PROBLEM

The photoacoustic problem was considered in the two-dimensional rectangular region (W—width and H—height) illustrated by Figure 6.9.

The mathematical model for the pressure wave, p, inside the domain is given by [145]:

$$\left(\nabla^2 - \frac{1}{v_s^2}\frac{\partial^2}{\partial t^2}\right)p(x,y,t) = -\frac{\beta}{c_p}\frac{\partial H_h(x,y,t)}{\partial t} \quad \text{in} \quad -\frac{W}{2} < x < \frac{W}{2}, \quad -\frac{H}{2} < y < \frac{H}{2}, \text{ for } t > 0 \quad (6.3.1)$$

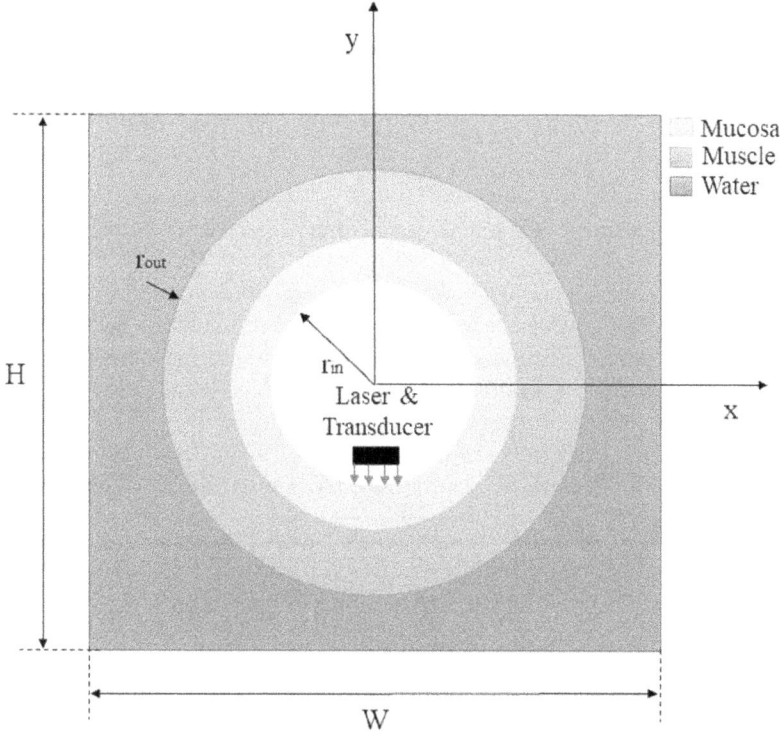

FIGURE 6.9 Photoacoustic problem for detection of inflamed regions in the bowel. (From Ref. [145].)

with v_s is the speed of sound, β is the coefficient of thermal expansion and c_p is the specific heat. The source term of equation (6.3.1) is the time derivative of the local energy deposited per unit volume by the photoacoustic laser pulse. It is assumed that the laser pulse used for the photoacoustic measurements is instantly absorbed by the tissues, which is consistent with the thermal confinement condition. Thus, the effects of the laser pulse can be taken into account in terms of the initial pressure distribution, $p_0(x,y,T)$ at $t = 0$, with $T \equiv T(x,y)$, and equation (6.3.1) is rewritten as:

$$\left(\nabla^2 - \frac{1}{v_s^2} \frac{\partial^2}{\partial t^2} \right) p(x,y,t) = 0 \quad \text{in} \quad -\frac{W}{2} < x < \frac{W}{2}, \quad -\frac{H}{2} < y < \frac{H}{2}, \quad \text{for } t > 0 \qquad (6.3.2a)$$

with initial conditions:

$$p(x,y,t) = p_0(x,y,T) \quad \text{at } t = 0, \text{ in} \quad -\frac{W}{2} < x < \frac{W}{2}, \quad -\frac{H}{2} < y < \frac{H}{2} \qquad (6.3.2b)$$

$$\frac{\partial p}{\partial t} = 0 \quad \text{at } t = 0, \text{ in} \quad -\frac{W}{2} < x < \frac{W}{2}, \quad -\frac{H}{2} < y < \frac{H}{2} \qquad (6.3.2c)$$

The boundary conditions imposed at the surfaces $x = \pm\frac{W}{2}$ and $y = \pm\frac{H}{2}$ were considered to simulate an infinite domain by using the so-called perfect match layers [156]. Once assuming that the volume expansion due to the absorbed energy is insignificant, the initial photoacoustic pressure can be written as [154–156]:

$$p_0(x,y,T) = \mu_a(x,y)\Gamma(x,y,T)\Phi(x,y) \qquad (6.3.3)$$

where $\mu_a(x,y)$ and $\Phi(x,y)$ are the absorption coefficient and laser fluence rate, respectively, and $\Gamma(x,y,T)$ is the Grüneisen parameter [145–156], which is defined by:

$$\Gamma(x,y,T) = \frac{v_s^2(x,y,T)\beta(x,y,T)}{c_p(x,y,T)} \qquad (6.3.4)$$

Since v_s, β and c_p are functions of the local temperature, the Grüneisen parameter was directly related to $T(x,y)$.

The photoacoustic problem examined in this work was focused on the cross section of the bowel presented in Figure 6.9. Perfect contact was assumed between the tissue layers, which were considered circular and centered in the rectangular region. The bowel inner and outer radiuses are r_{in} and r_{out}, respectively. Water was considered in the space between the muscle outer surface and the rectangular domain boundaries.

For the estimation of the temperature in the intestine wall by inverse photoacoustic analysis, the laser and the ultrasound transducer were located inside the intestine lumen in perfect contact with the mucosa. The Beer-Lambert law was used to model the laser propagation in the bowel wall, that is [145],

$$\Phi(x,y) = E_{0,i}e^{-\mu_{\mathrm{eff},i}\left(\sqrt{x^2+y^2}-r_i\right)} \qquad (6.3.5)$$

where $\mu_{\mathrm{eff},i}$ and $E_{0,i}$ are the effective attenuation coefficient and the incoming laser fluence at layer i, respectively. For $i = 1$, $E_{0,i} = E_0$ at $r_i = r_{in}$. The effective attenuation coefficient was derived from the diffusion approximation for radiation propagation and is strongly dependent on the laser wavelength. The effective attenuation coefficient is given by [145]:

$$\mu_{\mathrm{eff}} = \sqrt{3\mu_a\left(\mu_a + \mu_s'\right)} \qquad (6.3.6)$$

where μ_a is the absorption coefficient of light and μ_s' represents the reduced scattering coefficient, given in terms of the anisotropy factor, g, and the scattering coefficient, μ_s, by:

$$\mu_s' = \mu_s\left(1 - g\right) \qquad (6.3.7)$$

Here, the laser and the ultrasound transducer were located at the tip of a shaft that rotated inside a catheter, considered transparent to laser and ultrasound, in the bowel lumen. Since the duration of the laser pulse propagation was much shorter than the time between laser pulses, each pulse could be considered to propagate along a specific direction, as will be apparent below. Moreover, the directional characteristics of the laser and of the ultrasound transducer, with limited aperture angles, naturally made the problem two-dimensional as formulated above. Three-dimensional analyses could be performed by sequentially solving the two-dimensional problem at several longitudinal positions along the intestine.

INVERSE PROBLEM

The inverse problem examined in this work involved the estimation of the temperature field in the region presented by Figure 6.9 by using acoustic wave measurements taken by the transducer located at the rotating shaft inside the catheter.

Consider $\mathbf{T}$ as the vector with the temperatures at each one of the n finite volumes used for the discretization of the spatial region, while $\mathbf{P}_t$ is the vector of the m transient ultrasound pressure measurements. Let $\Pi(\mathbf{T})$ be the solution of the direct problem given by equations (6.3.2)-(6.3.7), at the measurement location, which is obtained with the temperature distribution, $T(x,y)$. The measurement errors are assumed as additive.

The mathematical model given by equation (6.3.2) is linear with respect to $p_0(x,y,T)$ and, in accordance with equation (6.3.4), the initial pressure is itself linear with respect to the Grüneisen parameter. Assuming also a linear dependence between the Grüneisen parameter and temperature, then the solution of the direct problem is linear with respect to the temperature $T(x,y)$. In summary, in discrete form we can write [145]:

$$\mathbf{P}_t = \mathbf{K}_A \mathbf{T} + \mathbf{K}_B + \boldsymbol{\varepsilon} \tag{6.3.8}$$

where $\mathbf{K}_A$ and $\mathbf{K}_B$ describe the discrete solution of the direct problem and $\boldsymbol{\varepsilon}$ is the vector of measurement errors.

We assume that the measurement errors are Gaussian with zero mean and known covariance matrix and independent of the local temperature. By also assuming a Gaussian prior distribution (equation 6.1.1), the inverse problem is solved in this work by the minimization of the MAP objective function given by equation (3.1.24). Due to the linearity of the inverse problem, the posterior density is also Gaussian with covariance matrix given by equation (3.2.7).

RESULTS

For the results presented below, the bowel inner and outer radiuses were taken as $r_{in} = 1.5$ mm and $r_{out} = 2.5$ mm, respectively. The mucosa and the muscle were assumed with wall thicknesses of 0.4 and 0.6 mm, respectively, and the computational domain was taken as $H = W = 6$ mm. In order to generate the photoacoustic pressure waves for the solution of the inverse problem that follows, the internal surface of the bowel mucosa was considered irradiated by pulses of a flat monochromatic laser beam, with energy of 1.9 mJ at the wavelength of 532 nm. The pulse duration was taken as 9 ns. The laser and the ultrasound transducer, which were located at the tip of a shaft inside a catheter in the intestine lumen, rotated with a frequency of 10 Hz. For each rotation, either 10 or 20 laser pulses provided initial pressures for specific sectors of the circular region, which sequentially covered 360°. Each sector corresponded to an angular view of the ultrasound transducer. The width of the laser beam was considered to be 1 mm for the pulse repetition frequency of 100 Hz and 0.5 mm for the pulse frequency of 200 Hz. Normal and inflamed tissues were assumed at the temperatures of 37°C and 38°C, respectively, while the water surrounding the intestine muscle was at 37°C. The ultrasound transducer was assumed to operate at 40 MHz and with an omnidirectional receiving radiation pattern [145].

Figure 6.10 shows the initial pressure generated by the first laser pulse for a region without inflammation, which reached 25.5 kPa in the mucosa, with a laser pulse frequency of 100 Hz. The initial pressure decayed inside the muscle surrounding the mucosa, due to the high attenuation of the laser pulse in both tissues. The photoacoustic pressure at the transducer location tended to zero and became negligible in about 4 μs after the laser pulse, as illustrated by Figure 6.11 that shows the pressure variation at the measurement location after the first laser pulse. Therefore, the inverse problem for estimating the temperature in the region with the photoacoustic technique, using the ultrasound pulses generated inside the intestine lumen, could be solved independently for each laser pulse.

Simulated measurements, obtained by adding Gaussian errors to the direct problem solution for the ultrasound pressure at the transducer location, were used in the inverse analysis. These errors were simulated with zero means and standard deviations of 1% of the exact maximum pressure, which was obtained from the solution of the direct problem for two situations: (i) Uniform inflammation in the mucosa; and (ii) three inflamed regions in the mucosa. MAP estimates were obtained by considering either one of the prior covariance matrices given by equations (6.1.2) or (6.1.3). For all cases, the means and the standard deviations of the priors for all pixels were supposed equal to 37°C and 1.5°C, respectively. Therefore, the whole region was supposed at a uniform temperature corresponding

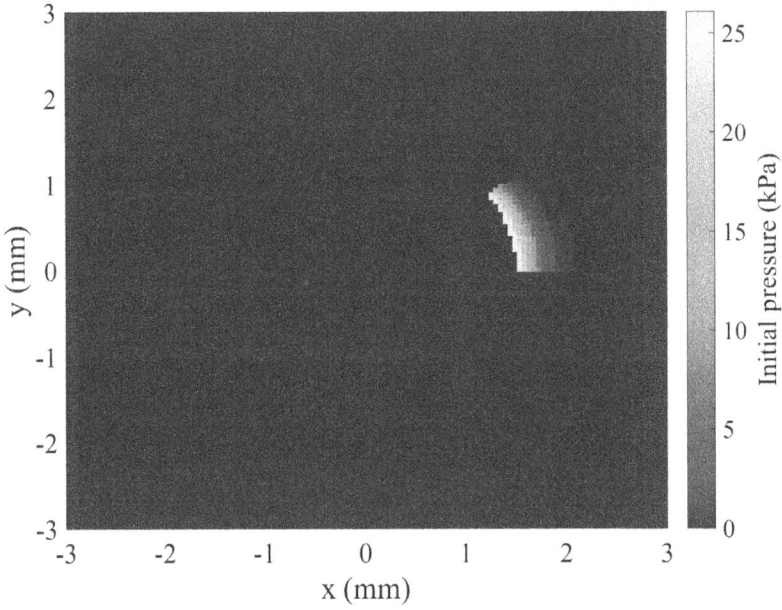

FIGURE 6.10 Initial pressure resulting from the first laser pulse in a region without inflammation with a laser pulse frequency of 100 Hz. (From Ref. [145].)

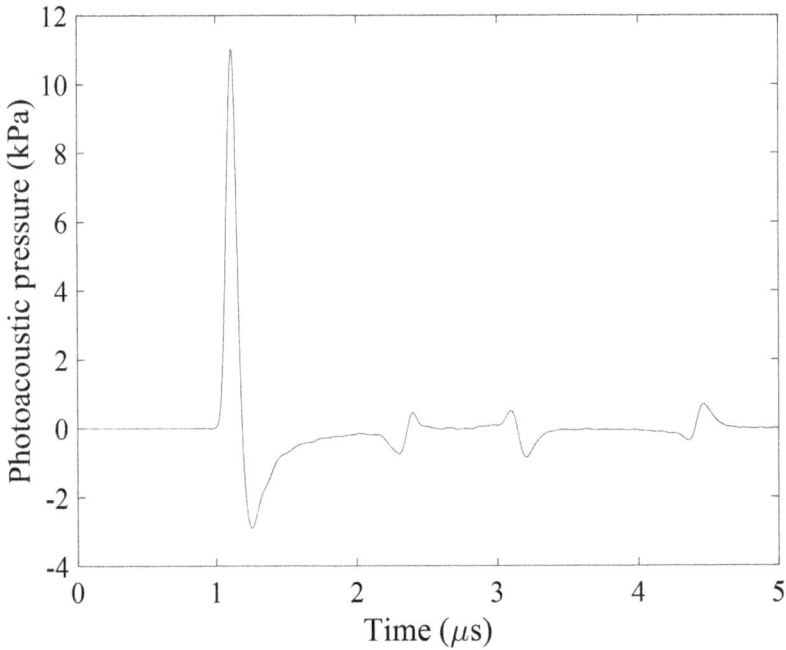

FIGURE 6.11 Ultrasound pressure at the transducer after the first laser pulse in a region without inflammation with a laser pulse frequency of 100 Hz. (From Ref. [145].)

to the normal tissue before the experiments, that is, the domain was *a priori* assumed without any inflamed region for the solution of the inverse problem. Furthermore, in order to challenge the identification of inflamed regions with the photoacoustic technique, a quite large standard deviation of 1.5°C was assigned to the prior distribution. By following previous works available in the literature, the parameters of the Matérn prior covariance matrix, that is, the characteristic length scale and the smoothness parameter, were set to $l = 0.0465$ mm (the pixel size) and $\alpha = 1.5$, respectively [145].

A comparison of the Gaussian marginal posterior distributions obtained for nine selected pixels in the inflamed mucosa is presented by Figure 6.12 for the uniform inflammation in the mucosa. The solid lines in this figure correspond to the distributions obtained with the uncorrelated prior (see equation 6.1.2), while the dashed lines are related to the distributions obtained with the Matérn covariance matrix given by equation (6.1.3). For all pixels, the standard deviations obtained with the uncorrelated prior were larger than those obtained with the correlated prior. Therefore, uncertainties of the temperatures estimated with the Matérn covariance matrix for the prior were smaller than those obtained with the independent prior. Furthermore, the marginal posterior distributions obtained with the Matérn covariance matrix were in fact centered around the exact temperature value of 38°C, while those corresponding to the uncorrelated prior were not. The posterior means obtained with the uncorrelated prior for pixels 1, 7 and 8 were practically identical to the prior means and inflammation was thus falsely not detected at these points.

After examining the situation with the inflammation in the whole mucosa, a case with three inflamed regions is now considered (see Figure 6.13) by employing a laser pulse frequency of 100 Hz. Figure 6.13 also shows the ten sectors for which independent inverse problems were sequentially solved, since the photoacoustic ultrasound pulses received by the transducers decayed much faster than the time between the laser pulses. Since the inverse problems for each sector were independent, the sizes and locations of the inflamed regions were selected so that they would not be completely within one single sector, in order to challenge the method. Based on the analysis performed previously for the situation with inflammation in the whole mucosa, only the results obtained with the Matérn prior covariance matrix are now presented for the case with three inflamed regions.

Figure 6.14 presents the estimated mean temperatures in the domain with a laser pulse frequency of 100 Hz, where the three inflamed regions could be accurately detected (see also Figure 6.13). The temperature distribution presented by Figure 6.14 consists of the composition of the estimations obtained

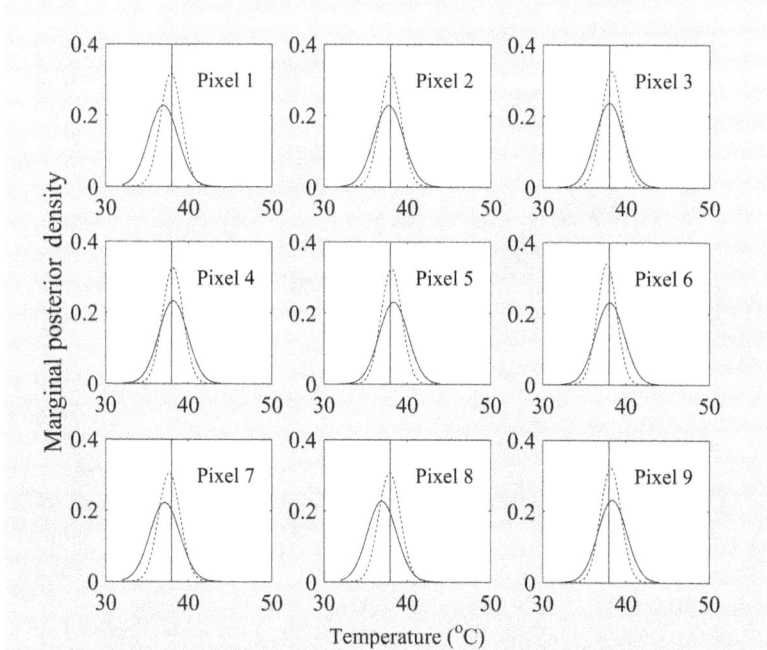

FIGURE 6.12 Marginal posterior distributions for nine pixels in the inflamed mucosa obtained with the uncorrelated prior (solid line) and with the correlated prior (dashed line). (From Ref. [145].)

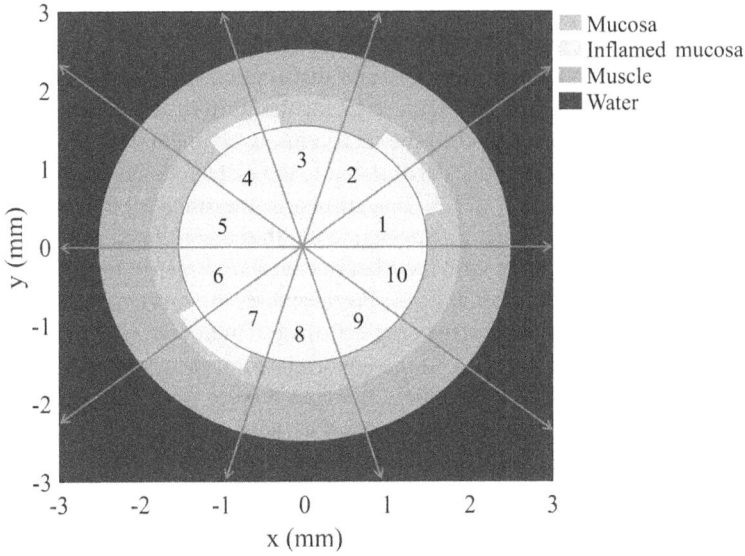

FIGURE 6.13 Case with three small inflamed regions in the mucosa illustrating the sectors corresponding to a laser pulse frequency of 100 Hz. (From Ref. [145].)

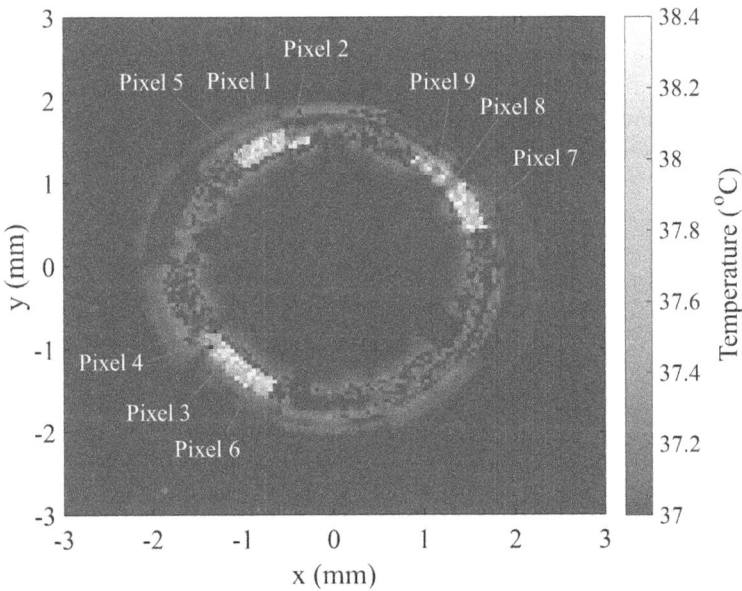

FIGURE 6.14 Temperatures estimated with a frequency of 100 Hz—case with three inflamed regions. (From Ref. [145].)

with each laser pulse, that is, the temperatures independently estimated in each of the sectors shown by Figure 6.13. Thus, Figure 6.14 exhibits discontinuities at the edges between adjacent sectors and the borders of the inflamed regions were not very well identified. The results presented by this figure were improved by increasing the laser pulse frequency, which then provided measurements taken in smaller sectors to compose the whole domain. The results obtained with a laser frequency of 200 Hz are presented in Figure 6.15. A comparison of Figures 6.14 and 6.15 reveals that the identification of the inflamed regions improved significantly when the laser frequency was increased from 100 to 200 Hz. In special, the borders of the inflamed regions became sharper with the frequency increase.

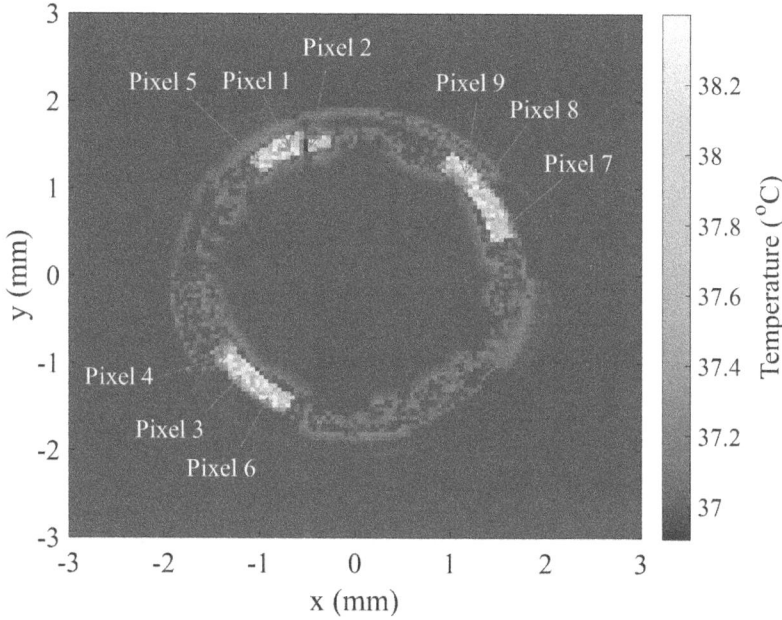

FIGURE 6.15 Temperatures estimated with a frequency of 200 Hz—case with three inflamed regions. (From Ref. [145].)

6.4 DETECTION OF CONTACT FAILURES BY USING INTEGRAL TRANSFORMED MEASUREMENTS

This section presents the solution of an inverse heat conduction problem aiming at the detection of contact failures in layered composites through the estimation of the contact conductance between the layers [157]. As discussed in Chapter 4, due to the computational demand of the MCMC method, its use with large dimensional problems is often prohibitive. Therefore, we applied here a data compression scheme. The temperatures measured with an infrared camera were spatially compressed through the integral transformation with eigenfunctions related to the actual physical problem. Only a few transformed modes were then used in the inverse analysis and the forward model was formulated directly in terms of the transformed (compressed) temperatures. The data compression applied in this work not only reduced the computational time required for the MCMC method, but also provided regularization for the inverse problem. Conceptually, the integral transform data compression scheme falls within the broader class of orthogonal decomposition methods, such as POD—Proper Orthogonal Decomposition, Principal Component Analysis, Karhunen-Loeve decomposition and Truncated Singular Value Decomposition [158–164].

DIRECT PROBLEM

The physical problem considered here involved heat conduction through a plate with two layers heated through its top surface by a heat flux $q(x,y,t)$ (see Figure 6.16). The bottom surface of the plate was thermally insulated and heat transfer was assumed negligible through its lateral surfaces. The plate was initially at a uniform temperature, T_0, and the physical properties of each layer were assumed homogeneous and not dependent on temperature. The length and width of the plate were a and b, respectively, while its thickness was denoted by c. A spatially distributed contact resistance between the two adjacent layers was modeled by a contact conductance $h_c(x,y)$.

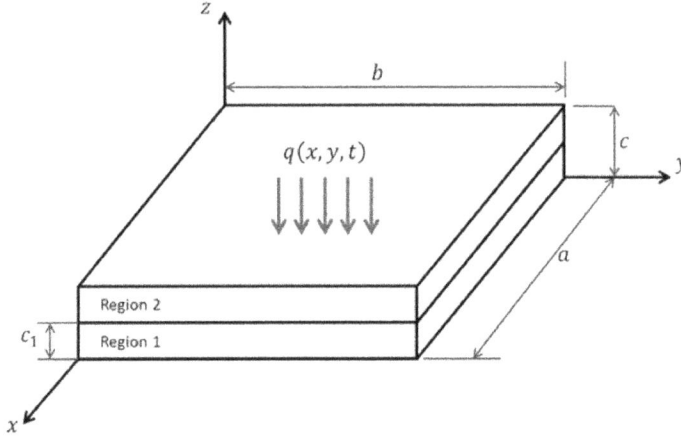

FIGURE 6.16 Physical problem. (From Ref. [157].)

The mathematical problem is written in dimensionless form by using the following variables:

$$\theta(X,Y,Z,\tau) = \frac{T(x,y,z,t) - T_0}{T_0} \qquad q^*(X,Y,\tau) = q(x,y,t)\frac{c}{k_{ref}T_0} \qquad (6.4.1a,b)$$

$$k^* = \frac{k}{k_{ref}} \qquad \alpha^* = \frac{\alpha}{\alpha_{ref}} \qquad (6.4.1c,d)$$

$$X = \frac{x}{c} \qquad Y = \frac{y}{c} \qquad Z = \frac{z}{c} \qquad Z_1 = \frac{c_1}{c} \qquad A = \frac{a}{c} \qquad B = \frac{b}{c} \qquad (6.4.1e\text{-}j)$$

$$\tau = \frac{\alpha_{ref}t}{c^2} \qquad Bi_c(X,Y) = \frac{h_c(x,y)c}{k_{ref}} \qquad (6.4.1k,l)$$

and then we obtain:

$$\frac{\partial\theta_1}{\partial\tau}(X,Y,\tau) = \alpha_1^*\nabla^2\theta_1 \quad \text{in } 0 < X < A, \ 0 < Y < B, \ 0 < Z < Z_1, \ \text{for} \quad \tau > 0 \qquad (6.4.2a)$$

$$\frac{\partial\theta_2}{\partial\tau}(X,Y,\tau) = \alpha_2^*\nabla^2\theta_2 \quad \text{in } 0 < X < A, \ 0 < Y < B, Z_1 < Z < 1, \ \text{for,} \quad \tau > 0 \qquad (6.4.2b)$$

$$\frac{\partial\theta_1}{\partial Z} = 0 \quad \text{at } Z = 0, \ \text{in } 0 < X < A, \ 0 < Y < B, \ \text{and } \tau > 0 \qquad (6.4.2c)$$

$$k_1^*\frac{\partial\theta_1}{\partial Z} = k_2^*\frac{\partial\theta_2}{\partial Z} \quad \text{at } Z = Z_1, \ \text{in } 0 < X < A, \ 0 < Y < B, \ \text{and } \tau > 0 \qquad (6.4.2d)$$

$$k_1^*\frac{\partial\theta_1}{\partial Z} = Bi_c(X,Y)[\theta_2 - \theta_1] \quad \text{at } Z = Z_1, \text{in } 0 < X < A, \ 0 < Y < B, \ \text{and } \tau > 0 \qquad (6.4.2e)$$

$$k_2^*\frac{\partial\theta_2}{\partial Z} = q^*(X,Y,\tau) \quad \text{at } Z = 1, \ \text{in } 0 < X < A, \ 0 < Y < B, \ \text{and } \tau > 0 \qquad (6.4.2f)$$

$$\frac{\partial\theta_1}{\partial X} = \frac{\partial\theta_2}{\partial X} = 0 \quad \text{at } X = 0, 0 < Y < B, 0 < Z < 1, \ \text{and } \tau > 0 \qquad (6.4.2g)$$

$$\frac{\partial \theta_1}{\partial X} = \frac{\partial \theta_2}{\partial X} = 0 \text{ at } X = A, 0 < Y < B, 0 < Z < 1, \text{ and } \tau > 0 \qquad (6.4.2h)$$

$$\frac{\partial \theta_1}{\partial Y} = \frac{\partial \theta_2}{\partial Y} = 0 \text{ at } Y = 0, 0 < X < A, 0 < Z < 1, \text{ and } \tau > 0 \qquad (6.4.2i)$$

$$\frac{\partial \theta_1}{\partial Y} = \frac{\partial \theta_2}{\partial Y} = 0 \text{ at } Y = B, \ 0 < X < A, \ 0 < Z < 1, \text{ and } t > 0 \qquad (6.4.2j)$$

$$\theta_1 = \theta_2 = 0 \text{ at } \tau = 0, \text{ in } 0 < X < A \ \ 0 < Y < B \ \ 0 < Z < 1 \qquad (6.4.2k)$$

where the contact interface is located at $Z = Z_1$.

The direct problem associated with the formulation given by equations (6.4.2a-k) involves the determination of the temperature fields $\theta_1(X,Y,Z,\tau)$ and $\theta_2(X,Y,Z,\tau)$. The direct problem was solved by using a hybrid analytical-numerical approach based on the generalized integral transform technique [165–169] and finite differences [86]. Problem (6.4.2) was integral transformed along the X and Y directions by using the following transform—inversion formulae pair [56]:

$$\text{Transform}: \ \tilde{\bar{\theta}}_{1,2}\left(\beta_i,\gamma_j,Z,\tau\right) = \int_{X=0}^{A}\int_{Y=0}^{B} \bar{\phi}_i\,\bar{\varphi}_j\,\theta_{1,2}(X,Y,Z,\tau)\,dY\,dX \qquad (6.4.3a)$$

$$\text{Inversion formula}: \ \theta_{1,2}(X,Y,Z,\tau) = \sum_{i=0}^{\infty}\sum_{j=0}^{\infty} \bar{\phi}_i\,\bar{\varphi}_j\,\tilde{\bar{\theta}}_{1,2}\left(\beta_i,\gamma_j,Z,\tau\right) \qquad (6.4.3b)$$

where the normalized eigenfunctions were obtained from Ref. [56]:

$$\bar{\phi}_i = \frac{\cos(X\beta_i)}{\sqrt{N_i}} \quad \bar{\varphi}_j = \frac{\cos(Y\gamma_j)}{\sqrt{N_j}} \quad \text{for } i = 0,\dots,\infty \text{ and } j = 0,\dots,\infty \qquad (6.4.4a,b)$$

with normalization integrals

$$N_i = A \quad \text{for } i = 0, \quad N_i = \frac{A}{2} \quad \text{for } i = 1,\dots,\infty \qquad (6.4.5a)$$

$$N_j = B \quad \text{for } j = 0, \quad N_j = \frac{B}{2} \quad \text{for } j = 1,\dots,\infty \qquad (6.4.5b)$$

and eigenvalues

$$\beta_i = \frac{i\pi}{A} \quad \text{for } i = 0,\dots,\infty \qquad (6.4.6a)$$

$$\gamma_j = \frac{j\pi}{B} \quad \text{for } j = 0,\dots,\infty \qquad (6.4.6b)$$

Hence, the following system of coupled equations for $i = 0,\dots,\infty$ and $j = 0,\dots,\infty$ resulted from the transformation of Problem (6.4.2):

$$\frac{1}{\alpha_1^*}\frac{\partial \tilde{\bar{\theta}}_1\left(\beta_i,\gamma_j,Z,\tau\right)}{\partial \tau} = \frac{\partial^2 \tilde{\bar{\theta}}_1}{\partial Z^2} - \left(\beta_i^2 + \gamma_j^2\right)\tilde{\bar{\theta}}_1 \quad \text{for } \tau > 0, \text{ in } 0 < Z < Z_1 \qquad (6.4.7a)$$

$$\frac{1}{\alpha_2^*}\frac{\partial\bar{\bar{\theta}}_2\left(\beta_i,\gamma_j,Z,\tau\right)}{\partial\tau} = \frac{\partial^2\bar{\bar{\theta}}_2}{\partial Z^2} - \left(\beta_i^2 + \gamma_j^2\right)\bar{\bar{\theta}}_2 \text{ for } \tau > 0, \text{ in } Z_1 < Z < 1 \qquad (6.4.7\text{b})$$

$$k_1^*\frac{\partial\bar{\bar{\theta}}_1}{\partial Z} = 0 \text{ at } Z = 0 \text{ and } \tau > 0 \qquad (6.4.7\text{c})$$

$$k_2^*\frac{\partial\bar{\bar{\theta}}_2}{\partial Z} = \bar{\bar{d}}_{i,j} \text{ at } Z = 1 \text{ and } \tau > 0 \qquad (6.4.7\text{d})$$

$$k_1^*\frac{\partial\bar{\bar{\theta}}_1}{\partial Z} = k_2^*\frac{\partial\bar{\bar{\theta}}_2}{\partial Z} \text{ at } Z = Z_1 \text{ and } \tau > 0 \qquad (6.4.7\text{e})$$

$$k_1^*\frac{\partial\bar{\bar{\theta}}_{1,(i,j)}}{\partial Z} = \sum_{m=0}^{\infty}\sum_{u=0}^{\infty}A_{i,j,m,u}\left[\bar{\bar{\theta}}_{2,(m,u)} - \bar{\bar{\theta}}_{1,(m,u)}\right] \text{ at } Z = Z_1 \text{ and } \tau > 0 \qquad (6.4.7\text{f})$$

$$\bar{\bar{\theta}}_1^* = \bar{\bar{\theta}}_2^* = 0 \text{ at } \tau = 0 \text{ in } 0 < Z < 1 \qquad (6.4.7\text{g})$$

where

$$\bar{\bar{d}}_{i,j} = \int\limits_{X=0}^{A}\int\limits_{Y=0}^{B} q^*(X,Y,\tau)\bar{\phi}_i\,\bar{\varphi}_j\,dY\,dX \qquad (6.4.8\text{a})$$

$$A_{i,j,m,u} = \int\limits_{X=0}^{A}\int\limits_{Y=0}^{B} \bar{\phi}_i\,\bar{\phi}_m\,\bar{\varphi}_j\,\bar{\varphi}_u\,Bi_c(X,Y)\,dY\,dX \qquad (6.4.8\text{b})$$

The system of infinite coupled partial differential equations (6.4.7a-g) for the transformed fields $\bar{\bar{\theta}}_1\left(\beta_i,\gamma_j,Z,\tau\right)$ and $\bar{\bar{\theta}}_2\left(\beta_i,\gamma_j,Z,\tau\right)$ was then discretized implicitly with finite differences along the Z direction [86]. The numerical solution was obtained by truncating the infinite system to a finite number of transform modes. The number of modes, as well as the finite difference grid size, was selected so that the computed fields $\theta_1(X,Y,Z,\tau)$ and $\theta_2(X,Y,Z,\tau)$ were within a user prescribed error tolerance.

INVERSE PROBLEM

The focus of this work was the detection of contact failures between layers 1 and 2 (see Figure 6.16) by identifying the dimensionless contact conductance $Bi_c(X,Y)$. For perfect contact, $Bi_c(X,Y)$ is sufficiently large to characterize temperature continuity at the interface, while a contact failure is detected by values of the contact conductance that tend to zero. Simulated and real transient temperature measurements obtained with an infrared camera over the top surface $Z = 1$ were used for the identification of $Bi_c(X,Y)$.

In this work, the dimensionless contact conductance $Bi_c(X,Y)$ was modeled as piecewise constant in each pixel of a grid with center points (X_I,Y_J), where $X_I = I\Delta X$, $Y_J = J\Delta Y$, $I = 1, ..., I_f, J = 1, ..., J_f$, and with grid spacing given by $\Delta X = A/I_f$ and $\Delta Y = B/J_f$. The total number of estimated points, which covered the spatial domain $0 < X < A$ and $0 < Y < B$, was then $M = I_f J_f$. The applied heat flux was analogously modeled as piecewise constant on a similar discretization of the top surface. Hence, the vector of unknown parameters for the inverse analysis is given by:

$$P^T = \left[Bi_{c1}, Bi_{c2}, ..., Bi_{cM}, q_1^*, q_2^*, ..., q_M^*, \alpha_1^*, \alpha_2^*, k_1^*, k_2^*\right] \qquad (6.4.9)$$

The priors used for the parameters will be discussed below. The vector containing the measured temperatures is written as:

$$\psi^T = \left(\vec{\psi}_1, \vec{\psi}_2, \dots, \vec{\psi}_{k_{max}}\right) \tag{6.4.10}$$

where $\vec{\psi}_k$ contains the measured temperatures of each of the M grid elements at time t_k, $k = 1, \dots, k_{max}$, that is,

$$\vec{\psi}_k = \left(\psi_{k1}, \psi_{k2}, \dots, \psi_{kM}\right) \quad \text{for} \quad k = 1, \dots, k_{max} \tag{6.4.11}$$

so that we have $D = M k_{max}$ measurements in total.

Temperature measurements obtained with an infrared camera have errors that can be modeled as Gaussian with zero mean and constant standard deviation σ (see Figure 1.6). Therefore, the *likelihood function* can be expressed as:

$$\pi(\Psi|\mathbf{P}) = (2\pi\sigma^2)^{-D/2} \exp\left\{-\frac{1}{2}\frac{[\Psi - \Theta(\mathbf{P})]^T[\Psi - \Theta(\mathbf{P})]}{\sigma^2}\right\} \tag{6.4.12}$$

where $\Theta(\mathbf{P})$ is the solution of the direct problem given by equations (6.4.2a-k) with vector $\mathbf{P}$ given by equation (6.4.9).

The inverse problem was solved with the Metropolis-Hastings algorithm presented in section 4-2. The solution of the direct problem, $\Theta(\mathbf{P})$, needs to be computed for all states of the Markov chain, at each position and time that a measurement is available. Typically, the number of states required for the Markov chain to generate samples that appropriately represent the posterior distribution is very large. Therefore, if the computational time for the solution of the direct problem is large, the application of the MCMC method for the solution of the inverse problem (acquiring an adequate number of samples with the MCMC chain) might not be feasible within a reasonable computational time. The use of fast reduced mathematical models for the solution of the direct problem in the inverse analysis, instead of the complete model that accurately represents the physics of the problem, can be formally treated within the Bayesian framework by modeling the approximation errors as Gaussian random variables and modifying the likelihood, such as in the Approximation Error Model (AEM) [17,127–131]. Sampling techniques like the delayed acceptance Metropolis-Hastings (DAMH) algorithm [126] have also been developed in order to expedite the application of MCMC methods with the use of reduced models. The use of the AEM and the DAMH algorithm to speed-up the inverse problem solution obtained with MCMC is presented in the next section.

Besides the three-dimensional nature of the heat conduction problem in this section, the large computational times for the solution of the present inverse problem result from the number of measurements made available by the infrared camera, which can provide experimental data with high spatial resolution and high frequency. Therefore, instead of applying model reduction and using either the AEM or the DAMH algorithm, a data compression approach was applied in order to reduce the computational work needed for the calculation of the likelihood function.

Data compression was performed by transforming the experimental data (temperatures at each pixel recorded by the infrared camera) with the same integral transform that was used in the forward model, equation (6.4.3a), that is,

$$\tilde{\bar{\psi}}(\beta_i,\gamma_j,\tau) = \int_{X=0}^{A}\int_{Y=0}^{B} \bar{\phi}_i\,\bar{\phi}_j\,\psi(X,Y,\tau)\,dY\,dX \tag{6.4.13}$$

Such as for $\tilde{\bar{\theta}}_1(\beta_i,\gamma_j,Z,\tau)$ and $\tilde{\bar{\theta}}_2(\beta_i,\gamma_j,Z,\tau)$, the transformed measured data, $\tilde{\bar{\psi}}(\beta_i,\gamma_j,\tau)$, were ordered with increasing eigenvalue $\beta_i^2 + \gamma_j^2$. The number of transformed modes was selected as the same used for the solution of Problem (6.4.7a-g) by also taking into account that the inversion of equation (6.4.13) with:

$$\psi(X,Y,\tau) = \sum_{i=0}^{\infty}\sum_{j=0}^{\infty} \bar{\phi}_i\,\bar{\varphi}_j\,\tilde{\psi}(\beta_i,\gamma_j,\tau) \tag{6.4.14}$$

must be represented up to a desired accuracy by the truncated series.

Since the transformation given by equation (6.4.13) is linear, and using the fact that the standard deviation of the measurements is constant, the covariances of the transformed measurements $\tilde{\psi}(\beta_i,\gamma_j,\tau)$ are then given by:

$$\tilde{\sigma}^2(\beta_i,\gamma_j,\tau) = \sigma^2 \int_{X=0}^{A}\int_{Y=0}^{B} \bar{\phi}_i\,\bar{\varphi}_j\, dY\, dX \tag{6.4.15}$$

With the transformed measurements obtained with equation (6.4.13), $\tilde{\mathbf{\Psi}}$, and with the transformed estimated temperatures obtained with the solution of the direct problem given by equation (6.4.7a-g), at $Z = 1$ and each time that a measurement is available, $\tilde{\mathbf{\Theta}}(\mathbf{P})$, the likelihood function can be rewritten in the transformed domain as:

$$\pi(\tilde{\mathbf{\Psi}}|\mathbf{P}) = (2\pi)^{-\tilde{D}/2}\left|\tilde{\mathbf{W}}\right|^{-1/2}\exp\left\{-\frac{1}{2}[\tilde{\mathbf{\Psi}} - \tilde{\mathbf{\Theta}}(\mathbf{P})]^T\,\tilde{\mathbf{W}}^{-1}[\tilde{\mathbf{\Psi}} - \tilde{\mathbf{\Theta}}(\mathbf{P})]\right\} \tag{6.4.16}$$

where $\tilde{\mathbf{W}}$ is the covariance matrix of $\tilde{\mathbf{\Psi}}$ with elements given by equation (6.4.15).

Vectors $\tilde{\mathbf{\Psi}}$ and $\tilde{\mathbf{\Theta}}(\mathbf{P})$ were computed at a limited number of transformed modes, $\tilde{D}$, much smaller than the actual number of measurements, D, thus resulting in substantial reduction of computational times required for the solution of the inverse problem, as will be apparent below.

RESULTS

For all results presented below, the prior for the dimensionless contact conductance was taken as a uniform improper distribution in the form

$$\pi[Bi_c(X,Y)] = \begin{cases} 1, & Bi_c(X,Y) \geq 0 \\ 0, & Bi_c(X,Y) < 0 \end{cases} \tag{6.4.17}$$

where the non-negativity constraint for the contact conductance was imposed. An upper bound for the prior of $Bi_c(X,Y)$ was not imposed because $Bi_c(X,Y) \to \infty$ in the regions of perfect contact. On the other hand, from the practical point of view, values of $Bi_c(X,Y)$ larger than ten already characterized perfect contact for the cases examined below, because the temperature difference across the interface became negligible, as given by equation (6.4.2e).

Simulated measurements were used for the verification of the inverse problem procedure based on measurements in the transformed domain and the non-informative prior given by equation (6.4.17) [157]. For the cases with actual measurements, the thermophysical properties of polymethyl methacrylate, which was used for manufacturing plates with known contact failures for controlled experiments, were measured. The thermal diffusivity and thermal conductivity were $1.31 \times 10^{-7}\,\mathrm{m}^2/\mathrm{s}$ and $0.22\,\mathrm{W/m\,K}$, respectively [157]. For the estimation procedures presented below, uncertainties in these parameters were taken into account through their priors, which were modeled as Gaussian distributions centered at these measured values, with standard deviations set to $1.31 \times 10^{-8}\,\mathrm{m}^2/\mathrm{s}$ and $0.022\,\mathrm{W/m\,K}$, respectively.

Before addressing the identification of the contact failures, the heat flux imposed on the top surface (equation 6.4.2f) needed to be estimated [157]. Then, for the estimation of the contact conductance $Bi_c(X,Y)$, the prior for $\left[q_1^*, q_2^*, ..., q_M^*\right]$ was given in the form of a Gaussian distribution with mean and covariance matrix obtained from the heat flux estimated in another controlled experiment.

The proposal density was given by the following nonsymmetrical random walk process:

$$\mathbf{P}^* = \mathbf{P}^{(t-1)} + \omega(\delta\mathbf{1} + \mathbf{N}) \qquad (6.4.18)$$

where $\mathbf{1}$ is a vector with all elements equal to one and $\mathbf{N}$ is a vector with elements taken from a Gaussian distribution with zero mean and unitary standard deviation. This nonsymmetrical proposal density was used in order to avoid a large rejection of candidates $\mathbf{P}^*$, especially in the regions of contact failure where the expected value of $Bi_c(X,Y)$ is zero, since this function needs to satisfy the non-negativity constraint given by equation (6.4.17). For the results presented below, which were obtained with this proposal density, the acceptance rate of candidate states was around 27%.

The first mode of the transformed measurements is presented in Figure 6.17a, while other selected modes are presented in Figure 6.17b, for a plate with a circular contact failure. These figures show that the magnitude of the first mode was significantly larger than of the other modes and that the 50th mode was practically zero. As a verification of the transformation process, the measured temperatures were then recovered by using 50 transformed modes through the application of the inverse formula given by equation (6.4.14). The actual measurements and the temperatures recovered by using 50 modes were in excellent agreement. In fact, the temperatures recovered with the integral transformation/inversion were smoother than the actual measurements; hence, the measurement errors of high frequency in space and time were filtered out with the integral transformation and inversion based on the most important modes.

The estimation of the contact conductance $Bi_c(X,Y)$ in a plate with a circular failure, obtained with 50 modes, is presented in Figures 6.18a and b. The circular region with the contact failure is represented in Figure 6.18a. While Figure 6.18a presents the means of the samples of the Markov chains after the burn-in period, Figure 6.18b presents the standard deviations of these samples, for each pixel. Despite the non-informative prior and the small number of pixels used to recover this function, Figures 6.18a and b show that the position of the contact failure could be accurately detected with small uncertainties. Note that the contact conductance was approximately zero within the region of the contact failure and large in the region of perfect contact (see Figure 6.18a), while the standard deviations were at least two

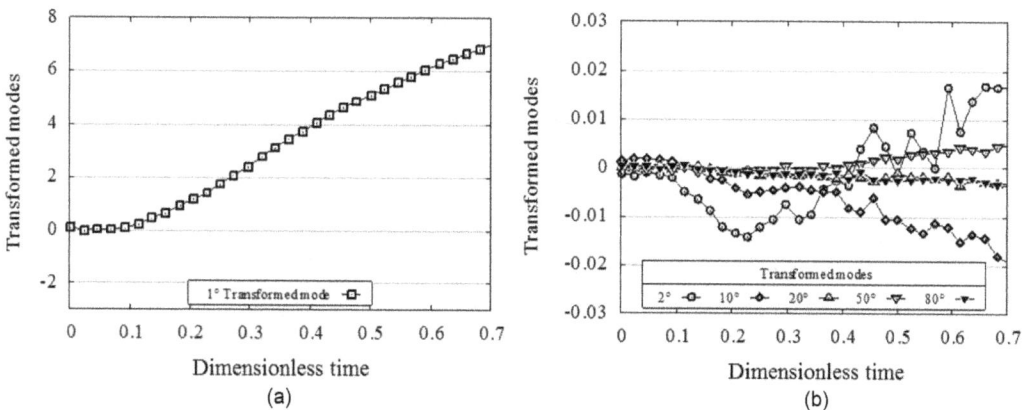

FIGURE 6.17 Transformed dimensionless measured temperature modes for a plate with circular failure: (a) first mode, (b) other selected modes. (From Ref. [157].)

order of magnitudes smaller that the estimated means of the contact conductance (see Figure 6.18b). Hence, despite the fact that the measurements were compressed through the integral transformation in the spatial domain, the transformed modes were capable of retaining the information about the spatial variation of $Bi_c(X,Y)$. Furthermore, the estimated contact conductance was smooth and did not exhibit oscillations, even at the edge of the contact failure. Such a result revealed that the number of modes used for the likelihood function in the transformed domain (equation 6.4.16) served as regularization of the inverse ill-posed problem by filtering the high frequencies (noise) that would be amplified if more modes would be accounted for in the inverse analysis. Indeed, the solution of the inverse problem obtained with 200 modes was completely unstable, as shown by Figure 6.18c. It should be noted that the non-informative uniform prior used here did not provide regularization, but only the non-negativity constraint, for the inverse problem solution. On the other hand, if the solution of the inverse problem is considered with the likelihood given by equation (6.4.12), Gaussian or Total Variation priors are required for appropriate regularization as shown in [170,171].

The Markov chains at a point of contact failure ($x = y = 0.02$ m) and at a point of perfect contact ($x = y = 0.005$ m) obtained with 50 modes in the plate with a circular failure are shown by Figure 6.19a. After starting at a condition of perfect contact, $Bi_c(X,Y) = 15$, the chains reached

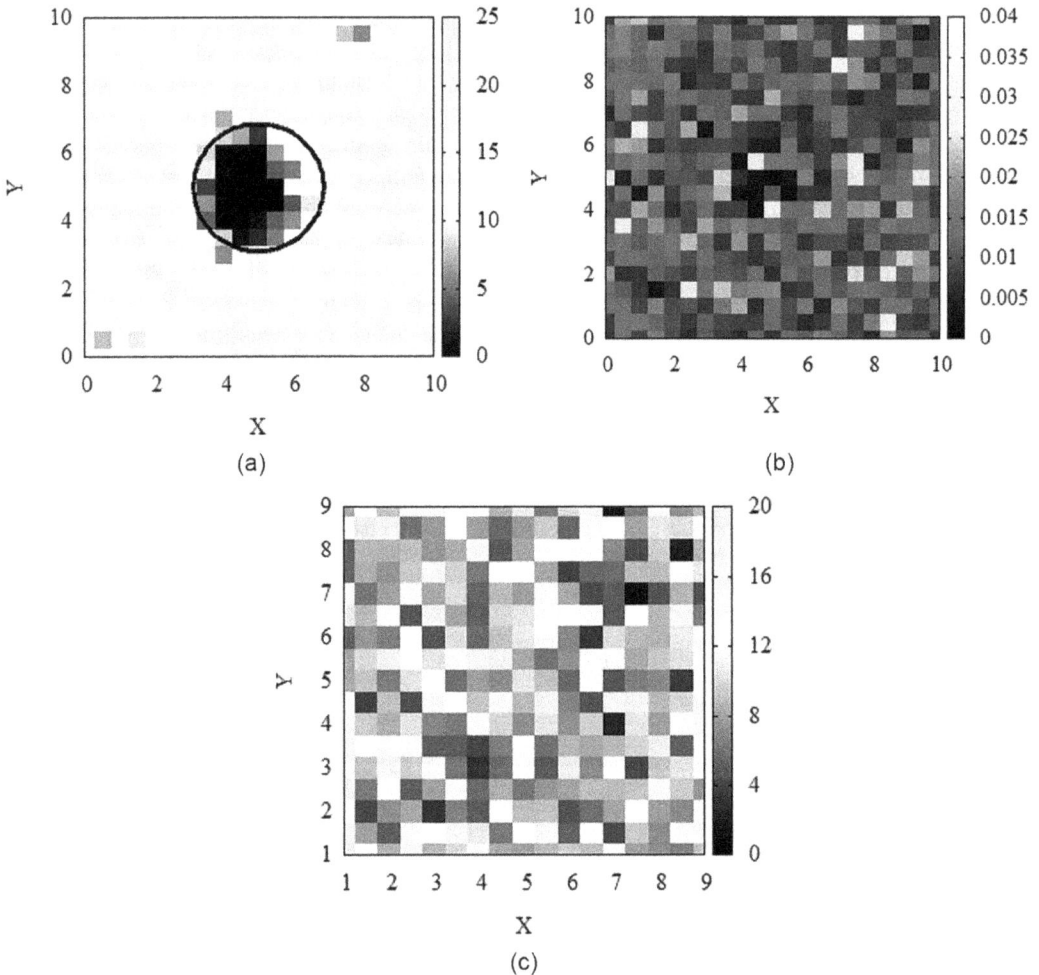

FIGURE 6.18 Dimensionless contact conductance for a plate with circular failure: (a) estimated means obtained with 50 modes, (b) estimated standard deviations obtained with 50 modes, (c) estimated means obtained with 200 modes. (From Ref. [157].)

equilibrium in about 8000 states and converged toward 12 and 0 at the pixels of perfect contact and contact failure, respectively. The histograms of the samples of the Markov chains, between 10,000 and 30,000 states, at the pixels located at $x = y = 0.02$ m and $x = y = 0.005$ m are presented in Figures 6.19b and c, respectively. Notice in Figure 6.19b the skewed histogram at the point of contact failure, caused by the non-negativity constraint for $Bi_c(X,Y)$, represented by the prior given by equation (6.4.17). The histogram at the point of perfect contact presented by Figure 6.19c shows that no samples were generated at regions of small $Bi_c(X,Y)$ at this point of perfect contact. Both histograms exhibit small posterior variances, as expected from the analysis of Figure 6.18b.

The temperature residuals (differences between measurements and temperatures calculated with the estimated parameters) are presented in Figure 6.20. The residuals were small, but correlated, despite the accurate predictions of the contact failures. Such behavior was due to the fact that the inverse problem was solved in the transformed temperature domain with a reduced number of modes.

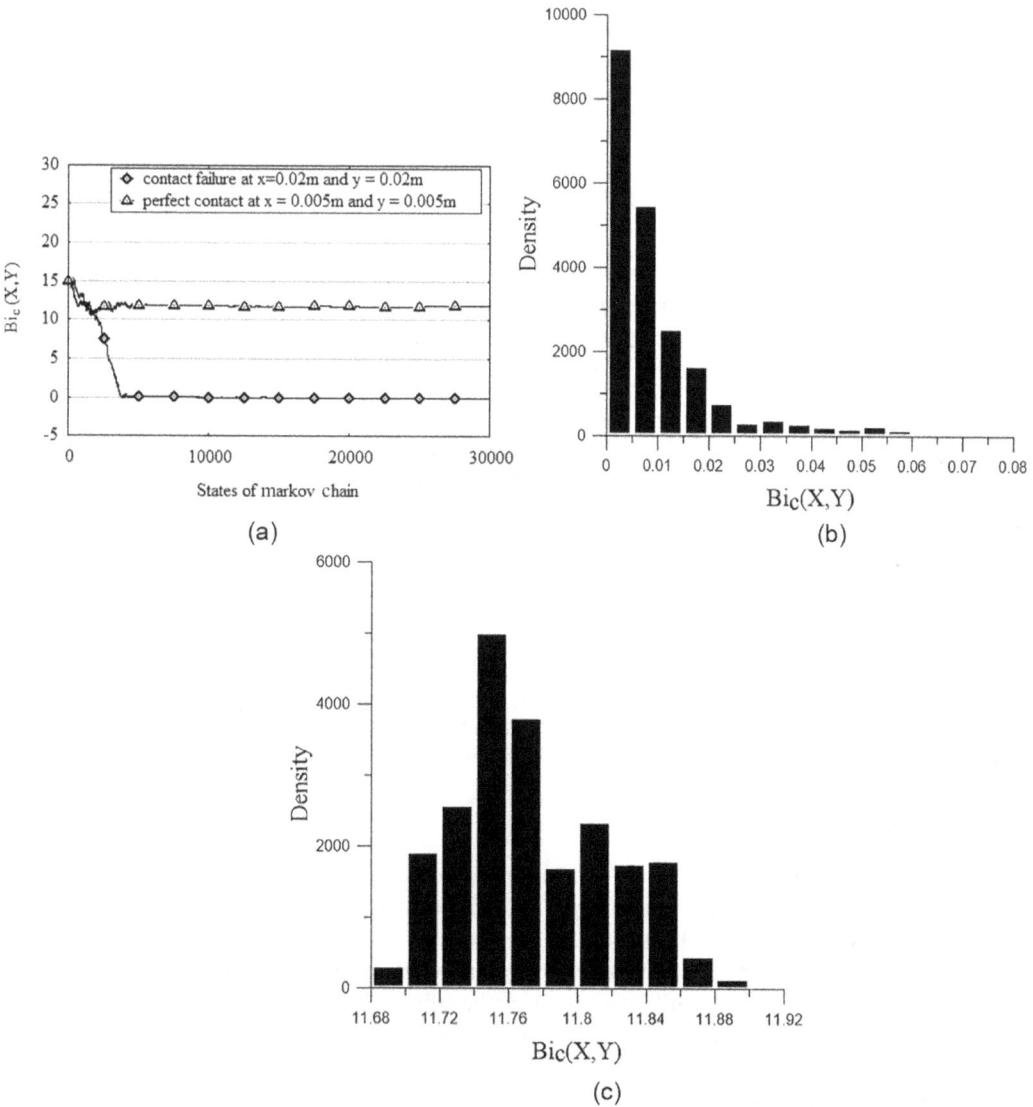

(a)

(b)

(c)

FIGURE 6.19 (a) Markov chains of the estimated dimensionless contact conductance obtained with 50 modes, (b) histogram for a point of contact failure at $x = y = 0.02$ m, (c) histogram for a point of perfect contact at $x = y = 0.005$ m. (From Ref. [157].)

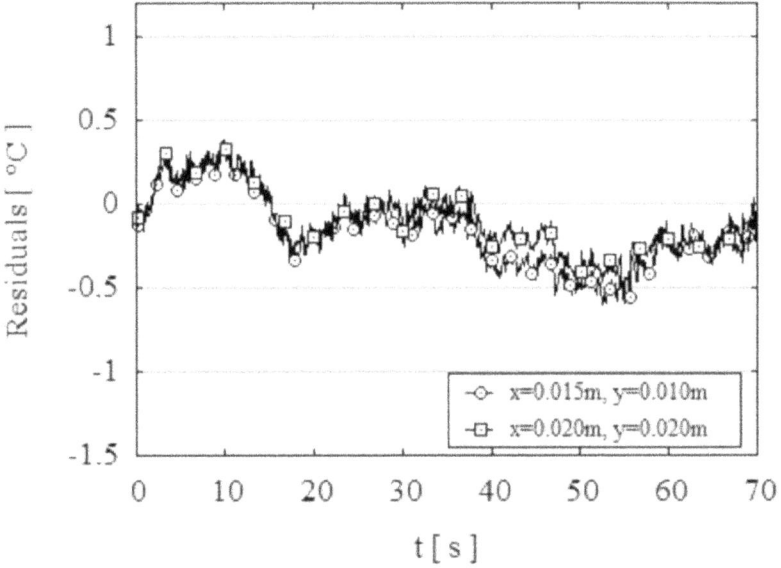

FIGURE 6.20 Temperature residuals at selected points. (From Ref. [157].)

For the experiments with duration of 90 s used in this work, the infrared camera provided 2,916,000 measurements. On the other hand, with the use of the likelihood function in the transformed domain with 50 modes, 40,500 transformed measurements were used in the inverse analysis, thus providing a data compression of 98.6%.

6.5 ACCELERATED BAYESIAN INFERENCE FOR THE ESTIMATION OF SPATIALLY VARYING HEAT FLUX

This section aims at the acceleration of inverse heat transfer problem solutions by using the MCMC method. The physical problem involved a spatially varying heat flux, which can reach very large magnitudes in small regions, such as in the heating imposed by high-power lasers [172]. The solution of the inverse problem was based on a reduced model, which consisted of an improved lumped formulation of a linearized version of the original nonlinear problem. Two different priors were considered for the sought heat flux, including a total variation density and a Gaussian density, which were described in Section 6.1. The DAMH algorithm [126] and the enhanced AEM [17,127–131] were applied with the objective to improve the accuracy of the inverse problem solution obtained with a reduced model.

DIRECT PROBLEM

The physical problem involved three-dimensional transient heat conduction in a plate with temperature-dependent thermal conductivity, $k(T)$, and volumetric heat capacity, $C(T)$, initially at a uniform temperature, T_0. The thickness of the plate, c, was assumed to be much smaller than its width, a, and length, b, so that heat transfer was neglected through the lateral boundaries. The plate bottom surface was assumed insulated, while a non-uniform heat flux $q(x,y)$ was imposed on its top surface.

The mathematical formulation of this problem is given by [172]:

$$C(T_c)\frac{\partial T_c(x,y,z,t)}{\partial t} = \frac{\partial}{\partial x}\left[k(T_c)\frac{\partial T_c}{\partial x}\right] + \frac{\partial}{\partial y}\left[k(T_c)\frac{\partial T_c}{\partial y}\right] + \frac{\partial}{\partial z}\left[k(T_c)\frac{\partial T_c}{\partial z}\right] \text{ in } 0 < x < a,$$

$$0 < y < b, 0 < z < c, \text{for } t > 0$$

(6.5.1a)

$$\frac{\partial T_c}{\partial x} = 0 \text{ at } x = 0 \text{ and } x = a, 0 < y < b, 0 < z < c, \text{ for } t > 0 \qquad (6.5.1\text{b,c})$$

$$\frac{\partial T_c}{\partial y} = 0 \text{ at } y = 0 \text{ and } y = b, 0 < x < a, 0 < z < c, \text{ for } t > 0 \qquad (6.5.1\text{d,e})$$

$$\frac{\partial T_c}{\partial z} = 0 \text{ at } z = 0, 0 < x < a, 0 < y < b, \text{ for } t > 0 \qquad (6.5.1\text{f})$$

$$k(T_c)\frac{\partial T_c}{\partial z} = q(x,y) \text{ at } z = c, 0 < x < a, 0 < y < b, \text{ for } t > 0 \qquad (6.5.1\text{g})$$

$$T_c = T_0 \text{ for } t = 0, \text{ in } 0 < x < a, 0 < y < b, 0 < z < c \qquad (6.5.1\text{h})$$

where the subscript c refers to the complete model, defined here as the one that perfectly reproduces all the phenomena of the physical problem, which was assumed to be exact within the accuracy of the numerical method used for its solution.

The average temperature across the thickness of the plate was defined by:

$$\bar{T_c}(x,y,t) = \frac{1}{c}\int_{z=0}^{c} T_c(x,y,z,t)\,dz \qquad (6.5.2)$$

The reduced model considered here involved partial lumping [168,172–179] along the z direction, since the plate was assumed to be thin, that is, $c \ll a$ and $c \ll b$. In addition, the reduced model was based on a linear approximation of Problem (6.5.1a-h), where the physical properties were evaluated at a temperature T^*. At this temperature, representative values C^* and k^* were obtained for the volumetric heat capacity and thermal conductivity within the plate, respectively. The solution of the linearized version of Problem (6.5.1a-h) is denoted $T(x,y,z,t)$.

By operating on equations (6.5.1a-e) and (6.5.1h) with $\frac{1}{c}\int_{z=0}^{c}(\cdot)\,dz$, making use of the boundary conditions (6.5.1f) and (6.5.1g), at $z = 0$ and $z = c$, respectively, and assuming constant thermophysical properties, the following problem was obtained for the computation of the approximate average temperature $\bar{T}(x,y,t)$:

$$C^*\frac{\partial \bar{T}(x,y,t)}{\partial t} = \frac{\partial}{\partial x}\left[k^*\frac{\partial \bar{T}}{\partial x}\right] + \frac{\partial}{\partial y}\left[k^*\frac{\partial \bar{T}}{\partial y}\right] + \frac{q(x,y)}{c} \text{ in } 0 < x < a, 0 < y < b, \text{ for } t > 0 \qquad (6.5.3\text{a})$$

$$\frac{\partial \bar{T}}{\partial x} = 0 \text{ at } x = 0 \text{ and } x = a, 0 < y < b, \text{ for } t > 0 \qquad (6.5.3\text{b,c})$$

$$\frac{\partial \bar{T}}{\partial y} = 0 \text{ at } y = 0 \text{ and } y = b, 0 < x < a, \text{ for } t > 0 \qquad (6.5.3\text{d,e})$$

$$\bar{T} = T_0 \text{ for } t = 0, \text{ in } 0 < x < a, 0 < y < b \qquad (6.5.3\text{f})$$

where

$$\bar{T}(x,y,t) = \frac{1}{c}\int_{z=0}^{c} T(x,y,z,t)\,dz \qquad (6.5.4)$$

Surface temperature measurements were assumed available for the inverse analysis. Therefore, from the first level of approximation, where the temperature $\bar{T}(x,y,t)$ was obtained as a solution of Problem (6.5.3a-f) and used in place of the actual average temperature $\bar{T}_c(x,y,t)$, several reduced models can be derived. These reduced models differ in terms of how $\bar{T}(x,y,t)$ relates to the surface temperatures at $z = 0$ and $z = c$.

In the *classical lumped formulation* [168,172–179], temperature gradients across the thickness of the plate are fully neglected. Therefore, the surface temperature becomes equal to the average temperature, that is,

$$T(x,y,0,t) = T(x,y,c,t) = \bar{T}(x,y,t) \tag{6.5.5a,b}$$

In the *improved lumped formulation* based on the coupled integral equations approach [168], the temperature gradients across the thickness of the plate are not neglected, but taken into account in an approximate form, which can involve different degrees of accuracy. Here, we use the so-called $H_{1,1}/H_{0,0}$ approximation [168], where the Hermite's formulae $H_{1,1}$ and $H_{0,0}$ are applied to approximate the average temperature $\bar{T}(x,y,t)$ and the integral of the temperature gradient along the z direction, respectively. The $H_{1,1}$ formula (corrected trapezoidal rule), as applied to the definition of the approximate average temperature (6.5.4), is given by [168]:

$$\bar{T}(x,y,t) \approx \frac{1}{2}\left[T(x,y,0,t) + T(x,y,c,t)\right] + \frac{c}{12}\left[\frac{\partial T}{\partial z}\bigg|_{z=0} - \frac{\partial T}{\partial z}\bigg|_{z=c}\right] \tag{6.5.6a}$$

The $H_{0,0}$ formula (trapezoidal rule) is now applied to the integral of the temperature gradient [168], that is:

$$\int_{z=0}^{c} \frac{\partial T(x,y,z,t)}{\partial z}dz = T(x,y,c,t) - T(x,y,0,t) \approx \frac{c}{2}\left[\frac{\partial T}{\partial z}\bigg|_{z=0} + \frac{\partial T}{\partial z}\bigg|_{z=c}\right] \tag{6.5.6b}$$

Equations (6.5.6a,b) are then solved, together with boundary conditions (6.5.1f,g), to yield the following relations between the surface temperatures and the approximate average temperature:

$$T(x,y,0,t) = \bar{T}(x,y,t) - \frac{c}{6k^*}q(x,y) \tag{6.5.7a}$$

$$T(x,y,c,t) = \bar{T}(x,y,t) + \frac{c}{3k^*}q(x,y) \tag{6.5.7b}$$

We note from equations (6.5.5) and (6.5.7) that the computational efforts for the classical and for the improved lumped formulations are practically the same. Differences between these efforts are limited to algebraic operations required to compute $T(x,y,0,t)$ and $T(x,y,c,t)$ with equations (6.5.7a,b), which can actually be done as post-processing to the solution of Problem (6.5.3a-f). On the other hand, much more accurate solutions are indeed obtained with the improved lumped formulation, since the temperature gradients are not neglected as in the classical lumped formulation.

INVERSE PROBLEM

The inverse problem was concerned with the estimation of the boundary heat flux $q(x,y)$, by using transient temperature measurements taken at the bottom surface, $z = 0$. The measurements were assumed to be taken with an infrared camera.

For the inverse analysis, the unknown function $q(x,y)$ was spatially discretized at the plate surface. The discretized flux was considered uniform over the cells with center points (x_i, y_j), where $x_i = i\Delta x$, $y_j = j\Delta y$, $i = 1, ..., I, j = 1, ..., J$, and with grid spacing given by $\Delta x = a/I$ and $\Delta y = b/J$. Therefore, the function was estimated in terms of its local values, which were suitably arranged in the vector of unknown parameters **P**. Simulated transient temperature measurements were also assumed to be available over this same grid. The measurement errors were additive, Gaussian, with zero means and known covariance matrix **W**, so that the likelihood function was given by equation (3.1.6).

Two different prior densities were examined for the spatially distributed heat flux: the total variation non-informative prior given by equation (6.1.9), and a *Gaussian* prior (equation 6.1.1) based on a model of very low accuracy that still took into account the physics of the problem (see Ref. [172] for more details). Since the reduced model, given by the improved lumped formulation of equations (6.5.7a,b), does not exactly reproduce the solution of the complete model given by equations (6.5.1a-h), the DAMH algorithm and the enhanced AEM (see Section 4.7) were applied. The DAMH algorithm was used with the TV prior, while the AEM approach was applied with the Gaussian prior. The AEM approach cannot be applied with an improper prior like TV, because the samples of the approximation error are generated by sampling from the prior distribution.

RESULTS

The plate was assumed to be made of stainless steel with dimensions $a = 0.12\,\mathrm{m}$, $b = 0.12\,\mathrm{m}$, $c = 0.003\,\mathrm{m}$. The plate was assumed to be initially at the uniform temperature $T_0 = 300$ K and the components $q(x_i, y_j)$ of the discretized heat flux were sought over a grid with $\Delta x = \Delta y = 0.005\,\mathrm{m}$, that is, $I = J = 24$ [172]. The constant thermophysical properties used in the reduced model were obtained at $T^* = 600$ K. In general, the direct problem solution with the complete model took around 7.2 s, while the solution with the reduced model in the form of the improved lumped approach given by equations (6.5.7a,b) took around 0.09 s of computational time. This clearly demonstrates the necessity of a reduced model for the solution of the inverse problem with the MCMC method in the present case. In fact, if we consider 10^5 states of the Markov chain, the solution of the direct problem alone with the complete model would require at least 8 days of computation. On the other hand, with the reduced model the required computational time decreased to around 2.5 h.

Three different heat flux functions were examined, involving sharp discontinuities and heat fluxes of large magnitudes; they were selected in order to challenge the inverse problem solution procedure with each prior distribution, as well as with either the DAMH algorithm or the AEM approach. These functions are denoted as *Flux A* (with two small heating spots), *Flux B* (with one single small heating spot) and *Flux C* (with one single large heating spot). They are given by [172]:

$$\text{Flux } A: \quad q(x_i, y_j) = \begin{cases} 10^7\,\mathrm{W/m^2}, & \text{for } 8 \le i \le 10 \quad \text{and} \quad 8 \le j \le 10 \\ 10^7\,\mathrm{W/m^2}, & \text{for } 18 \le i \le 20 \quad \text{and} \quad 18 \le j \le 20 \\ 0 & \text{elsewhere} \end{cases} \quad (6.5.8a)$$

$$\text{Flux } B: q(x_i, y_j) = \begin{cases} 10^7\,\mathrm{W/m^2}, & \text{for } 8 \le i \le 10 \quad \text{and} \quad 8 \le j \le 10 \\ 0, & \text{elsewhere} \end{cases} \quad (6.5.8b)$$

$$\text{Flux } C: q(x_i, y_j) = \begin{cases} 10^7\,\mathrm{W/m^2} & \text{for } 8 \le i \le 15 \quad \text{and} \quad 8 \le j \le 15 \\ 0 & \text{elsewhere} \end{cases} \quad (6.5.8c)$$

where $1 \le i \le I = 24$ and $1 \le j \le J = 24$.

For the solution of the inverse problem, the simulated temperature measurements were given at $z = 0$, also over the grid with $\Delta x = \Delta y = 0.005\,\text{m}$, at every $\Delta t = 0.01\,\text{s}$. The simulated measurements were generated from the solution of the complete model given by equations (6.5.1a-h). Uncorrelated errors with a Gaussian distribution, zero mean and two different levels of a constant standard deviation were then added to the solution of the complete model to simulate actual measurements. The two standard deviations of the simulated measurements were 0.02 and 1.25 K. The lower value of the standard deviation is typical of accurate infrared cameras used in laboratory measurements. The larger value of the standard deviation was selected to challenge the techniques for field applications where low-accuracy infrared cameras are commonly used.

A comparison of the estimated means and the exact applied heat fluxes is presented in Figures 6.21 and 6.22, for measurements with standard deviations of $\sigma = 0.02$ K and $\sigma = 1.25$ K, respectively. The results presented in Figure 6.21 were obtained with the AEM approach, while the results presented in Figure 6.22 were obtained with the DAMH algorithm. Such figures illustrate the accuracy of these two techniques as applied to the solution of the present inverse problem. For the results obtained with measurements of small standard deviation, we note that both the locations and the magnitudes of the applied heat fluxes were extremely well estimated, as shown by Figure 6.21. An analysis of the results presented in Figure 6.22, which were obtained with measurements containing errors of standard deviation $\sigma = 1.25$ K by using the DAMH algorithm, shows that this technique was capable of accurately estimating the regions of the applied heat fluxes. Although oscillations can be observed within the regions where the heating was imposed, the estimated heat fluxes were stable and practically null, as expected, in regions where heat transfer was neglected (outside the heating spots).

PROBLEMS

6.1 Consider the following heat conduction problem in dimensionless form:

$$\frac{\partial T}{\partial t} = \frac{\partial^2 T}{\partial x^2} \text{ in } 0 < x < 1 \text{ for } t > 0$$

$$\frac{\partial T}{\partial x} = 0 \text{ at } x = 0 \text{ for } t > 0$$

$$\frac{\partial T}{\partial x} = q(t) \text{ at } x = 1 \text{ for } t > 0$$

$$T = 0 \text{ for } t = 0 \text{ in } 0 < x < 1$$

Solve the inverse problem of estimating the boundary heat flux $q(t)$ by using the Metropolis-Hastings algorithm. Consider a parameterization of the continuous heat flux function, $q(t)$, in terms of N heat flux components $q(t_i) \equiv P_i$, supposed constant in $t_i - \Delta t/2 < t < t_i + \Delta t/2$, where Δt is the time interval between two consecutive measurements. Use for the inverse analysis 100 equally spaced transient measurements in $0 < t \leq 1$ of a sensor located at $x_{\text{meas}} = 0$. In order to generate the simulated measurements, utilize the following functional forms:

$$q(t) = 1 + t$$

$$q(t) = 1 + t + t^2$$

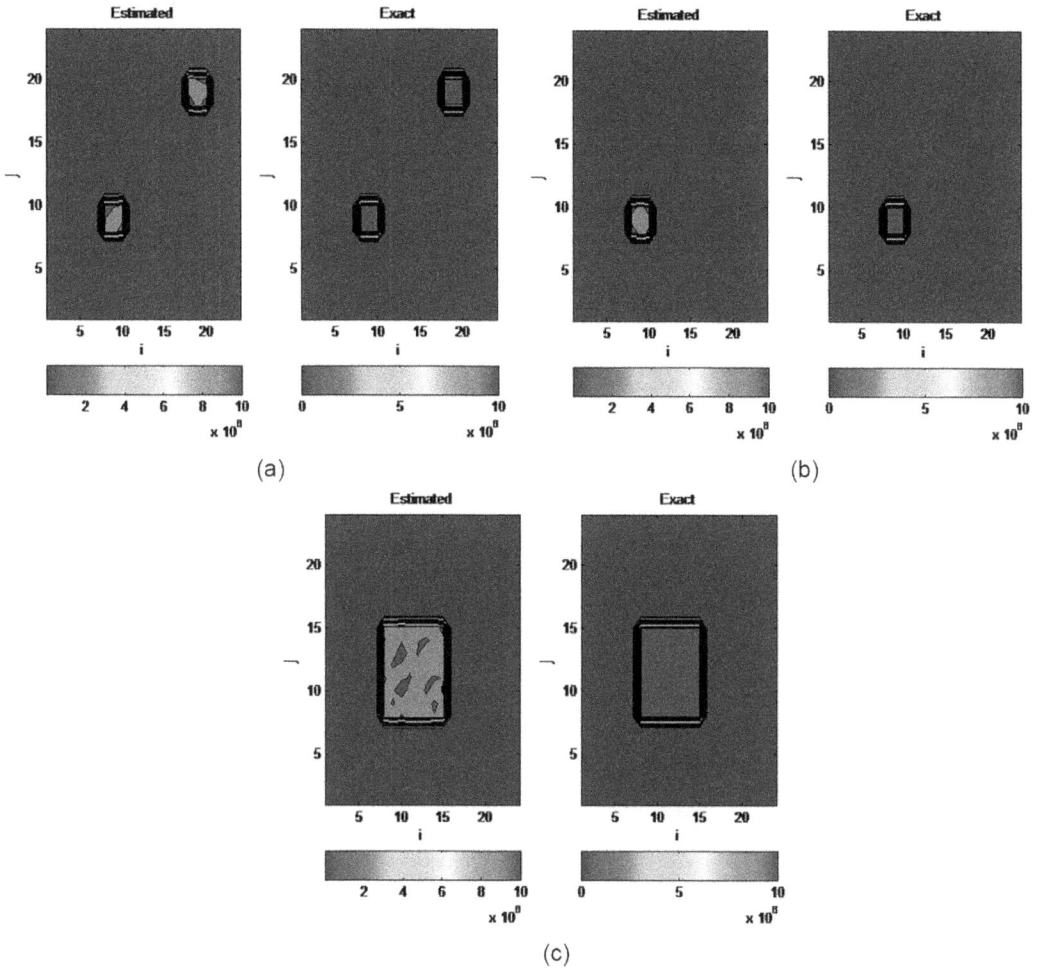

FIGURE 6.21 Estimated and exact heat fluxes—AEM approach and $\sigma = 0.02$ K. (a) Flux A; (b) Flux B; (c) Flux C. (From Ref. [172].)

$$q(t) = \begin{cases} 1, & t \leq 0.3 \quad \text{and} \quad t \geq 0.7 \\ 2, & 0.3 < t < 0.7 \end{cases}$$

$$q(t) = \begin{cases} 1, & t \leq 0.3 \text{ and } t \geq 0.7 \\ 5t - 0.5, & 0.3 < t \leq 0.5 \\ -5t + 4.5, & 0.5 < t < 0.7 \end{cases}$$

Use as prior distribution for the unknown heat flux the Markov random field given by equation (6.1.4) with the matrix **D** given by equation (6.1.5a). Use a Rayleigh distribution for the hyperparameter γ, that is, the posterior distribution is given by equation (6.1.12). Examine the effects of the center of the Rayleigh distribution, γ_0, on the inverse problem solution.

6.2 Repeat Problem 6.1, but now utilize the TV prior given by equation (6.1.9), instead of the Gaussian Markov random field.

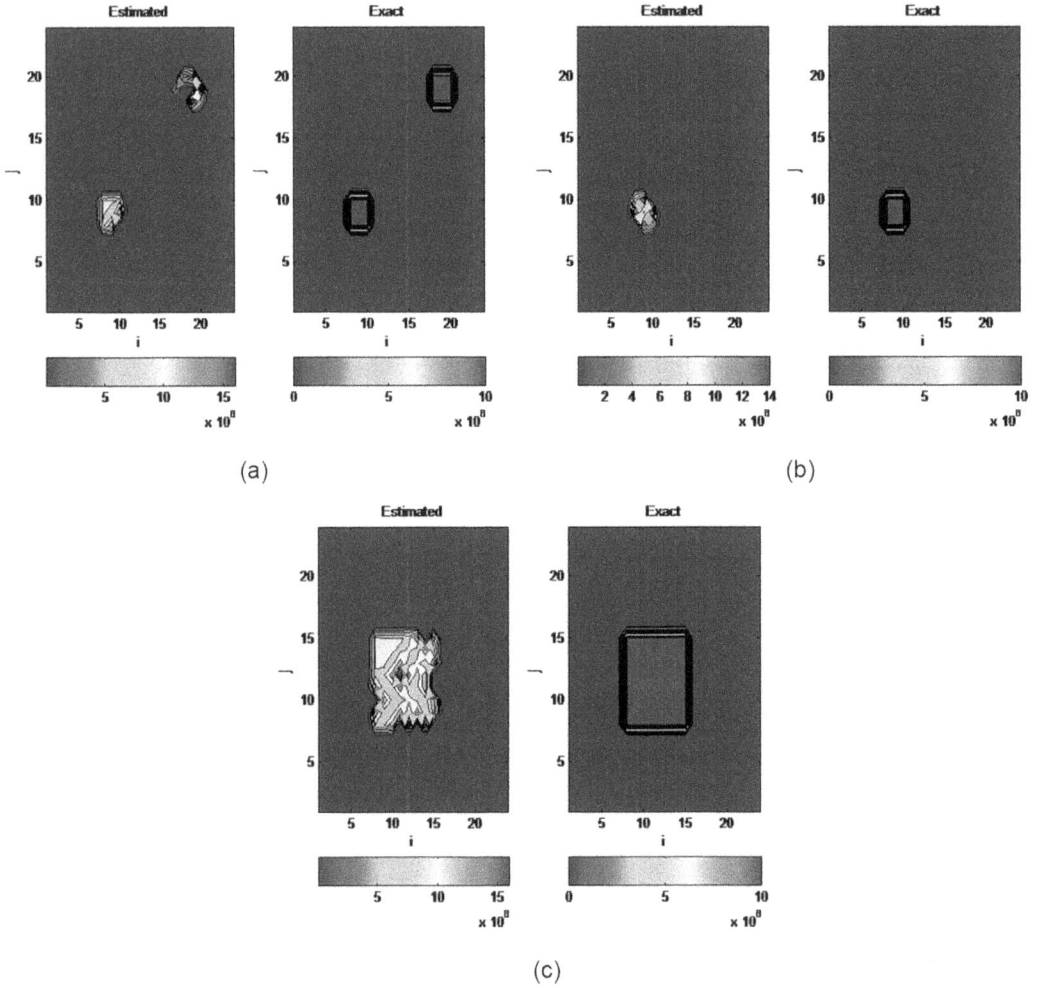

FIGURE 6.22 Estimated and exact heat fluxes—DAMH algorithm and $\sigma = 1.25$ K. (a) Flux A; (b) Flux B; (c) Flux C. (From Ref. [172].)

Part III

State Estimation

7 State Estimation
Kalman Filter

State estimation problems, also designated as nonstationary inverse problems [17], are of great interest in innumerable practical applications. In such kind of problems, the available measured data are used together with prior knowledge about the phenomena of interest and the measurement devices in order to sequentially produce estimates of the desired dynamic variables. This is accomplished in such a manner that the error is statistically minimized [52]. Hence, state estimation problems deal with the combination of model predictions containing uncertainties and measurements that are also intrinsically uncertain, in order to obtain more accurate estimations of the system variables.

State estimation problems are solved with the so-called Bayesian filters [17,52–55]. The most widely known Bayesian filter method is the Kalman filter [17,52–55,180–182]. However, the application of the Kalman filter is limited to linear models with additive Gaussian noises. Extensions of the Kalman filter were developed in the past for less restrictive cases by using linearization and sampling techniques [17,54]. Similarly, Monte Carlo methods have been developed in order to represent the posterior density in terms of random samples and associated weights. Such Monte Carlo methods, usually denoted as particle filters among other designations found in the literature, do not require the restrictive hypotheses of the Kalman filter. Hence, particle filters can be applied to nonlinear models with non-Gaussian errors [17,53–55,180–191].

This third part of the book is devoted to the solution of state estimation problems. The Kalman filter, as well as a modified version of its recursive procedure, is presented in this chapter. Chapter 8 is focused on the particle filter method. Practical applications in heat transfer illustrate the Kalman filter and the particle filter in these two chapters, which are referred herein as Techniques VI and VII, respectively.

7.1 STATE ESTIMATION PROBLEM

State estimation problems are written in terms of two stochastic processes, namely, the *evolution model* and the *observation model* [17,53–55,180–191]. The vector $\mathbf{x}_k$, referred to as the *state vector*, contains all the state variables that describe the system at a given time instant t_k, $k = 1, 2,...$ Moreover, for the solution of the state estimation problem, the vector $\mathbf{Y}_k^{\text{meas}}$ of measurements is considered available.

The *evolution model* is given by the mathematical formulation of the problem under analysis and accounts for all phenomena that influence the dynamic evolution of the state vector $\mathbf{x}_k$. Similarly, the *observation model* is the mathematical formulation of the phenomena involved in the measurement process and relates the measured variables $\mathbf{Y}_k$ to the state vector $\mathbf{x}_k$. The evolution model and the observation model can be, respectively, written in the following general forms [17,53–55,180–191]:

$$\text{Evolution model}: \mathbf{x}_k = \mathbf{f}_k\left(\mathbf{x}_{k-1},\mathbf{P},\mathbf{v}_k\right), \quad k = 1,2,... \tag{7.1.1a}$$

$$\text{Observation model}: \mathbf{Y}_k = \mathbf{g}_k\left(\mathbf{x}_k,\mathbf{P},\mathbf{n}_k\right), \qquad k = 1,2,... \tag{7.1.1b}$$

where $\mathbf{f}_k(.)$ and $\mathbf{g}_k(.)$ are known vector functions, $\mathbf{Y}_k$ is the prediction of the measurements $\mathbf{Y}_k^{\text{meas}}$ obtained with the observation model, $\mathbf{P}$ is a vector containing all the non-dynamic parameters of both models, while $\mathbf{v}_k$ and $\mathbf{n}_k$ represent the noises in the state evolution model and in the observation model, respectively.

The probability density $\pi(\mathbf{x}_0 \mid \mathbf{Y}_0^{\text{meas}}, \mathbf{P}) = \pi(\mathbf{x}_0)$ at the initial state $t = t_0$ is considered known. Thus, the state estimation problem aims at *sequentially* obtaining information about $\mathbf{x}_k$ based on the state evolution model (7.1.1a), on the measurement model (7.1.1b) and on the set of measured data $\mathbf{Y}_{1:k}^{\text{meas}} = \{\mathbf{Y}_i^{\text{meas}}, i = 1, \ldots, k\}$ [17,53–55,180–191].

The evolution and observation models given by equations (7.1.1a,b) include the following assumptions [17,180]: (i) The state vector $\mathbf{x}_k$ follows a Markovian process; (ii) the prediction of the measured data $\mathbf{Y}_k$ follows a Markovian process with respect to the history of $\mathbf{x}_k$ and (iii) the state vector $\mathbf{x}_k$ depends on the past observations only through its own history. Mathematically, these assumptions are, respectively, written as:

$$\pi(\mathbf{x}_k \mid \mathbf{x}_0, \mathbf{x}_1, \ldots, \mathbf{x}_{k-1}, \mathbf{P}) = \pi(\mathbf{x}_k \mid \mathbf{x}_{k-1}, \mathbf{P}) \tag{7.1.2a}$$

$$\pi(\mathbf{Y}_k \mid \mathbf{x}_0, \mathbf{x}_1, \ldots, \mathbf{x}_k, \mathbf{P}) = \pi(\mathbf{Y}_k \mid \mathbf{x}_k, \mathbf{P}) \tag{7.1.2b}$$

$$\pi(\mathbf{x}_k \mid \mathbf{x}_{k-1}, \mathbf{Y}_{1:k-1}, \mathbf{P}) = \pi(\mathbf{x}_k \mid \mathbf{x}_{k-1}, \mathbf{P}) \tag{7.1.2c}$$

In addition to the above hypotheses for the evolution and observation models, it is assumed heretofore that the noise vectors $\mathbf{v}_i$ and $\mathbf{v}_j$, as well as $\mathbf{n}_i$ and $\mathbf{n}_j$, are independent for $i \neq j$ and also independent of the initial state $\mathbf{x}_0$. The vectors $\mathbf{v}_i$ and $\mathbf{n}_j$ are also mutually independent for all i and j [17].

Different kinds of state estimation problems can be considered [17]:

  i. The *prediction problem*, concerned with the determination of $\pi(\mathbf{x}_k \mid \mathbf{Y}_{1:k-1}^{\text{meas}}, \mathbf{P})$;
 ii. The *filtering problem*, concerned with the determination of $\pi(\mathbf{x}_k \mid \mathbf{Y}_{1:k}^{\text{meas}}, \mathbf{P})$;
iii. The *fixed-lag smoothing problem*, concerned with the determination of $\pi(\mathbf{x}_k \mid \mathbf{Y}_{1:k+p}^{\text{meas}}, \mathbf{P})$, where $p \geq 1$ is the fixed lag;
 iv. The *whole-domain smoothing problem*, concerned with the determination of $\pi(\mathbf{x}_k \mid \mathbf{Y}_{1:I}^{\text{meas}}, \mathbf{P})$, where $\mathbf{Y}_{1:I}^{\text{meas}} = \{\mathbf{Y}_i^{\text{meas}}, i = 1, \ldots, I\}$ is the complete set of measurements.

Only the filtering problem is considered in Part III of this book, where the posterior probability density $\pi(\mathbf{x}_k \mid \mathbf{Y}_{1:k}^{\text{meas}}, \mathbf{P})$ is sequentially obtained with Bayesian filters in two steps [17,53–55,180–191]: *prediction and update*. From the distribution at the initial time, $\pi(\mathbf{x}_0 \mid \mathbf{Y}_0^{\text{meas}}, \mathbf{P}) = \pi(\mathbf{x}_0)$, a prior information for the state variables at the subsequent time, t_1, is *predicted* by using the state evolution model given by equation (7.1.1a). Then, the measurements $\mathbf{Y}_1^{\text{meas}}$ are used to *update* this prior, together with the observation model given by equation (7.1.1b), in the likelihood function (the statistical model for the measurement errors), so that the posterior $\pi(\mathbf{x}_1 \mid \mathbf{Y}_1^{\text{meas}}, \mathbf{P})$ is obtained. This process is then repeated for the estimation of state variables at future times as the measured data is sequentially assimilated.

7.2   TECHNIQUE VI: THE KALMAN FILTER

The application of the Kalman filter requires that the evolution and observation models given by equations (7.1.1a,b) be linear and with additive Gaussian noises. It is also assumed that the noises in such models have zero means and known covariance matrices $\mathbf{Q}$ and $\mathbf{R}$, respectively. In addition, it is assumed that the distribution of the state variables at the initial time, $\pi(\mathbf{x}_0 \mid \mathbf{Y}_0^{\text{meas}}, \mathbf{P}) = \pi(\mathbf{x}_0)$, is Gaussian. Therefore, the posterior density $\pi(\mathbf{x}_k \mid \mathbf{Y}_{1:k}^{\text{meas}}, \mathbf{P})$ at t_k, $k = 1, 2, \ldots$, is Gaussian and the Kalman filter results in the *optimal solution* to the state estimation problem [17,52,54,180–182].

With the above hypotheses, the evolution and observation models can be written, respectively, as:

$$\mathbf{x}_k = \mathbf{F}_k \mathbf{x}_{k-1} + \mathbf{s}_k + \mathbf{v}_k \tag{7.2.1a}$$

$$Y_k = H_k x_k + n_k \qquad (7.2.1b)$$

where $\mathbf{F}$ and $\mathbf{H}$ are known matrices that can vary in time but do not depend on $\mathbf{x}$. While $\mathbf{F}$ gives the linear evolution of the state vector $\mathbf{x}$, $\mathbf{H}$ provides the linear relation between the state vector and the measurements in the observation model. In equation (7.2.1a), $\mathbf{s}$ is a known vector of inputs for the evolution model.

The prediction and update steps of the Kalman filter are given by [17,52,54,180–182]:

Prediction:

$$\mathbf{x}_k^- = \mathbf{F}_k \hat{\mathbf{x}}_{k-1} + \mathbf{s}_k \qquad (7.2.2a)$$

$$\mathbf{V}_k^- = \mathbf{F}_k \mathbf{V}_{k-1} \mathbf{F}_k^T + \mathbf{Q}_k \qquad (7.2.2b)$$

Update:

$$\mathbf{K}_k = \mathbf{V}_k^- \mathbf{H}_k^T \left(\mathbf{H}_k \mathbf{V}_k^- \mathbf{H}_k^T + \mathbf{R}_k \right)^{-1} \qquad (7.2.3a)$$

$$\hat{\mathbf{x}}_k = \mathbf{x}_k^- + \mathbf{K}_k (\mathbf{Y}_k^{\text{meas}} - \mathbf{H}_k \mathbf{x}_k^-) \qquad (7.2.3b)$$

$$\mathbf{V}_k = (\mathbf{I} - \mathbf{K}_k \mathbf{H}_k) \mathbf{V}_k^- \qquad (7.2.3c)$$

The matrix $\mathbf{K}_k$ given by equation (7.2.3a) is called Kalman's gain matrix. In equation (7.2.3b), $\mathbf{K}_k$ multiplies the difference between the measurements, $\mathbf{Y}_k^{\text{meas}}$, and the output of the observation model, $\mathbf{H}_k \mathbf{x}_k^-$, to update the state variables predicted by equation (7.2.2a). Notice above that, after predicting the state variable $\mathbf{x}$ and its covariance matrix $\mathbf{V}$ with equations (7.2.2a,b), *a posteriori* estimates for such quantities are obtained in the update step with equations (7.2.3b,c). The superscripts " – " and "^" indicate predicted quantities and the estimated state vector, respectively.

Example 7.1

Consider transient linear heat conduction in a medium with axial symmetry in cylindrical coordinates given by the following dimensionless formulation:

$$\frac{\partial \theta(R,\tau)}{\partial \tau} = \frac{\partial^2 \theta(R,\tau)}{\partial R^2} + \frac{1}{R} \frac{\partial \theta(R,\tau)}{\partial R} \qquad 0 \le R < 1, \tau > 0 \qquad (7.2.4a)$$

$$\frac{\partial \theta(R,\tau)}{\partial R} + Bi\,\theta(R,\tau) = 0 \qquad R = 1, \tau > 0 \qquad (7.2.4b)$$

$$\theta(R,0) = 1 \qquad 0 \le R < 1, \tau = 0 \qquad (7.2.4c)$$

Estimate the temperature field in the region $0 \le R < 1$ from limited temperature data available at its surface by solving a state estimation problem with the Kalman filter.

Solution: The heat conduction Problem (7.2.4) was discretized by explicit finite differences [86,192,193] by using N nodes in the region $0 \le R < 1$. The resulting system of algebraic equations was thus written in the form given by equation (7.2.1a), with the state vector $\mathbf{x}$ containing the dimensionless temperatures at each of the N nodes in the region, $\mathbf{s} = 0$ and

$$
\mathbf{F} = \begin{bmatrix}
(1-4B) & 4B & & & & \\
\left(B - \dfrac{B}{4}\right) & (1-2B) & \left(B + \dfrac{B}{4}\right) & & & \\
& \ddots & \ddots & \ddots & & \\
& \ddots & \ddots & \ddots & \ddots & \\
& & \left(B - \dfrac{B}{2(N-1)}\right) & (1-2B) & \left(B + \dfrac{B}{2(N-1)}\right) \\
& & & 2B & C
\end{bmatrix} \tag{7.2.5a}
$$

where

$$
B = \frac{\Delta \tau}{(\Delta R)^2} \qquad C = \left(1 - 2B - 2B\,Bi\,\Delta R + \frac{B\,Bi\,\Delta R}{N}\right) \tag{7.2.5b,c}
$$

$\Delta \tau$ and ΔR are the time step and spatial grid spacing, respectively.

The matrix $\mathbf{H}$ in this case is a mapping that selects the temperature at the measurement location from the vector $\mathbf{x}$, since the state variable is itself measured at the surface of the cylindrical region.

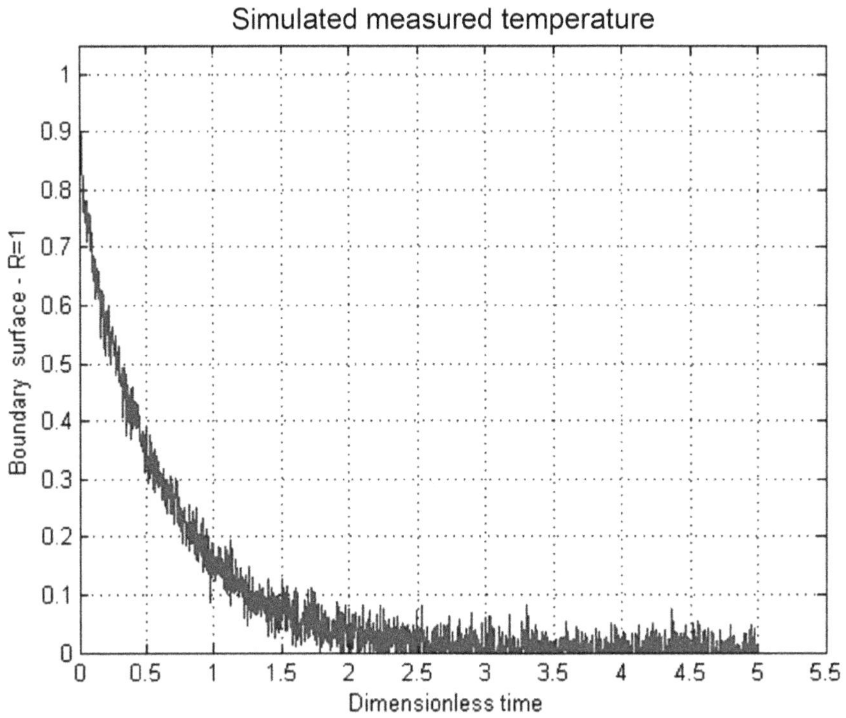

FIGURE 7.1 Simulated measured temperatures. (From Ref. [192,193].)

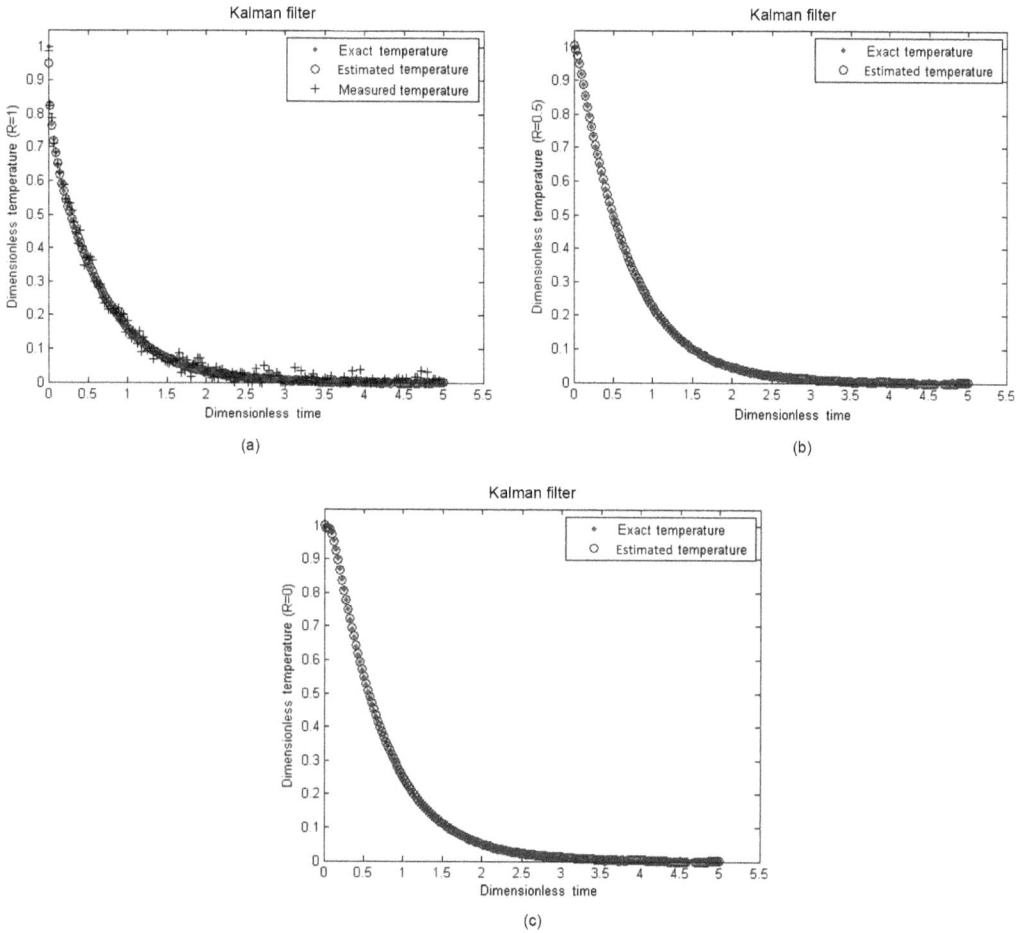

FIGURE 7.2   Temperatures at (a) $R = 1$, (b) $R = 0.5$ and (c) $R = 0$—standard deviation of the evolution model errors of 0.5°C. (From Ref. [192,193].)

Figure 7.1 presents the simulated measured data. The transient measurements were simulated at the boundary surface considering additive Gaussian noise with constant standard deviation of 3°C. Figure 7.2 presents the temperatures estimated at $R = 1$, $R = 0.5$ and $R = 0$ with standard deviation of the evolution model errors of 0.5°C. Note that the estimated temperatures are in very good agreement with the exact ones. However, for the situation in which the standard deviation of the evolution model errors was 5°C, the estimated temperatures tended to follow the measurements more closely. This fact can be observed in Figure 7.3a. On the other hand, since measurements inside the domain were not available, the estimations were dominated by the evolution model in this region. Figures 7.3b and c show the behavior of the estimated temperatures at $R = 0.5$ and $R = 0$, respectively.

7.3   ESTIMATION OF A TRANSIENT BOUNDARY HEAT FLUX THAT VARIES OVER THE SURFACE

The same inverse problem considered in Section 6.5 is examined here, but with an increased difficulty level, since now the unknown heat flux is supposed to vary in time, as well as spatially over the heated surface [194]. The inverse problem is solved in this section using Technique VI with

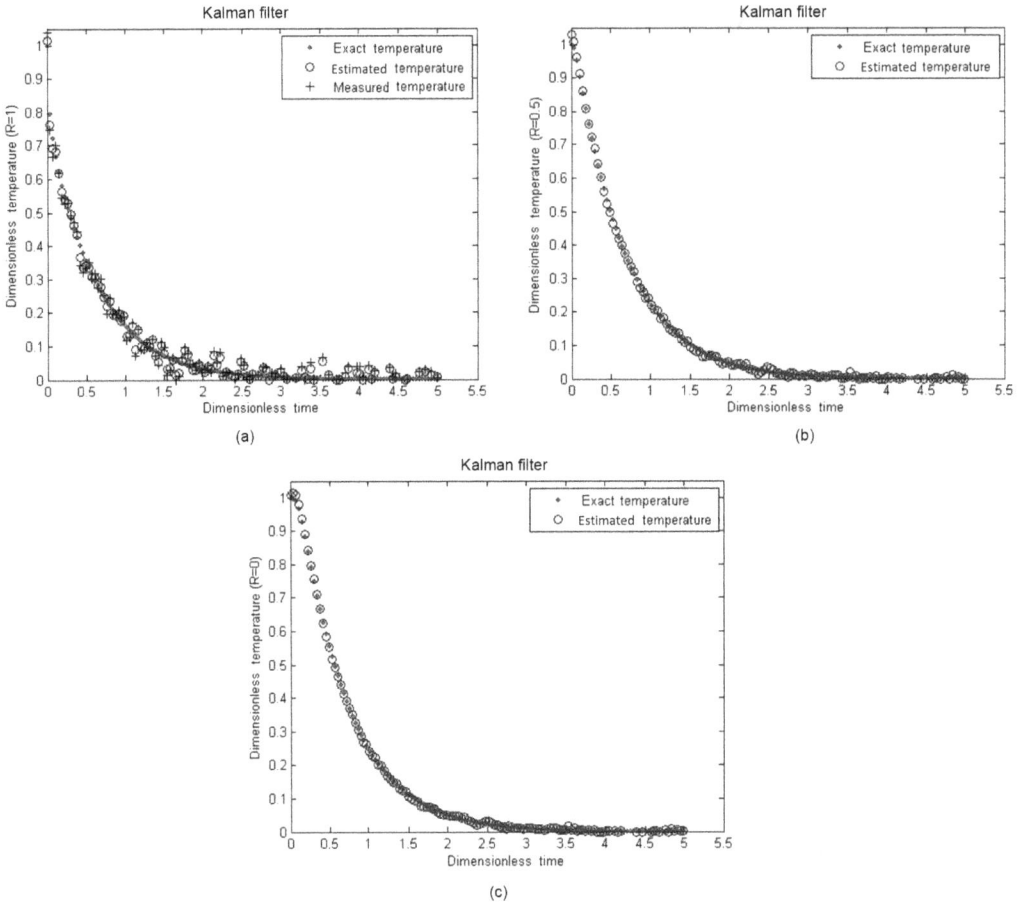

FIGURE 7.3   Temperatures at (a) $R = 1$, (b) $R = 0.5$ and (c) $R = 0$—standard deviation of the evolution model errors of 5°C. (From Ref. [192,193].)

the reduced models described in Section 6.5. The approximation error model approach is used to cope with the modeling errors.

The state vector $\mathbf{x}_k$ is presented in equation (7.3.1), where $\overline{\mathbf{T}}_k$ contains the values of the mean temperatures in the z direction at the center of each control volume of the numerical grid along the x and y directions, while $\mathbf{q}_k$ contains the heat flux values at the same locations, at each time t_k. The observations $\mathbf{Y}_k^{\mathrm{meas}}$ used for the estimation of the state vector $\mathbf{x}_k$ are given by the measured temperatures at the surface $z = 0$ at the center of the same control volumes used for $\overline{\mathbf{T}}_k$ and $\mathbf{q}_k$. The total number of unknowns in $\mathbf{x}_k$ is thus $2IJ$, where I is the number of finite volumes in the x direction and J is the number of finite volumes in the y direction.

$$\mathbf{x}_k = \begin{bmatrix} \overline{\mathbf{T}}_k \\ \mathbf{q}_k \end{bmatrix} \tag{7.3.1}$$

The matrix $\mathbf{F}_k$, with size $2IJ \times 2IJ$, is built by joining four smaller matrices of size $IJ \times IJ$, that is [194],

$$\mathbf{F}_k = \begin{bmatrix} \mathbf{A}_k & \mathbf{B}_k \\ \mathbf{0} & \mathbf{I} \end{bmatrix} \tag{7.3.2}$$

where $\mathbf{A}_k$ and $\mathbf{B}_k$ are matrices that result from the discretization of the reduced model; $\mathbf{A}_k$ accounts for heat diffusion in the domain, $\mathbf{B}_k$ considers the effect of the heat flux on the temperatures, while $\mathbf{0}$ and $\mathbf{I}$ correspond to a matrix with zero elements and the identity matrix, respectively.

A random walk model is used for the evolution of the heat flux in the form [194]:

$$\mathbf{q}_k = \mathbf{q}_{k-1} + \sigma_q \omega_k \tag{7.3.3}$$

where ω is a Gaussian vector with zero mean and identity covariance matrix, while the standard deviation of the random walk is σ_q.

The matrix $\mathbf{H}_k$ of the observation model is of size $IJ \times 2IJ$ and is also built by combining two matrices of size $IJ \times IJ$:

$$\mathbf{H}_k = \begin{bmatrix} \mathbf{I} & \mathbf{C} \end{bmatrix} \tag{7.3.4}$$

Matrix $\mathbf{C}$ depends if the measurement model is based on the classical or on the improved lumped system analysis (see Section 6.5), that is,

$$\mathbf{C} = \mathbf{0} \quad \text{for the classical lumped analysis} \tag{7.3.5a}$$

$$\mathbf{C} = -\frac{c}{6k^*}\mathbf{I} \quad \text{for the improved lumped analysis} \tag{7.3.5b}$$

For the solution of the state estimation problem, evolution and observation models that are supposed to accurately represent the phenomena involved in the analysis would possibly be considered if their numerical solution could be calculated within a feasible computational time. Here, we write such *complete models* in terms of general nonlinear functions such as in (7.1.1a,b), that is,

$$\mathbf{x}_k = \mathbf{f}^*(\mathbf{x}_{k-1}, \mathbf{P}, \mathbf{v}_k^*) \tag{7.3.6a}$$

$$\mathbf{Y}_k = \mathbf{g}^*(\mathbf{x}_k, \mathbf{P}, \mathbf{n}_k^*) \tag{7.3.6b}$$

On the other hand, if the computational times are too large for the use of the complete models given by equations (7.3.6a,b), an approximate evolution model, $\mathbf{f}(\mathbf{x}_{k-1}, \mathbf{P}, \mathbf{v}_k)$, and an approximate observation model, $\mathbf{g}(\mathbf{x}_k, \mathbf{P}, \mathbf{n}_k)$, could be considered for the inverse analysis. The solution of the state estimation problem with the approximate models is faster, but errors between the complete and approximate models must be taken into account in the analysis.

By adding and subtracting $\mathbf{f}(\mathbf{x}_{k-1}, \mathbf{P}, \mathbf{v}_k)$ to equation (7.3.6a) and $\mathbf{g}(\mathbf{x}_k, \mathbf{P}, \mathbf{n}_k)$ to equation (7.3.6b), we can write:

$$\mathbf{x}_k = \mathbf{f}(\mathbf{x}_{k-1}, \mathbf{P}, \mathbf{v}_k) + [\mathbf{f}^*(\mathbf{x}_{k-1}, \mathbf{P}, \mathbf{v}_k^*) - \mathbf{f}(\mathbf{x}_{k-1}, \mathbf{P}, \mathbf{v}_k)] \tag{7.3.7a}$$

$$\mathbf{Y}_k = \mathbf{g}(\mathbf{x}_k, \mathbf{P}, \mathbf{n}_k) + [\mathbf{g}^*(\mathbf{x}_k, \mathbf{P}, \mathbf{n}_k^*) - \mathbf{g}(\mathbf{x}_k, \mathbf{P}, \mathbf{n}_k)] \tag{7.3.7b}$$

or, alternatively,

$$\mathbf{x}_k = \mathbf{f}(\mathbf{x}_{k-1}, \mathbf{P}, \mathbf{v}_k) + \mathbf{e}_k \tag{7.3.8a}$$

$$\mathbf{Y}_k = \mathbf{g}(\mathbf{x}_k, \mathbf{P}, \mathbf{n}_k) + \boldsymbol{\zeta}_k \tag{7.3.8b}$$

where $\mathbf{e}_k = [\mathbf{f}^*(\mathbf{x}_{k-1}, \mathbf{P}, \mathbf{v}_k^*) - \mathbf{f}(\mathbf{x}_{k-1}, \mathbf{P}, \mathbf{v}_k)]$ and $\boldsymbol{\zeta}_k = [\mathbf{g}^*(\mathbf{x}_k, \mathbf{P}, \mathbf{n}_k^*) - \mathbf{g}(\mathbf{x}_k, \mathbf{P}, \mathbf{n}_k)]$ are the approximation errors for the evolution and observations models, respectively.

In order to develop the *approximation error model* approach [17,127–131], one can assume, for example, $\mathrm{cov}(\mathbf{x}_k, \mathbf{e}_k) = \mathrm{cov}(\mathbf{x}_k, \boldsymbol{\zeta}_k) = 0$, that is, the sought state variables and the approximation errors are independent, which is usually referred to as the *enhanced error model* (see Section 4.7). For the application of the approximation error model approach [17,127–131] to the present linear state estimation problem, the evolution and observations models given by equation (7.3.8a,b) are then rewritten as:

$$\mathbf{x}_k = \mathbf{F}_k \mathbf{x}_{k-1} + \mathbf{s}_k + \mathbf{v}_k + \mathbf{e}_k \tag{7.3.9a}$$

$$\mathbf{Y}_k = \mathbf{H}_k \mathbf{x}_k + \mathbf{n}_k + \boldsymbol{\zeta}_k \tag{7.3.9b}$$

where $\mathbf{F}$ and $\mathbf{H}$ are the matrices or the approximate models.

Then the recursive equations of the Kalman filter become [194]:

Prediction:

$$\mathbf{x}_k^- = \mathbf{F}_k \hat{\mathbf{x}}_{k-1} + \mathbf{s}_k + \mathrm{E}(\mathbf{e}_k) \tag{7.3.10a}$$

$$\mathbf{V}_k^- = \mathbf{F}_k \mathbf{V}_{k-1} \mathbf{F}_k^T + \mathbf{Q}_k + \mathrm{cov}(\mathbf{e}_k) \tag{7.3.10b}$$

Update:

$$\mathbf{K}_k = \mathbf{V}_k^- \mathbf{H}_k^T \left[\mathbf{H}_k \mathbf{V}_k^- \mathbf{H}_k^T + \mathbf{R}_k + \mathrm{cov}(\boldsymbol{\zeta}_k) \right]^{-1} \tag{7.3.11a}$$

$$\hat{\mathbf{x}}_k = \mathbf{x}_k^- + \mathbf{K}_k \left[\mathbf{Y}_k^{\mathrm{meas}} - \mathbf{H}_k \mathbf{x}_k^- - \mathrm{E}(\boldsymbol{\zeta}_k) \right] \tag{7.3.11b}$$

$$\mathbf{V}_k = \left(\mathbf{I} - \mathbf{K}_k \mathbf{H}_k \right) \mathbf{V}_k^- \tag{7.3.11c}$$

Therefore, for the proper application of these recursive equations, the expected values and covariance matrices of the modeling errors must be quantified. Here, the approximation error statistics were calculated simultaneously to the estimation process by sequentially calculating the system's state variables with both the reduced and the complete models [194].

The inverse problem was solved by using simulated measurements obtained from the solution of the complete model with a reference heat flux on a sufficiently fine grid. Thus, the simulated measurements and the estimates from the inverse problem were obtained using different mathematical models and different grid sizes to avoid an inverse crime [17]. The measurement errors were modeled as uncorrelated and with a constant standard deviation, σ_y. Thus, to simulate the noisy data, a Gaussian vector with zero mean and covariance matrix $\sigma_y^2 \mathbf{I}$ was added to the temperatures obtained from the solution of the complete model, where $\mathbf{I}$ stands for the identity matrix. The measurements were simulated by solving the complete model on a grid with $768 \times 768 \times 64$ volumes and a time step of $\Delta t = 10^{-4}$ s. The inverse problem was solved on a 24×24 grid with time step $\Delta t = 10^{-2}$ s. The initial temperature was considered as 300 K. The standard deviation of the simulated measurements was selected as $1\,°C$. The standard deviation of the random walk model for the evolution of the heat flux was 5×10^4 W/m^2, while the standard deviation of the evolution model for temperatures was $1\,°C$.

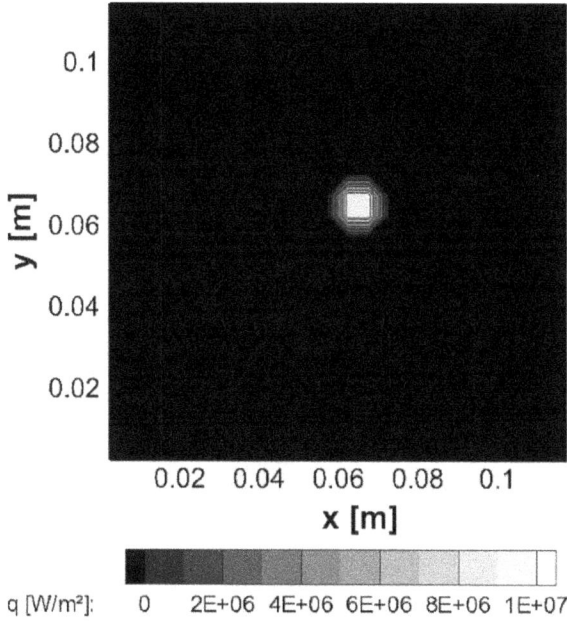

FIGURE 7.4   Exact heat flux on the coarse grid at time $t = 2.0$ s. (From Ref. [194].)

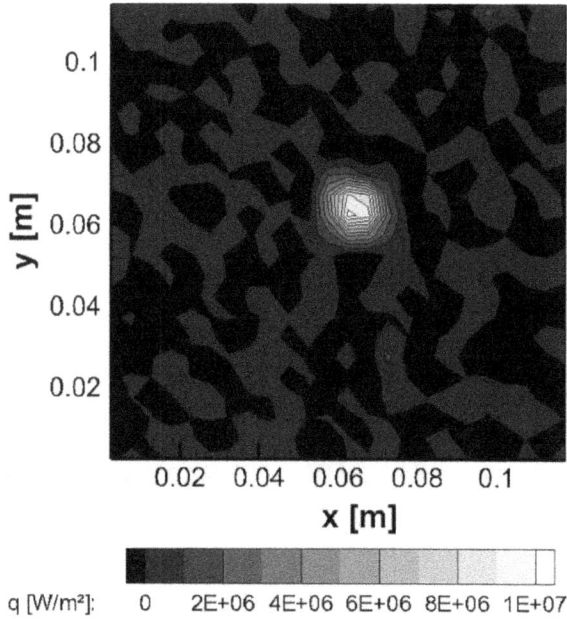

FIGURE 7.5   Estimated heat flux on the coarse grid at time $t = 2.0$ s. (From Ref. [194].)

The heat flux used to generate the simulated measurements at time $t = 2.0$ s is presented for the coarse grid in Figure 7.4, while Figure 7.5 shows the heat flux estimated with the Kalman filter and the approximation error model approach (equations 7.3.10 and 7.3.11) at this same time instant. The improved lumped formulation was used as the reduced model (see Section 6.5). A comparison of the time evolution of the estimated state variables with their exact values is shown for point $(x, y, z) = (6.25, 6.25, 0.00)$ cm

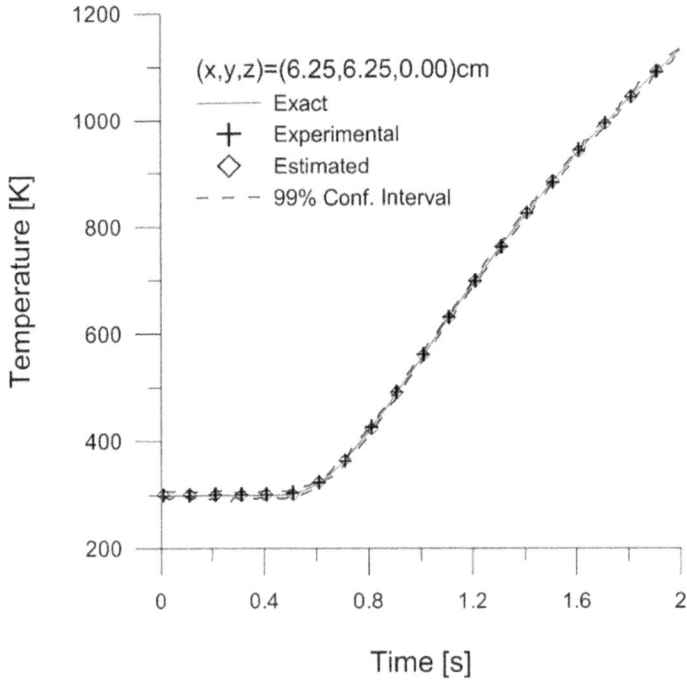

FIGURE 7.6 Evolution in time of the reference and estimated temperatures at $(x, y) = (6.25, 6.25, 0.00)$ cm. (From Ref. [194].)

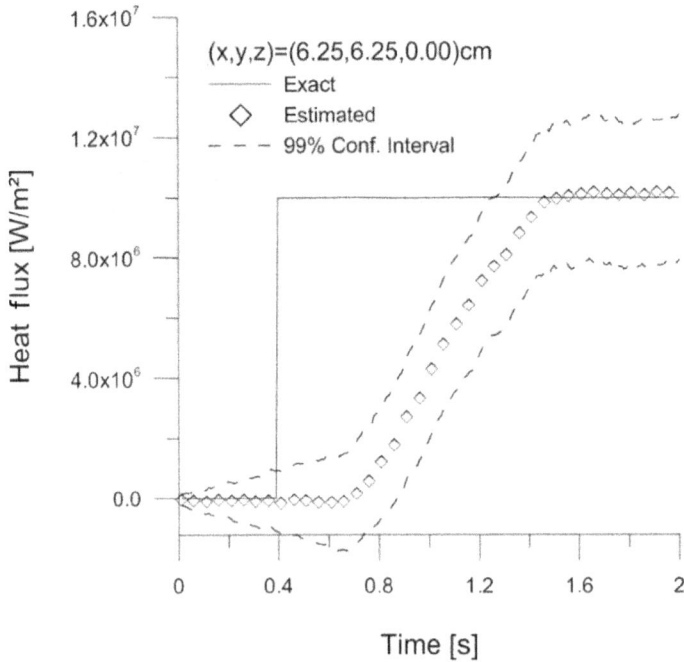

FIGURE 7.7 Evolution in time of the reference and estimated heat flux at $(x, y) = (6.25, 6.25, 0.00)$ cm. (From Ref. [194].)

in Figures 7.6 and 7.7. The results presented in these figures show excellent agreement between exact and estimated temperature and heat flux, respectively. The results presented in Figures 7.5 and 7.7 show that the AEM approach successfully compensated for the approximation errors due to the reduced model, thus resulting in a very good agreement between estimated and exact heat fluxes. The time variation of the estimated heat flux was smooth, but lagged with respect to the exact function. This behavior was due to the standard deviation of the random walk model for the heat flux given by equation (7.3.3). A larger standard deviation would allow faster responses of the Kalman filter, but with larger variances due to the ill-posed character of the inverse problem.

7.4   THE STEADY-STATE KALMAN FILTER

The recursive equations of the Kalman filter require floating point operations of the order D^3 per time step, where D is the size of the observation vector [195]. For linear time-invariant systems, where the evolution and observation matrices are time-invariant, as well as the covariance matrices of the evolution and observation noises are constant, that is,

$$\mathbf{F}_k = \mathbf{F} \quad \mathbf{H}_k = \mathbf{H} \quad \mathbf{Q}_k = \mathbf{Q} \quad \mathbf{R}_k = \mathbf{R} \tag{7.4.1a-d}$$

the Kalman gain matrix, as well as the prior and posterior error covariance matrices, present an asymptotic behavior [195]. Thus,

$$\mathbf{K}_k \approx \mathbf{K}_\infty \quad \text{and} \quad \mathbf{V}_k^- \approx \mathbf{V}_k \approx \mathbf{V}_\infty \tag{7.4.2}$$

In this case, the so-called *steady-state Kalman filter* [195] can be used in order to reduce the computational effort to floating point operations of the order D^2 per time step. Although this version of the Kalman filter is not optimal, it has been successfully applied to inverse problems of practical interest and allows for real-time solutions due to the reduced computational cost [196–198].

The implementation of the steady-state Kalman filter includes the following discrete algebraic Riccati equation, which needs to be solved only once for $\mathbf{V}_\infty$ [195–198]:

$$\mathbf{V}_\infty = \mathbf{F}\mathbf{V}_\infty\mathbf{F}^T - \mathbf{F}\mathbf{V}_\infty\mathbf{H}^T \left[\mathbf{H}\mathbf{V}_\infty\mathbf{H}^T + \mathbf{R}\right]^{-1} \mathbf{H}\mathbf{V}_\infty\mathbf{F}^T + \mathbf{Q} \tag{7.4.3}$$

The solution of equation (7.4.3) is very time consuming. However, it can be performed offline, that is, before the application of the Kalman filter, because it does not depend on the measured data and on the sequentially estimated state variables. After solving (7.4.3) and obtaining $\mathbf{V}_\infty$, the asymptotic Kalman gain matrix can be calculated as follows [195–198]:

$$\mathbf{K}_\infty = \mathbf{V}_\infty\mathbf{H}^T \left[\mathbf{H}\mathbf{V}_\infty\mathbf{H}^T + \mathbf{R}\right]^{-1} \tag{7.4.4}$$

The Kalman filter then reduces to the recursive application of the following single equation [195–198]:

$$\hat{\mathbf{x}}_k = \left[\mathbf{I} - \mathbf{K}_\infty\mathbf{H}\right]\mathbf{F}\hat{\mathbf{x}}_{k-1} + \mathbf{K}_\infty\mathbf{Y}_k^{\text{meas}} \tag{7.4.5}$$

We note that the problem examined in Section 7.3 can be straightforwardly solved with this version of the Kalman filter, because the observation and measurement models are time-invariant. In order to

TABLE 7.1

Parameters of the Heat Flux Used to Generate the Simulated Measurements [196]

Quantity	Value (mm)	Quantity	Value (mm)	Quantity	Value
$x_{1,1}$	30	$y_{1,1}$	30	q_1	5×10^6 W/m^2
$x_{1,2}$	50	$y_{1,2}$	50	t_1	0.4 s
$x_{2,1}$	90	$y_{2,1}$	90	q_2	1×10^7 W/m^2
$x_{2,2}$	100	$y_{2,2}$	100	t_2	0.6 s

illustrate the application of the steady-state Kalman filter to this problem, consider a test case where the simulated measurements were generated with the following equation for the heat flux [196]:

$$q(x,y,t) = \begin{cases} q_1 & \text{in } x_{1,1} \le x \le x_{1,2}, \quad y_{1,1} \le y \le y_{1,2} \text{ and } t \ge t_1 \\ q_2 & \text{in } x_{2,1} \le x \le x_{2,2}, \quad y_{2,1} \le y \le y_{2,2} \text{ and } t \ge t_2 \\ 0 & \text{otherwise} \end{cases} \qquad (7.4.6)$$

where the different parameters of $q(x,y,t)$ are presented in Table 7.1.

The simulated measurements and the heat flux used to generate the simulated measurements are shown in Figures 7.8a and b, respectively, at time $t = 2.0$ s. Figure 7.8b also highlights two control volumes, located inside the heated regions at $x = y = 40$ mm and $x = y = 95$ mm, where the time evolutions of exact and estimated quantities are compared below.

The contour plots of the estimated temperatures and the applied heat flux at $t = 2.0$ s are shown in Figures 7.9a and b, respectively. The results obtained with the steady-state Kalman filter show an excellent agreement between reference and estimated values (see also Figures 7.8a and b). The comparisons of the time evolutions of exact and estimated temperatures and heat fluxes at the selected control volumes ($x = y = 40$ mm and $x = y = 95$ mm) are presented in Figures 7.10 and 7.11, respectively. The agreement between estimated and exact temperatures is excellent, as shown by Figures

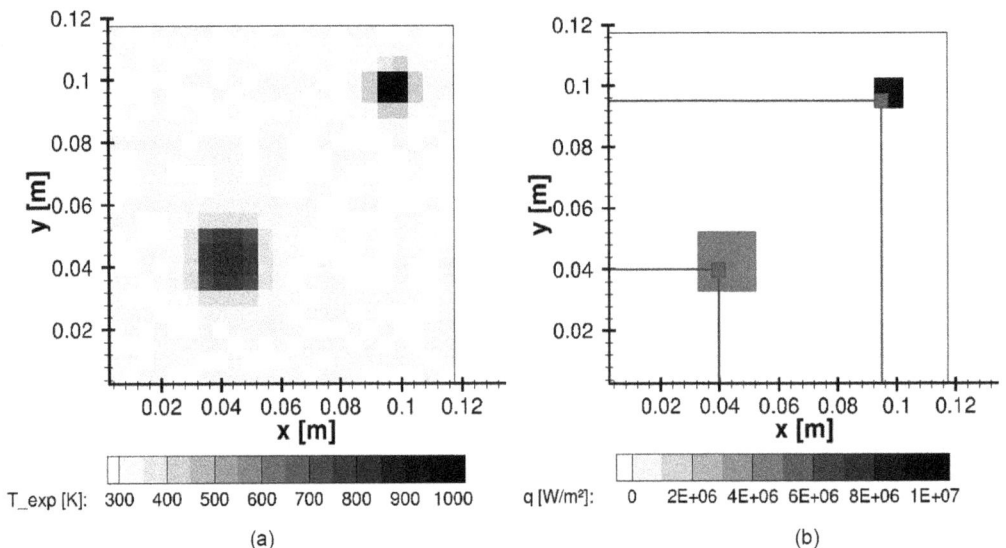

FIGURE 7.8   (a) Simulated measurements and (b) exact heat flux at $t = 2.0$ s. (From Ref. [196].)

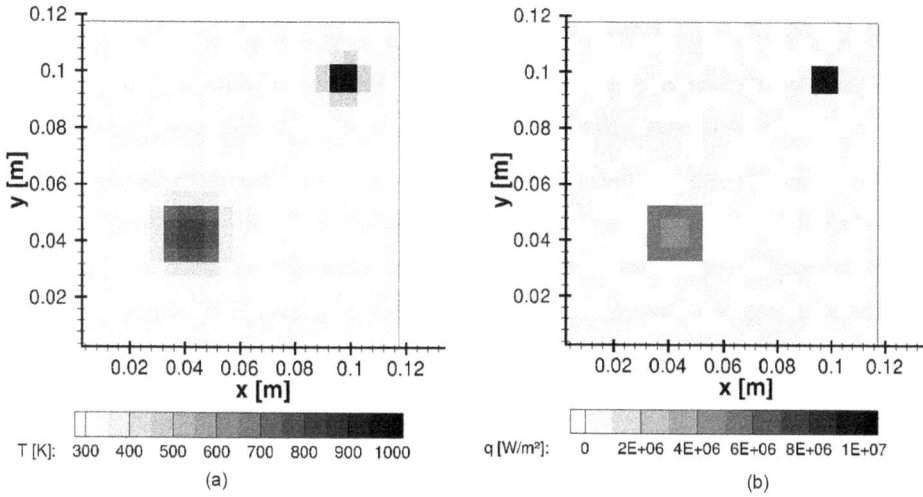

FIGURE 7.9 Estimated quantities at $t = 2.0$ s: (a) temperature; (b) heat flux. (From Ref. [196].)

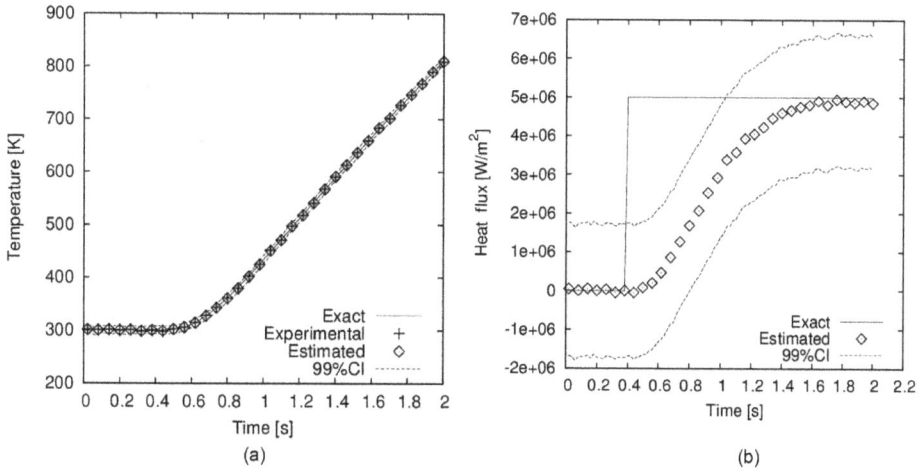

FIGURE 7.10 Time evolution of temperature (a) and heat flux (b) at $x = y = 40$ mm. (From Ref. [196].)

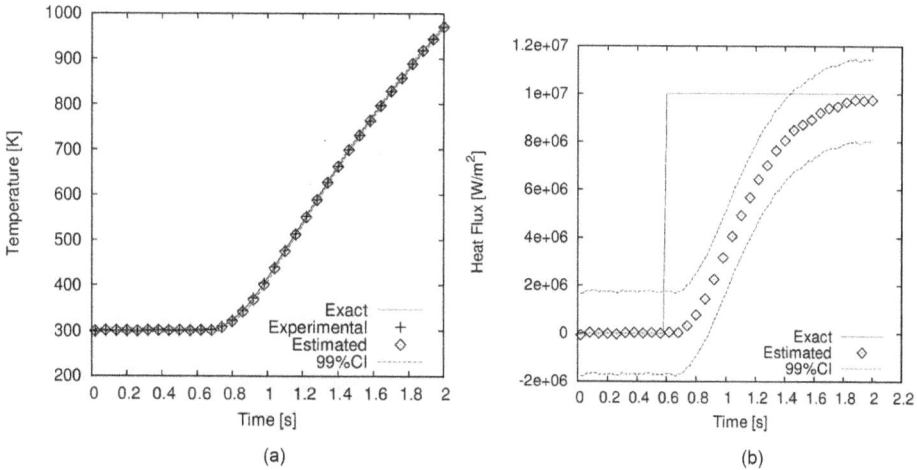

FIGURE 7.11 Time evolution of temperature (a) and heat flux (b) at $x = y = 95$ mm. (From Ref. [196].)

7.10a and 7.11a. Similarly, Figures 7.10b and 7.11b show that the exact heat fluxes were accurately identified after 1 s.

For this test case, the boundary heat flux could be estimated in real time, since the computational time was smaller than the actual duration of the simulated physical problem [196]. This is a significant improvement in comparison with the Kalman filter solution presented in Section 7.3, where the computational times were of the order of several minutes for a simulated experiment with duration of 2 s [194]. The steady-state Kalman filter has also been applied to solve state estimation problems related to the use of magnetic resonance for non-intrusive measurements of temperatures in internal regions of the human body [197,198].

PROBLEMS

7.1 Consider a slab of thickness L, initially at the uniform temperature T_0, which is subjected to a uniform heat flux $q(t)$ over one of its surfaces. The other surface exchanges heat by convection and linearized radiation with a medium at a temperature T_∞ with a heat transfer coefficient h. Temperature gradients are neglected inside the slab and a lumped formulation is used. The formulation for this problem is thus given by:

$$\frac{d\theta(t)}{dt} + m\theta(t) = \frac{mq(t)}{h} \quad \text{for} \ t > 0$$

$$\theta = \theta_0 \quad \text{for} \ t = 0$$

where

$$\theta(t) = T(t) - T_\infty, \ \ \theta_0 = T_0 - T_\infty, \ \ m = \frac{h}{\rho c L}$$

ρ is the density and c is the specific heat of the homogeneous material of the slab.

Let the heat flux $q(t) = q_0$ be constant and deterministically known. Use the Kalman filter to estimate the transient temperature in a slab made of aluminum ($\rho = 2707\,\text{kg/m}^3$, $c = 896\,\text{J/kg K}$), with thickness $L = 0.03\,\text{m}$, for $q_0 = 8000\,\text{W/m}^2$, $T_\infty = 20°\text{C}$, $h = 50\,\text{W/m}^2$ K and $T_0 = 50°\text{C}$. Measurements of the transient temperature of the slab are assumed available. These measurements contain additive, uncorrelated, Gaussian errors, with zero mean and a constant standard deviation σ_z. The errors in the state evolution model are also supposed to be additive, uncorrelated, Gaussian, with zero mean and a constant standard deviation, σ_θ. Examine the effects of σ_z and σ_θ on the inverse problem solution.

7.2 Repeat Problem 7.1 but now consider the heat flux $q(t) = q_0 f(t)$ with known q_0 but unknown time variation $f(t)$. Use a Gaussian random walk model for $f(t)$ like equation (7.3.3).

8 State Estimation
Particle Filter

The *particle filter method* is a sequential Monte Carlo technique for the solution of state estimation problems. The method was devised for the solution of nonlinear and/or non-Gaussian problems, for which the posterior distribution of the state variables is neither Gaussian nor analytical. By using the particle filter method, the posterior density of the state variables at a given time instant t_k, $k = 1$, 2,…, is represented by a set of random samples (particles) and their corresponding weights.

In the particle filter method, the *prediction* step involves the application of the evolution model to advance the direct problem solution by one time step for each particle. The measurements and the observation model are then used in the *update* step to compute the weights for each particle, based on the likelihood function. The estimates of the state variables are obtained with these particles and weights. The prediction and update steps are sequentially repeated in order to advance the solution of the state estimation problem in time, as new measurements are assimilated.

As the number of samples becomes large, this Monte Carlo characterization becomes an equivalent representation of the posterior probability function and the solution approaches the optimal Bayesian estimate [17,53–55,180–191]. Particle filter algorithms demand large computational times as any Monte Carlo method, but they are not restricted to linear and Gaussian problems like the Kalman filter discussed in the previous chapter. Although the key ideas for particle filters have been introduced in the 1950s, not until recently its use became practical because of limited computational resources. Furthermore, in the early implementations most of the particles would have negligible weights after few time steps, which resulted in a large computational time wasted to advance particles with small likelihoods [54].

Although several different versions of the particle filter method can be found in the literature, this chapter presents three algorithms that are at the same simple and robust. These three algorithms are presented below, in order of increasing complexity. They can cope with state estimation problems of practical interest as illustrated with examples in this chapter.

8.1 TECHNIQUE VII: THE SAMPLING IMPORTANCE RESAMPLING (SIR) ALGORITHM

We recall that the evolution model and the observation model are, respectively, given by:

$$\text{Evolution model: } \mathbf{x}_k = \mathbf{f}_k\left(\mathbf{x}_{k-1}, \mathbf{P}, \mathbf{v}_k\right), \quad k = 1,2,\dots \tag{8.1.1a}$$

$$\text{Observation model: } \mathbf{Y}_k = \mathbf{g}_k\left(\mathbf{x}_k, \mathbf{P}, \mathbf{n}_k\right), \quad k = 1,2,\dots \tag{8.1.1b}$$

In the particle filter method, the posterior probability density of the state vector $\mathbf{x}$ at time t_k, $\pi\left(\mathbf{x}_k \mid \mathbf{Y}_{1:k}^{\text{meas}}, \mathbf{P}\right)$, is represented by a set of N particles $\{\mathbf{x}_k^i, i = 1,\dots,N\}$ and by their normalized weights $\{w_k^i, i = 1,\dots,N\}$, where $\sum_{i=1}^{N} w_k^i = 1$. The posterior density is then discretely approximated by [17,53–55,180–191]:

$$\pi\left(\mathbf{x}_k \mid \mathbf{Y}_{1:k}^{\text{meas}}, \mathbf{P}\right) \approx \sum_{i=1}^{N} w_k^i \, \delta\left(\mathbf{x}_k - \mathbf{x}_k^i\right) \tag{8.1.2}$$

233

where δ (.) is the Dirac delta function and the weights are computed from [54]:

$$w_k^i \propto w_{k-1}^i \frac{\pi\left(\mathbf{Y}_k^{\mathrm{meas}}\middle|\mathbf{x}_k^i,\mathbf{P}\right)\pi\left(\mathbf{x}_k^i\middle|\mathbf{x}_{k-1}^i,\mathbf{P}\right)}{\varphi\left(\mathbf{x}_k^i\middle|\mathbf{x}_{k-1}^i,\mathbf{Y}_k^{\mathrm{meas}},\mathbf{P}\right)} \tag{8.1.3}$$

Most particle filter algorithms rely on the so-called importance sampling, where an *importance density* is used to generate the particles, instead of the exact posterior density that is not exactly known [54]. In equation (8.1.3), $\varphi\left(\mathbf{x}_k^i\middle|\mathbf{x}_{k-1}^i,\mathbf{Y}_k^{\mathrm{meas}},\mathbf{P}\right)$ is the importance density, which is assumed to be a Markovian process for $\mathbf{x}$. The optimal choice of the importance density that minimizes the variance of the importance weights is given by $\varphi\left(\mathbf{x}_k^i\middle|\mathbf{x}_{k-1}^i,\mathbf{Y}_k^{\mathrm{meas}},\mathbf{P}\right) = \pi\left(\mathbf{x}_k^i\middle|\mathbf{x}_{k-1}^i,\mathbf{Y}_k^{\mathrm{meas}},\mathbf{P}\right)$. However, for most practical problems this optimal choice is not analytical and a suboptimal importance density is taken as the transition prior, that is, $\varphi\left(\mathbf{x}_k^i\middle|\mathbf{x}_{k-1}^i,\mathbf{Y}_k^{\mathrm{meas}},\mathbf{P}\right) = \pi\left(\mathbf{x}_k^i\middle|\mathbf{x}_{k-1}^i,\mathbf{P}\right)$ [54]. Hence, equation (8.1.3) becomes:

$$w_k^i \propto w_{k-1}^i\, \pi\left(\mathbf{Y}_k^{\mathrm{meas}}\middle|\mathbf{x}_k^i,\mathbf{P}\right) \tag{8.1.4}$$

The simplest particle filter algorithm then consists of the following steps to advance the solution from time t_{k-1} to t_k, where the probability density $\pi(\mathbf{x}_0 \mid \mathbf{Y}_0^{\mathrm{meas}},\mathbf{P}) = \pi(\mathbf{x}_0)$ at the initial time $t = t_0$ is known:

1. *Predict* the prior for the state vector at time t_k, $\pi\left(\mathbf{x}_k^i\middle|\mathbf{x}_{k-1}^i,\mathbf{P}\right)$, with the evolution model given by equation (8.1.1a).
2. Calculate the normalized weight of each particle by using equation (8.1.4) and the total weight $w_k = \sum_{i=1}^{N} w_k^i$. In equation (8.1.4), the likelihood $\pi\left(\mathbf{Y}_k^{\mathrm{meas}}\middle|\mathbf{x}_k^i,\mathbf{P}\right)$ is computed with the measurements $\mathbf{Y}_k^{\mathrm{meas}}$ and with the observation model given by equation (8.1.1b).
3. *Update* the prior by approximating the posterior distribution with equation (8.1.2).
4. Repeat steps 1–3 in order to advance the solution of the state estimation problem in time, as new measurements are sequentially made available.

At each time t_k, the vector of means and the vector of standard deviations of the state variables can be, respectively, calculated with:

$$\bar{\mathbf{x}}_k = \sum_{i=1}^{N} w_k^i \mathbf{x}_k^i \tag{8.1.5a}$$

$$\sigma_k = \sqrt{\left(\frac{N}{N-1}\right)\sum_{i=1}^{N} w_k^i \left(\mathbf{x}_k^i - \bar{\mathbf{x}}_k\right)^2} \tag{8.1.5b}$$

The above steps represent the *sequential importance sampling* (SIS) algorithm of the particle filter [54]. The application of the SIS algorithm might result in the *degeneracy phenomenon*, where only few particles have significant weights after some time steps. Consequently, a large computational effort is spent to update particles with negligible contribution to the posterior density function given by equation (8.1.2).

TABLE 8.1

SIR Algorithm [53, 54]

Step 1

For $i = 1, \ldots, N$, draw new particles $\mathbf{x}_k^i$ from the prior density $\pi\left(\mathbf{x}_k \middle| \mathbf{x}_{k-1}^i, \mathbf{P}\right)$ and then use the likelihood density to calculate the corresponding weights $w_k^i = \pi\left(\mathbf{Y}_k^{\text{meas}} \middle| \mathbf{x}_k^i, \mathbf{P}\right)$.

Step 2

Calculate the total weight $T_w = \sum_{i=1}^{N} w_k^i$ and then normalize the particle weights, that is, for $i = 1, \ldots, N$ let $w_k^i = T_w^{-1} w_k^i$.

Step 3

Resample the particles as follows:

Construct the cumulative sum of weights (CSW) by computing $c_i = c_{i-1} + w_k^i$ for $i = 1, \ldots, N$, with $c_0 = 0$.

Let $i = 1$ and draw a starting point u_1 from the uniform distribution $U\left[0, N^{-1}\right]$.

For $j = 1, \ldots, N$:

    Move along the CSW by making $u_j = u_1 + N^{-1}(j-1)$

    While $u_j > c_i$ make $i = i + 1$.

    Assign sample $\mathbf{x}_k^j = \mathbf{x}_k^i$

    Assign sample weight $w_k^j = N^{-1}$

The degeneracy of the particles can be overcome with a *resampling step* in the application of the particle filter, where particles with small weights are discarded and particles with large weights are replicated [17,53–55,180–191]. After the resampling step, all particles are assigned with identical weights $w_k^i = 1/N$. Resampling can be performed if the number of particles with large weights falls below a certain threshold number. Alternatively, it can be performed at every instant t_k, such as in the *sampling importance resampling* (SIR) algorithm [53,54]. Such algorithm can be summarized in the steps presented in Table 8.1, as applied to the system evolution from time t_{k-1} to t_k. Note in Step 1 of the SIR algorithm that equation (8.1.4) reduces to $w_k^i \propto \pi\left(\mathbf{Y}_k^{\text{meas}} \middle| \mathbf{x}_k^i, \mathbf{P}\right)$, since resampling is applied every time step and equal weights are assigned to all particles after resampling (Step 3 in Table 8.1).

8.2 TECHNIQUE VII: THE AUXILIARY SAMPLING IMPORTANCE RESAMPLING (ASIR) ALGORITHM

Although the resampling step reduces the effects of degeneracy, it may lead to the loss of diversity of the particles $\mathbf{x}_k^i$, $i = 1, \ldots, N$. In this case, many particles become identical, since the resampling step replicates the particles with large original weights and then assigns equal weights to all particles.

Sample impoverishment can be reduced with the *auxiliary sampling importance resampling* (ASIR) algorithm of the particle filter [53–55]. In this algorithm, the resampling step is performed at one time instant backward, t_{k-1}, but with the measurements at time t_k [54]. The resampling step is based on some point estimate μ_k^i that characterizes $\pi\left(\mathbf{x}_k^i \middle| \mathbf{x}_{k-1}^i, \mathbf{P}\right)$, for example, the mean of the particles that represent this distribution.

The ASIR algorithm can be summarized in the steps presented in Table 8.2, as applied to the system evolution from t_{k-1} to t_k [53–55].

TABLE 8.2

ASIR Algorithm [53–55]

Step 1

For $i = 1, \ldots, N$, draw new particles $\mathbf{x}_k^i$ from the prior density $\pi(\mathbf{x}_k | \mathbf{x}_{k-1}^i, \mathbf{P})$ and then calculate some point characteristic μ_k^i of $\mathbf{x}_k^i$. Then, use the likelihood density to calculate the corresponding weights $w_k^i = w_{k-1}^i \, \pi\left(\mathbf{Y}_k^{\mathrm{meas}} | \mu_k^i, \mathbf{P}\right)$.

Step 2

Calculate the total weight $T_w = \displaystyle\sum_{i=1}^{N} w_k^i$ and then normalize the particle weights, that is, for $i = 1, \ldots, N$ let $w_k^i = T_w^{-1} \, w_k^i$.

Step 3

Resample the particles as follows:

Construct the CSW by computing $c_i = c_{i-1} + w_k^i$ for $i = 1, \ldots, N$, with $c_0 = 0$.

Let $i = 1$ and draw a starting point u_1 from the uniform distribution $\mathrm{U}\left[0, N^{-1}\right]$.

For $j = 1, \ldots, N$:

    Move along the CSW by making $u_j = u_1 + N^{-1}(j-1)$

    While $u_j > c_i$ make $i = i+1$

    Assign sample $\mathbf{x}_{k-1}^j = \mathbf{x}_{k-1}^i$ and $\mu_k^j = \mu_k^i$

    Assign parent $i^j = i$

Step 4

For $j = 1, \ldots, N$ draw particles $\mathbf{x}_k^j$ from the prior density $\pi\left(\mathbf{x}_k | \mathbf{x}_{k-1}^{i^j}, \mathbf{P}\right)$, using the parent i^j, and then use the likelihood density to calculate the correspondent weights $w_k^j = \pi\left(\mathbf{Y}_k^{\mathrm{meas}} | \mathbf{x}_k^j, \mathbf{P}\right) \big/ \pi\left(\mathbf{Y}_k^{\mathrm{meas}} | \mu_k^{i^j}, \mathbf{P}\right)$.

Step 5

Calculate the total weight $T_w = \displaystyle\sum_{j=1}^{N} w_k^j$ and then normalize the particle weights, that is, for $j = 1, \ldots, N$ let $w_k^j = T_w^{-1} \, w_k^j$.

8.3   TECHNIQUE VII: THE ALGORITHM OF LIU AND WEST

The above algorithms of the particle filter relied on known values of the model parameters $\mathbf{P}$. If these parameters are to be estimated simultaneously with the state variables, one possibility is to apply the SIR or ASIR algorithms by mimicking the parameters as state variables with an evolution model, for example, in the form of a random walk process, that is,

$$\mathbf{P}_k = \mathbf{P}_{k-1} + \mathbf{e}_k \qquad (8.3.1)$$

where $\mathbf{e}_k$ is a random vector with zero mean. The subscript k for $\mathbf{P}$ is to denote that the parameters will be sequentially estimated together with the state variables; it does not mean that the parameters are time dependent. Although such an approach can result on accurate estimates for the parameters, even for physically complicated nonlinear problems like in fire propagation (see the next section) [199], the simulation of parameters as state variables might degenerate the particles very fast [200,201].

On the other hand, the algorithm by Liu and West [200], which is based on the ASIR version of the particle filter, can be used for the estimation of the posterior probability distribution $\pi\left(\mathbf{x}_k, \mathbf{P} | \mathbf{Y}_{1:k}^{\mathrm{meas}}\right)$, where the vector $\mathbf{P}$ contains the random parameters of the model. The algorithm of Liu and West for the particle filter is based on West's hypothesis [202] of a Gaussian mixture for the vector of parameters $\mathbf{P}$, that is [202,203],

$$\pi\left(\mathbf{P} | \mathbf{Y}_{1:k-1}^{\mathrm{meas}}\right) \approx \sum_{i=1}^{N} w_{k-1}^i \, \mathrm{N}\left(\mathbf{m}_{k-1}^i, \eta^2 \mathbf{V}_{k-1}\right) \qquad (8.3.2)$$

where $N(\mathbf{a}, \mathbf{B})$ is a Gaussian density with mean $\mathbf{a}$ and covariance matrix $\mathbf{B}$, η is a smoothing parameter and $\mathbf{V}_{k-1}$ is the posterior covariance matrix at time t_{k-1}.

Equation (8.3.2) shows that the density $\pi\left(\mathbf{P}\middle|\mathbf{Y}_{1:k-1}^{\mathrm{meas}}\right)$ is a mixture of $N\left(\mathbf{m}_{k-1}^i, \eta^2\mathbf{V}_{k-1}\right)$ Gaussian distributions weighted by w_{k-1}^i. The kernel locations are specified by using the following shrinkage rule [200]:

$$\mathbf{m}_{k-1}^i = a\,\mathbf{P}_{k-1}^i + (1-a)\overline{\mathbf{P}}_{k-1} \tag{8.3.3}$$

where $a = \sqrt{1-\eta^2}$ and $\overline{\mathbf{P}}_{k-1}$ is the mean of $\mathbf{P}$ at time t_{k-1}. The shrinkage factor, a, is computed as [200]:

$$a = \frac{3\delta - 1}{2\delta} \tag{8.3.4}$$

where $0.95 < \delta < 0.99$.

Table 8.3 summarizes the basic steps of Liu and West's algorithm [200], as applied for the advancement of the particles from time t_{k-1} to t_k.

TABLE 8.3
Liu and West's Algorithm [200]

Step 1

Find the mean $\overline{\mathbf{P}}_{k-1}$ of the parameters $\mathbf{P}$ at time t_{k-1}.

Step 2

For $i = 1,\ldots,N$, compute $\mathbf{m}_{k-1}^i$ with equation (8.3.3), draw new particles $\mathbf{x}_k^i$ from the prior density $\pi\left(\mathbf{x}_k \mid \mathbf{x}_{k-1}^i, \mathbf{m}_{k-1}^i\right)$ and then calculate some characteristic $\boldsymbol{\mu}_k^i$ of $\mathbf{x}_k^i$. Then, use the likelihood density to calculate the corresponding weights $w_k^i = w_{k-1}^i\,\pi\left(\mathbf{Y}_k^{\mathrm{meas}}\middle|\boldsymbol{\mu}_k^i, \mathbf{m}_{k-1}^i\right)$.

Step 3

Calculate the total weight $T_w = \sum_{i=1}^{N} w_k^i$ and then normalize the particle weights, that is, for $i = 1, \ldots, N$ let $w_k^i = T_w^{-1}\,w_k^i$.

Step 4

Resample the particles as follows:

Construct the CSW by computing $c_i = c_{i-1} + w_k^i$ for $i = 1, \ldots, N$, with $c_0 = 0$.

Let $i = 1$ and draw a starting point u_1 from the uniform distribution $U\left[0, N^{-1}\right]$.

For $j = 1,\ldots,N$:

    Move along the CSW by making $u_j = u_1 + N^{-1}(j-1)$

    While $u_j > c_i$ make $i = i+1$

    Assign samples $\mathbf{x}_{k-1}^j = \mathbf{x}_{k-1}^i$, $\mathbf{m}_{k-1}^j = \mathbf{m}_{k-1}^i$ and $\boldsymbol{\mu}_k^j = \boldsymbol{\mu}_k^i$

    Assign parent $i^j = i$

Step 5

For $j = 1, \ldots, N$ draw samples $\mathbf{P}_k^j$ from $N(\mathbf{m}_{k-1}^{i^j}, \eta^2\mathbf{V}_{k-1})$ by using the parent i^j.

Step 6

For $j = 1, \ldots, N$ draw particles $\mathbf{x}_k^j$ from the prior density $\pi\left(\mathbf{x}_k\middle|\mathbf{x}_{k-1}^{i^j}, \mathbf{P}_k^j\right)$, using the parent i^j, and then use the likelihood density to calculate the correspondent weights $w_k^j = \pi\left(\mathbf{Y}_k^{\mathrm{meas}}\middle|\mathbf{x}_k^j, \mathbf{P}_k^j\right)\middle/\pi\left(\mathbf{Y}_k^{\mathrm{meas}}\middle|\boldsymbol{\mu}_k^{i^j}, \mathbf{m}_{k-1}^{i^j}\right)$.

Step 7

Calculate the total weight $T_w = \sum_{i=1}^{N} w_k^i$ and then normalize the particle weights, that is, for $i = 1,\ldots, N$ let $w_k^i = T_w^{-1}\,w_k^i$.

8.4 ESTIMATION OF THE FIRE FRONT IN REGIONAL SCALE WILDFIRE SPREAD

This section demonstrates the capability of particle filters for sequentially improving the simulation and forecast of wildfire propagation, as new fire front observations become available [199]. The SIR and the ASIR algorithms are compared, as applied to the sequential estimation of a dynamic variable and of vegetation parameters of the rate of fire spread (ROS) model, which are all treated as state variables.

The propagation of wildfires results from complex interactions between pyrolysis, combustion, heat transfer and flow dynamics, as well as atmospheric dynamics and chemistry, among other phenomena. These interactions occur over a wide range of scales: vegetation scales that characterize the biomass fuel; topographical scales that characterize the terrain and vegetation boundary layer; and meteorological micro-/meso-scales that characterize atmospheric conditions. Here, we adopt a regional-scale perspective and simulate a wildfire as a thin flame zone (i.e., as a front) that propagates normal to itself toward the unburnt vegetation. In this representation, the main quantity of interest is the ROS that is the local propagation speed of the front.

Based on Rothermel's model [204], the ROS was formulated as a semi-empirical function of a reduced number of parameters that locally characterize the vegetation (fuel) properties, the weather conditions and the terrain topography. The local ROS, denoted by Γ [m/s], can be written as [199]:

$$\Gamma \equiv \Gamma(x, y, t) = P\left(M_f, \Sigma, \mathbf{u}_w(x, y, t), \ldots\right)\delta(x, y) \qquad (8.4.1)$$

where δ [m] is the fuel depth (the vegetation layer thickness) and P [s^{-1}] is a function of the fuel moisture content M_f, the fuel particle surface-to-volume ratio Σ [m^{-1}] and the wind velocity (at mid-flame height) $\mathbf{u}_w$ [m/s]. Here, Σ, M_f and δ were treated as spatially uniform parameters. Note that $\mathbf{u}_w$ is spatially distributed along the fire front evolving on the two-dimensional horizontal plane (x,y). This variable results from the projection of the wind velocity vector on the direction normal to the contour lines of the progress variable. The propagation of the fire front with the ROS given by equation (8.4.1) was simulated using a standard level-set front-tracking technique [199]. A progress variable, denoted by c and also referred to as the level-set function, was introduced as a flame marker, where $c = 0$ in the unburnt vegetation, $c = 1$ in the burnt vegetation; and the flame front is identified by the two-dimensional contour line where $c = 0.5$. The transient spatial evolution of $c = c(x,y,t)$ was calculated as a solution of the following advection equation:

$$\frac{\partial c}{\partial t} = \Gamma |\nabla c| \qquad (8.4.2)$$

where Γ is the ROS [m/s] along the direction $\mathbf{n} = -\nabla c / |\nabla c|$ normal to the contour lines of the progress variable c.

The SIR and ASIR algorithms were applied to estimate the fire front position together with the estimation of some physical parameters involved in the evolution model formulation. Data were taken from an experiment involving a small-scale (4 m × 4 m) open-field grassland fire occurring under moderate wind velocity of 1 m/s. The fire spread was recorded during 350 s using an infrared camera; the resulting observations were the time-evolving positions of the fire front (see Figure 8.1), identified as the contour lines where the temperature reached 600 K, which was considered as the temperature of combustion ignition.

In the state estimation process, measurements of fire front locations were assimilated every 14 s, from $t = 64$ s to $t = 106$ s (the associated fronts are represented in black solid lines in Figure 8.1). Thus, the update step of the particle filter was successively performed at $t = 64$ s, 78 s, 92 s and 106 s. Each observed front was discretely represented with 200 markers, whose error standard deviation was estimated as 0.047 m (based on the spatial resolution of the infrared camera), as shown by Figure 8.2.

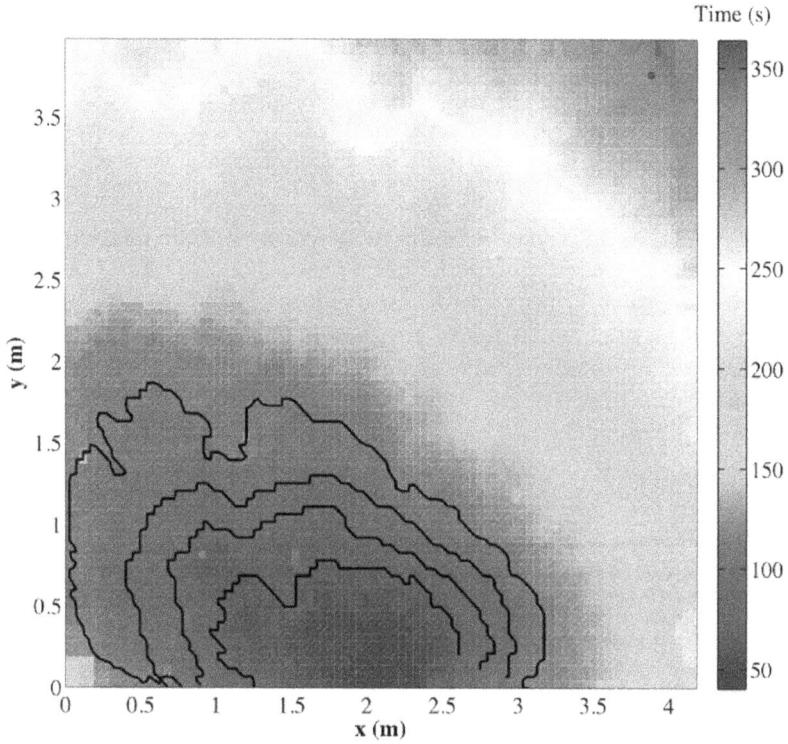

FIGURE 8.1   Arrival times of the fire front (in color) and observed fire fronts separated by 14 s (at $t = 64$ s, 78 s, 92 s, 106 s) in black solid lines. (From Ref. [199].)

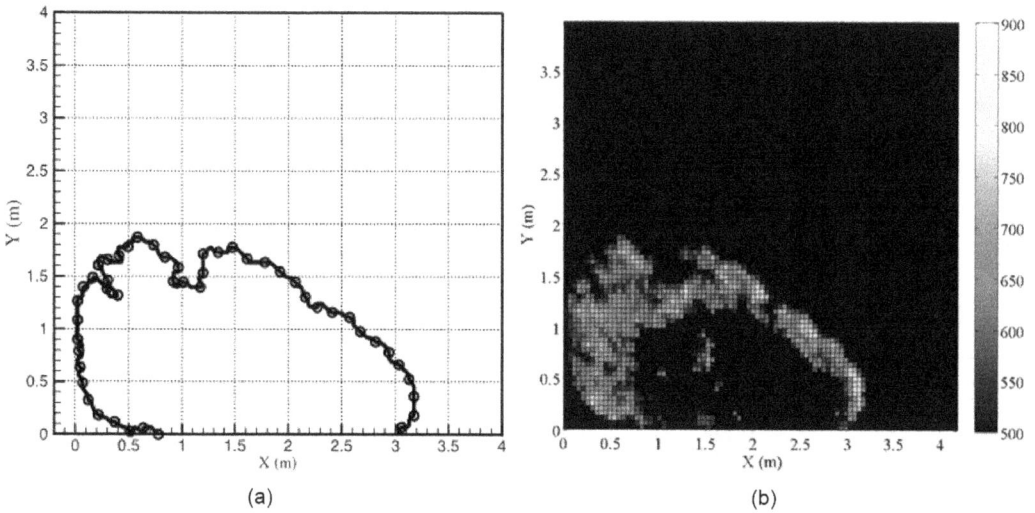

FIGURE 8.2   Extraction of the fire front location (a) from thermal-infrared imaging (b) at $t = 106$ s; the fire front was identified as the 600 K temperature contour line. (From Ref. [199].)

The objective of the present state estimation problem was to estimate the posterior distributions of the fire front location. Due to their importance and inherent uncertainties, the fuel moisture content, M_f, and the fuel particle surface-to-volume ratio, Σ (see equation 8.4.1), were also treated as state variables in this work and estimated through the application of the particle filter algorithms under analysis. These two parameters were assumed to be spatially uniform and with a random walk model given by:

$$M_f(t_k) = M_f(t_{k-1}) + \sigma_f R_f \qquad (8.4.3a)$$

$$\Sigma(t_k) = \Sigma(t_{k-1}) + \sigma_\Sigma R_\Sigma \qquad (8.4.3b)$$

where R_f and R_Σ are Gaussian random numbers, with zero mean and unitary standard deviation, while σ_f and σ_Σ are the standard deviations of the random walk models for M_f and Σ, respectively.

The state evolution model for the vector containing the values of the progress variables at the grid points used in the spatial discretization, $\mathbf{c}(t_k)$, was obtained from the discrete integration of equation (8.4.2). Uncertainties for $\mathbf{c}(t_k)$ were assumed to be additive, Gaussian, with zero mean and a constant standard deviation of 0.01.

Figures 8.3 and 8.4 present the measurements and the time-evolving locations of the fire fronts (from $t = 64$ s to $t = 106$ s), estimated with the SIR and ASIR algorithms with different numbers

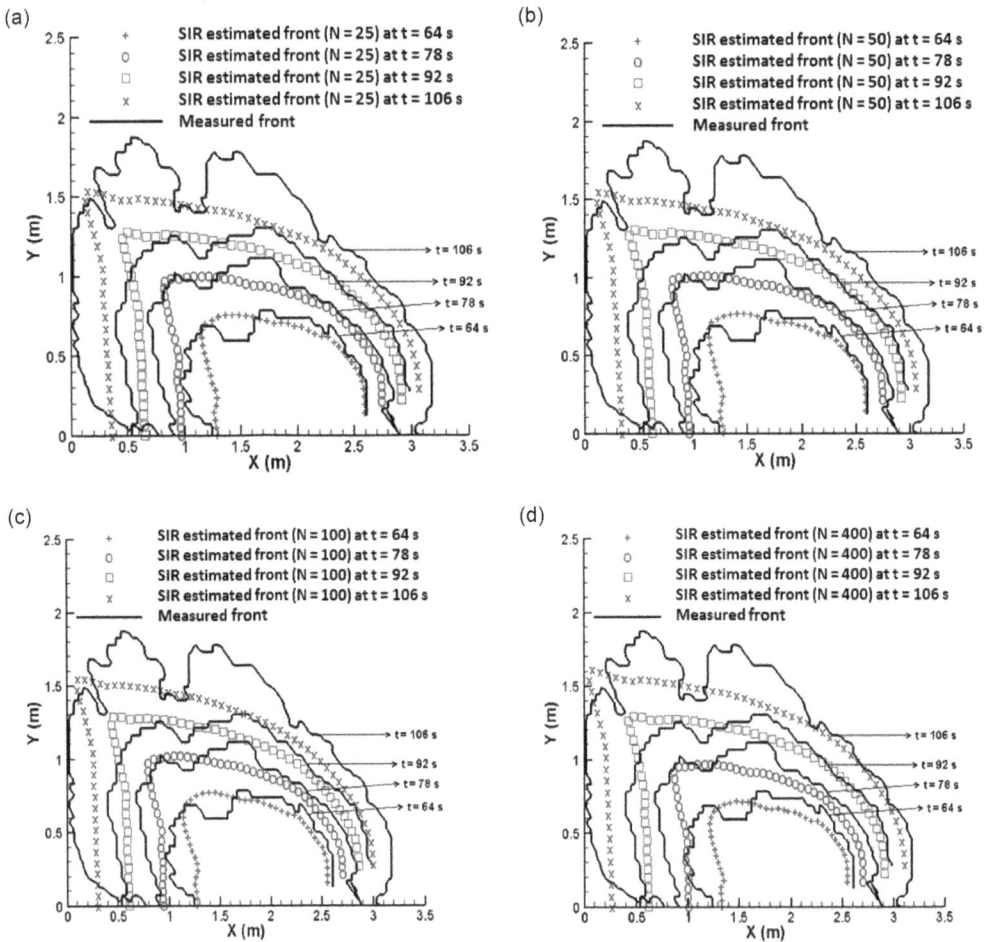

FIGURE 8.3 Comparison between simulated and measured fire front positions from $t=64$ s to $t=106$ s using the SIR filter, for: (a) 25 particles, (b) 50 particles, (c) 100 particles and (d) 400 particles. Observations are represented in solid lines; simulated fire fronts associated with the posterior PDF of the control vector are represented in symbols. (From Ref. [199].)

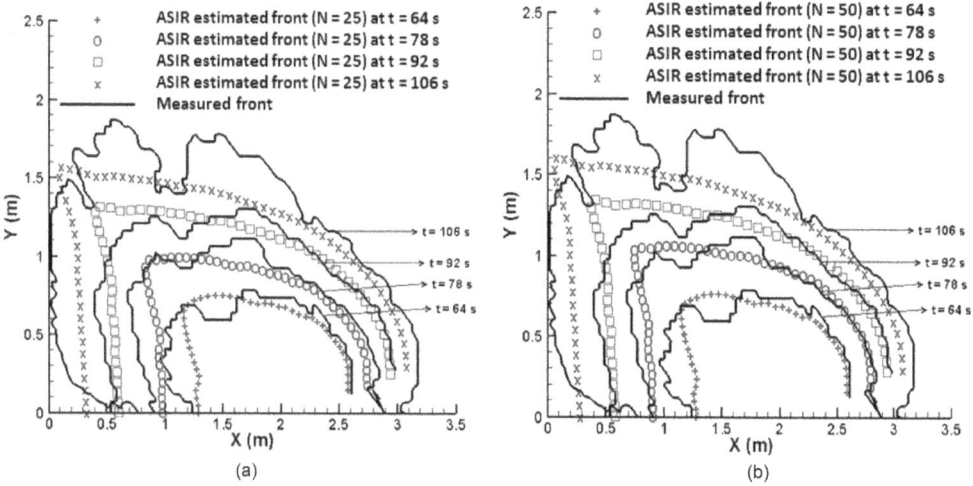

FIGURE 8.4 Comparison between simulated and measured fire front positions from $t = 64\,s$ to $t = 106\,s$ using the ASIR filter for: (a) 25 particles and (b) 50 particles. Observations are represented in solid lines; simulated fire fronts associated with the posterior PDF of the control vector are represented in symbols. (From Ref. [199].)

of particles, respectively. These results show that both algorithms were able to closely track the observed fire fronts. The SIR algorithm with N particles required the same computational time as the ASIR algorithm with $N/2$ particles, as expected from the analysis of Tables 8.1 and 8.2.

8.5 A COMPARISON OF PARTICLE FILTER ALGORITHMS IN BIOHEAT TRANSFER

A comparison of the three algorithms presented in Sections 8.1–8.3 of Technique VII is performed here, with focus on applications related to the hyperthermia treatment of cancer, with heating imposed either by a laser in the near-infrared range or by radiofrequency waves [205–210].

The use of heat for the treatment of cancer can be aimed at: (i) A mild temperature increase of the tumor, in order to make their cells more susceptible to the effects of other treatments, like radiotherapy or chemotherapy; or (ii) a large temperature increase of the tumor to kill their cells solely by the effects of heat. Within the medical community, these treatments are usually referred to as hyperthermia and thermal ablation, respectively. The hyperthermia treatment of cancer consists in raising tumor tissues to temperatures between 41°C and 47°C during a pre-specified period of time. Among other types of heating, electromagnetic energy sources in the radiofrequency and near-infrared ranges have been used to deliver energy to the target region, due to the biological windows of human tissues that exhibit small absorption. One major problem of the hyperthermia treatment of cancer is the lack of selectivity of the heating procedure. On the other hand, with recent advancements in nanotechnology, nanoparticles have been used as absorbing agents in the near-infrared and in the radiofrequency ranges, in order to provide localized thermal damage to the tumor, with minimal damage to the healthy cells [205–210].

We utilize the classical bioheat transfer model proposed by Pennes [211]. The mathematical formulation of the bioheat transfer problem in a domain Ω, with position-dependent thermophysical properties to account for different tissues or organs, and third-kind boundary conditions over the body surface Γ, is given by [205]:

$$\rho(\mathbf{r})c_p(\mathbf{r})\frac{\partial T(\mathbf{r},t)}{\partial t} = \nabla \cdot \left[k(\mathbf{r})\nabla T(\mathbf{r},t)\right] + Q(\mathbf{r}),\ \text{in } \Omega,\ \text{for } t > 0 \qquad (8.5.1a)$$

$$k(\mathbf{r})\nabla T(\mathbf{r},t)\cdot\mathbf{n}+h(\mathbf{r})T(\mathbf{r},t)=h(\mathbf{r})T_{\infty}(\mathbf{r}),\ \text{for}\ \mathbf{r}\in\Gamma,\ t>0 \qquad (8.5.1b)$$

$$T(\mathbf{r},t)=T_s(\mathbf{r})\ \text{in}\ \Omega,\ t=0 \qquad (8.5.1c)$$

where $\mathbf{r}$ is the position vector, $\mathbf{n}$ is the unit vector normal to the surface, $h(\mathbf{r})$ is the heat transfer coefficient at the surface of the body, $T_{\infty}(\mathbf{r})$ is the temperature of the surrounding media and $T_s(\mathbf{r})$ is the initial temperature distribution within the medium, supposed to be the steady-state temperature of the problem when the external heating is null.

The heat source is given by:

$$Q(\mathbf{r})=\rho_b c_{p,b}\omega_b(\mathbf{r})\big[T_b-T(\mathbf{r},t)\big]+Q_{\text{met}}(\mathbf{r})+Q_{\text{ext}}(\mathbf{r}) \qquad (8.5.2)$$

which includes the term resulting from the external heating for the hyperthermia heating, $Q_{\text{ext}}(\mathbf{r})$, as well as due to metabolism, $Q_{\text{met}}(\mathbf{r})$, and the effect of blood perfusion with a coefficient $\omega_b(\mathbf{r})$.

The first physical problem considered here involved the hyperthermia treatment of a subcutaneous tumor induced by an external collimated plane laser beam under constant illumination. The skin was represented as a nonhomogeneous cylindrical medium with five layers, where each layer corresponded to a specific tissue, namely: epidermis, dermis, fat, muscle and a tumor within the dermis (see Figure 8.5). The tumor was assumed to be loaded with gold nanorods in order to enhance the hyperthermia effects and to limit such effects to the tumor region. The heat transfer problem resulting from the laser irradiation of the medium was given by Pennes' model in two-dimensional cylindrical coordinates with axial symmetry. The internal surface (at $z=L_z$) was assumed to exchange heat with the deeper tissues with a heat transfer coefficient h_{int}, while the irradiated surface (at $z=0$) was assumed to be cooled by air in order to avoid overheating of the skin. At the external surface of the skin, at $z=0$, the heat transfer coefficient and the temperature of the surrounding medium were given, respectively, by h_c and T_c. Heat transfer was neglected through the lateral surfaces of the medium.

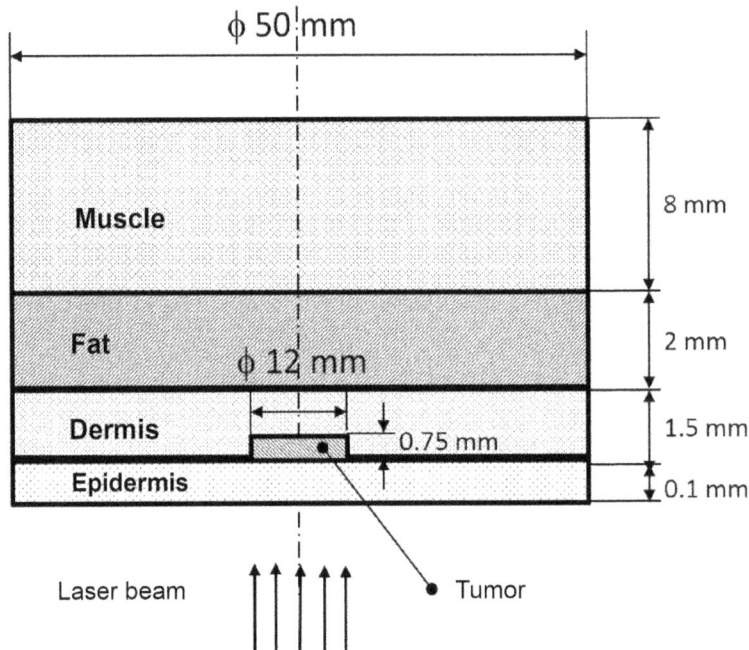

FIGURE 8.5   Physical problem for laser heating. (From Ref. [205].)

The laser radiation propagation in the skin was modeled with the δ-P1 diffusion approximation. The laser beam was assumed to be co-axial with the cylindrical skin model, so that the problem was formulated as two dimensional with axial symmetry. At the external surface of the skin, the incident laser radiation was assumed to be partially reflected (specular reflection), with reflection coefficient R_{sc}. The internal surface of the irradiated boundary was assumed to partially and diffusively reflect the incident radiation, with reflectivity characterized by Fresnel's coefficient A_1, while opacity was assumed for the remaining boundaries. The refractive indexes (n_t) of the different tissues were assumed constant and homogeneous. The mathematical formulation of the radiation problem within the δ-P1 approximation is given by [205]:

$$\nabla \cdot \left[-D(r,z)\nabla\Phi_s(r,z) + \frac{\sigma_s'(r,z)g'(r,z)}{\beta_{tr}(r,z)}\Phi_p(r,z)\hat{s}_c \right] + \kappa(r,z)\Phi_s(r,z)$$

$$= \sigma_s'(r,z)\Phi_p(r,z) \quad \text{in } 0 < r < L_r \text{ and } 0 < z < L_z \tag{8.5.3a}$$

$$-D(r,z)\nabla\Phi_s(r,z)\cdot\mathbf{n} + \frac{1}{2A_1}\Phi_s(r,z) = -\frac{\sigma_s'(r,z)g'(r,z)}{\beta_{tr}(r,z)}\Phi_p(r,z) \quad \text{at } z = 0, 0 < r < L_z \tag{8.5.3b}$$

$$\Phi_s(r,z) = 0 \quad \text{at } z = L_z, 0 < r < L_r \tag{8.5.3c}$$

$$\nabla\Phi_s(r,z)\cdot\mathbf{n} = 0 \quad \text{at } r = 0, 0 < z < L_z \tag{8.5.3d}$$

$$\Phi_s(r,z) = 0 \quad \text{at } r = L_r, 0 < z < L_z \tag{8.5.3e}$$

where

$$D = \frac{1}{3\beta_{tr}}; \quad \sigma_s' = (1-g^2)\sigma_s; \quad g' = \frac{g}{1+g}; \quad A_1 = (1+R_2)/(1-R_1); \quad \beta_{tr} = \kappa + \sigma_s(1-g) \tag{8.5.4a-e}$$

with g being the anisotropy factor of scattering, σ_s the scattering coefficient, while R_1 and R_2 are the first and second moments of Fresnel's reflection coefficient, respectively.

The collimated component of the fluence rate, $\Phi_p(r,z)$, followed the generalized Beer-Lambert's law, with the imposed laser flux given by:

$$E(r) = \begin{cases} E_0, & r \leq L_{tumor} \\ 0, & r > L_{tumor} \end{cases} \tag{8.5.5}$$

where L_{tumor} is the radius of the tumor. The total fluence rate was obtained by adding both diffuse and collimated components, that is,

$$\Phi(r,z) = \Phi_p(r,z) + \Phi_s(r,z) \tag{8.5.6}$$

and the heat source term resulting from the laser absorption is given by:

$$Q_{ext}(\mathbf{r}) = \kappa(r,z)\Phi(r,z) \tag{8.5.7}$$

The physical problem considered for radiofrequency heating involved a domain in two dimensions, consisting of a rectangle (healthy tissue—Ω_1), containing a circular domain (tumor—Ω_2), as shown by Figure 8.6. Heating was imposed by radiofrequency waves through the electrodes Ω_1' and Ω_2', which were maintained at the voltages U and zero with respect to ground, respectively.

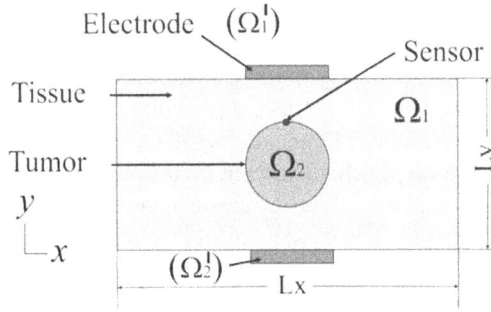

FIGURE 8.6 Physical problems for radiofrequency heating. (From Ref. [205].)

The remaining surfaces were electrically insulated. Heat generated inside the domain was propagated by conduction and by blood perfusion, as given by Pennes' model. The top and bottom boundaries exchange heat by convection, where h_c is the heat transfer coefficient and T_c is the temperature of the surrounding medium. The remaining boundaries of the domain were supposed insulated.

The electric potential within the domain can be obtained by solving the following Laplace's equation [205]:

$$\nabla \cdot [\varepsilon(x,y) \cdot \nabla \varphi(x,y)] = 0 \quad x, y \in \Omega_1 \cup \Omega_2 \tag{8.5.8a}$$

where φ is the potential and ε is the permittivity, which varies spatially depending on the tissue and tumor regions. The boundary conditions for equation (8.5.8a) were given by:

$$\varphi(x,y) = U \quad \text{at} \quad \Omega_1' \tag{8.5.8b}$$

$$\varphi(x,y) = 0 \quad \text{at} \quad \Omega_2' \tag{8.5.8c}$$

$$\nabla \varphi(x,y) \cdot \mathbf{n} = 0 \quad \text{elsewhere over the boundary} \tag{8.5.8d}$$

After solving Problem (8.5.8), the electric field strength $\mathbf{E}$ was obtained with:

$$\mathbf{E}(x,y) = -\nabla \varphi(x,y) \tag{8.5.9a}$$

and the intensity of the magnetic field $\mathbf{H}$ with:

$$|\mathbf{H}(x,y)| = \frac{1}{1 + N(\chi)} \frac{|\mathbf{E}(x,y)|}{\mu_0 \pi f R} \tag{8.5.9b}$$

where $N(\chi) = 1/3$ is the demagnetizing factor of the composite tissue, χ is the susceptibility of the magnetic nanoparticles that can be described in terms of complex susceptibility, $\chi = \chi' + i\chi''$, μ_0 is the dielectric constant permeability of free space $\mu_0 = 4\pi \times 10^{-7}$ Tm/A, f is the electromagnetic frequency and R is the radius of magnetic induction loop.

The heat source term in the healthy tissue resulting from the radiofrequency heating is given by

$$Q_{\text{ext}}(\mathbf{r}) = \frac{\sigma_1 |\mathbf{E}(x,y)|^2}{2} \quad \text{in } \Omega_1 \tag{8.5.10a}$$

while in the tumor, the heat source term including the contribution of the magnetic nanoparticles is obtained from:

$$Q_{ext}(\mathbf{r}) = (1-\theta)\frac{\sigma_2 |\mathbf{E}(x,y)|^2}{2} + \theta\left[\frac{9}{16}\frac{\chi''}{\mu_0 \pi f R^2}|\mathbf{E}(x,y)|^2\right] \quad \text{in } \Omega_2 \quad (8.5.10b)$$

where $\theta = n\pi r^2/A$ is the concentration of nanoparticles, r is the mean radius of the supposedly spherical nanoparticles, n is the number of nanoparticles, A is the area of the tumor and σ_2 is the electrical conductivity of the tumor tissue embedded with nanoparticles, which can be approximated by $1/\sigma_2 = (1-\theta)/\sigma_2' + \theta/\sigma_3$, where σ_2' and σ_3 are the electrical conductivity of tumor and nanoparticles, respectively. The permittivity of the tumor with nanoparticles was approximated by the permittivity of the tumor.

The inverse problems dealt with the estimation of the transient temperature field for each of the hyperthermia problems described above. The state evolution model for temperature was given by the numerical solution of the bioheat transfer problem given by equations (8.5.1a-c), in the domains presented by Figures 8.5 and 8.6, for laser or radiofrequency heating, respectively. In order to cope with uncertainties in the temperature evolution model, a Gaussian uncorrelated noise with zero mean and constant standard deviation was added to the solution of equations (8.5.1a-c).

For the application of the SIR and ASIR algorithms, uncertainties in the model parameters, $\mathbf{P}$, were taken into account through an additive Gaussian noise for the temperature evolution model as described above, as well as through an additive Gaussian noise for the heat source term resulting from the external heating. Therefore, for the application of the SIR and ASIR algorithms, the evolution model for this heat source term, required for the solution of problem (8.5.1), was taken in the form of a random walk given by:

$$Q_{ext,k}^i(\mathbf{r}) = Q_{ext,k-1}^i(\mathbf{r}) + \xi_k^i(\mathbf{r}) \qquad (8.5.11)$$

where i is the particle number and $\xi_k^i(\mathbf{r})$ is a Gaussian random variable with zero mean and a constant standard deviation. The subscript k in equation (8.5.11) does not represent a time evolution of $Q_{ext}(\mathbf{r})$, but the fact that it is treated as state variable for the application of the particle filter. The particles $Q_{ext,0}^i(\mathbf{r})$ were initially sampled from Gaussian distributions with means obtained from the deterministic solutions of either Problems (8.5.3) or (8.5.8), depending if the heating was imposed by the laser in the near-infrared range or by the radiofrequency waves, respectively. Thus, in the application of the SIR and ASIR algorithms, the radiation or electric problems were decoupled from the bioheat transfer problem, and needed to be solved only once, while the bioheat transfer problem was solved recursively within the filter with the heat source term given by equation (8.5.11) for each particle.

For the application of Liu and West's algorithm, uncertainties in the optical or electrical parameters (depending on the applied heating strategy), as well as uncertainties in the thermal parameters, were given by Gaussian mixtures (equation 8.3.2). Therefore, the radiation or electric problems were also solved, together with the bioheat transfer problem given by equations (8.5.1a-c), at each evolution step of the particle filter.

The state estimation problems were solved by using one single temperature measurement point inside the domain. Uncertainties in such measurements, as well as in the observation model, were taken as Gaussian, additive, uncorrelated, with zero mean and a constant standard deviation.

The performances of the three particle filter algorithms were compared in terms of computational time and accuracy of the estimated temperatures. The root mean square error between the estimated and exact temperatures was used for this purpose, which is given by:

$$\text{RMS} = \sqrt{\frac{\sum_{p=1}^{P}\left(T_{est,p} - T_{exa,p}\right)^2}{P}} \qquad (8.5.12)$$

where $T_{est,p}$ and $T_{exa,p}$, respectively, represent the estimated and exact temperatures at a position $\mathbf{r}_{i,j}$ and at a time instant t_k, and P is the total number of time steps and locations where the temperatures were compared. The RMS errors were reported in terms of their means and standard deviation values, obtained with 30 runs of each algorithm, in order to avoid any bias resulting from the simulated measurements.

For the hyperthermia treatment with laser heating, the state estimation problem was solved by assuming transient temperature measurements available from one single sensor located inside the tumor, taken at a rate of one measurement every 1 s. The measurement errors were assumed Gaussian with zero mean and a constant standard deviation of 0.5°C. For the application of the SIR and ASIR algorithms, the fluence rate was treated as a state variable with evolution model in the form of a random walk, with Gaussian noise of zero mean and a standard deviation of 1% of the deterministic value of the fluence rate at each position of the finite volume mesh. A constant standard deviation of 0.5°C was assumed for the evolution model of the temperature. On the other hand, in the case of Liu and West's algorithm, Gaussian prior probability densities were assumed for the optical and thermophysical properties, with zero means and standard deviations corresponding to ~2% of the means [205]. Figures 8.7b-d present the estimated temperatures at $t = 20$ s, obtained with the three different filters, for $N = 100$ particles. The exact temperature distribution at $t = 20$ s is presented in Figure 8.7a. This figure shows a good agreement between the estimated and exact temperature distributions. The accuracy of the estimated temperatures can also be verified in Figure 8.8,

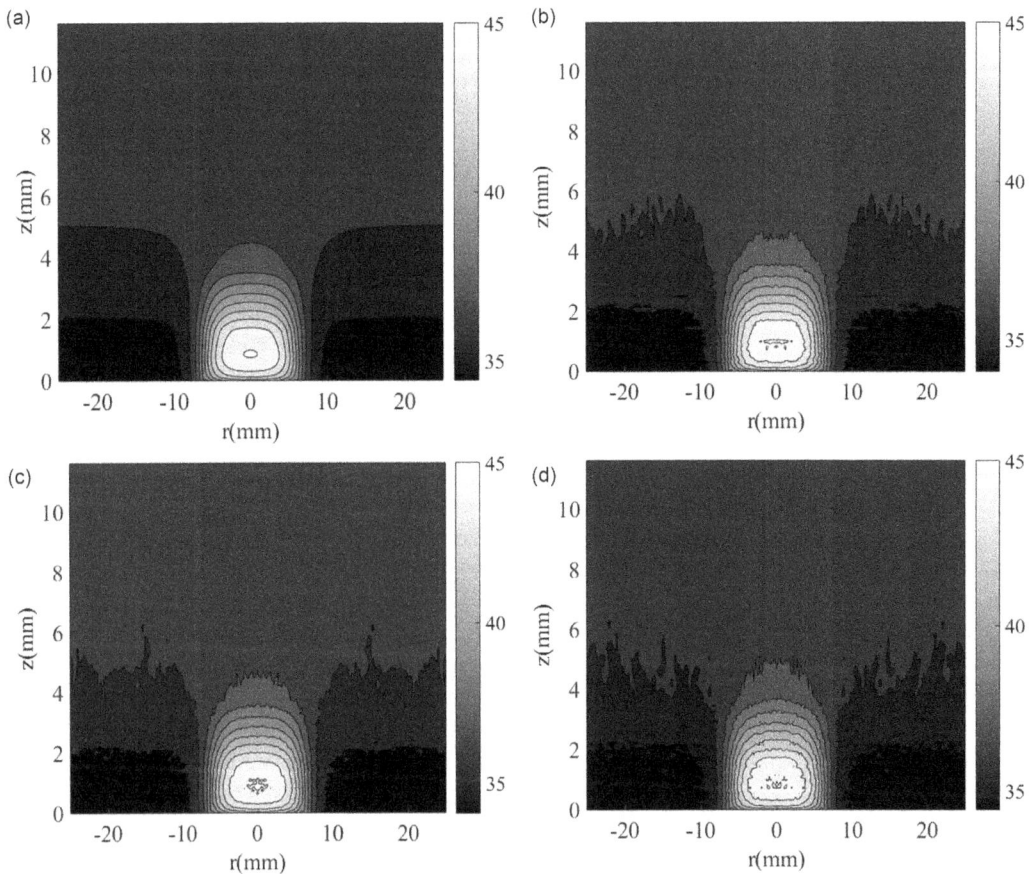

FIGURE 8.7 Comparison of estimated and exact temperature distributions at $t = 20$ s for $N = 100$ particles: (a) exact; (b) SIR; (c) ASIR; (d) Liu & West. (From Ref. [205].)

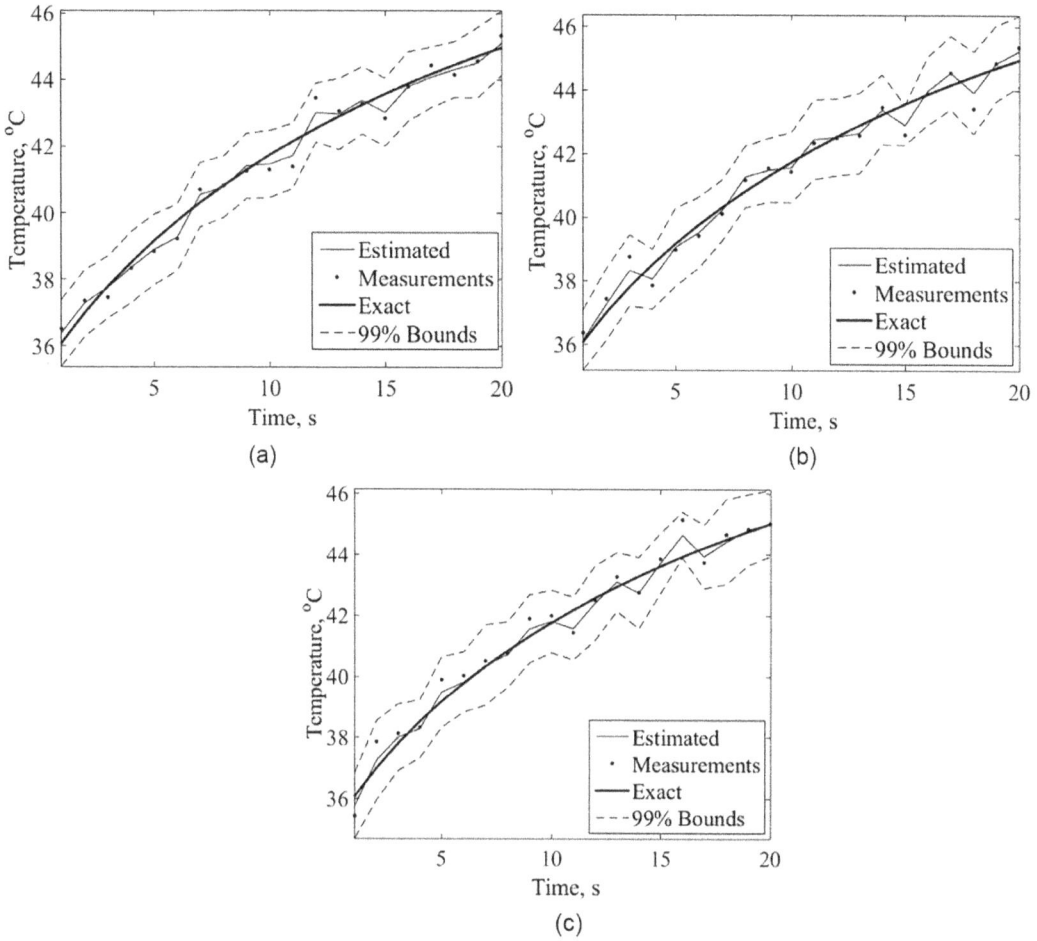

FIGURE 8.8 Comparison of the estimated and exact transient temperature variations with the temperature measurements at the sensor position for $N = 100$ particles: (a) SIR; (b) ASIR; (c) Liu & West. (From Ref. [205].)

where the transient variations of the estimated temperatures are compared with the exact ones at the sensor location, for $N = 100$ particles. The simulated transient temperature measurements were also included in these figures. Figures 8.7 and 8.8 show that the estimated transient variations of temperature obtained with the three different filters followed the exact ones, and that the estimated temperatures were generally closer to the exact temperatures than the simulated temperatures.

Table 8.4 shows the performance of the SIR, ASIR and Liu and West algorithms in terms of the computational time, as well as of mean and standard deviation values of the RMS errors of the temperature, for two different number of particles $N = 100$ and $N = 250$. Notice in this table that, as expected, the RMS error was reduced when the number of particles increased. Furthermore, it can be noticed that, for the same number of particles, the SIR filter presented the smallest RMS errors, while the largest RMS errors were obtained with Liu and West's filter. In addition, the SIR filter generated estimates with the smallest dispersions, thus, the estimates were closer to the exact values a larger number of runs than for the other filters. Differently from the SIR and ASIR algorithms, the cases run with Liu and West's algorithm contained uncertainties in the model parameters as well as in the evolution models; for this reason the RMS errors of Liu and West's algorithm were the largest. Anyhow, Figures 8.7 and 8.8 reveal that the three particle filter algorithms resulted in accurate estimations of the unknowns for this case, even with a small number of particles such as $N = 100$.

The particle filter algorithms were also compared in terms of the computational cost for the state estimation solution. Table 8.4 shows the computational time of one run of the SIR, ASIR and Liu and West algorithms, for $N = 100$ and $N = 250$ particles. It can be noticed in this table that the SIR filter presented the smallest computational cost, while that of the ASIR filter was approximately as twice as that of the SIR filter. This was due to the fact that in the ASIR filter the state evolution was performed twice (see Table 8.2). The computational cost of Liu and West's filter was several times higher than those of the SIR and ASIR filters, since both radiation and bioheat transfer problems were solved at each time step for each particle, as discussed above.

For the case of radiofrequency heating, a 2D rectangular domain of dimensions $L_x = 80$ mm and $L_y = 40$ mm was considered, while the tumor was assumed as a circle of radius $R = 10$ mm located at the center of the 2D rectangular domain. The lengths of both electrodes were considered of 20 mm, so that $\Omega_1' = \{-10 \text{ mm} \leq x \leq 10 \text{ mm}, y = 20 \text{ mm}\}$ and $\Omega_2' = \{-10 \text{ mm} \leq x \leq 10 \text{ mm}, y = -20 \text{ mm}\}$, where the x, y axes were supposed to be located at the tumor center (see Figure 8.6). The voltage applied over Ω_1' was $U = 10$ V. Temperature measurements of one single sensor were assumed available for the inverse analysis, located at the position $\{x = 10 \text{ mm}, y = 0\}$. The measurements were generated from the solution of the forward problem with the parameters specified above. Uncorrelated Gaussian errors with zero mean and a constant standard deviation of 1°C were added to the solution of the direct problem. The simulated measurements were supposed available every 20 s, during 900 s. The particle filter algorithms were applied with $N = 100$, 250 and 500 particles.

For the application of the SIR and ASIR algorithms, Gaussian uncorrelated noise with zero mean and a constant standard deviation of 1°C was added to the evolution model for temperature, which was solved with each sample of a Gaussian distribution for the electrical heat source; the means of such Gaussian distribution were obtained from the deterministic solution of the electric Problem (8.5.8), with standard deviations given by 10% of these values. For the application of Liu and West's algorithm, Gaussian uncorrelated noise with zero mean and a constant standard deviation of 1°C was also added to the solution of the bioheat transfer problem used for the evolution of temperatures. However, differently from the SIR and ASIR filters, the electric problem was also solved for each particle at each time step of Liu and West's algorithm, by considering Gaussian uncertainties in the electric parameters with zero means and standard deviations of 10% of the exact parameter values. The other model parameters were also supposed as Gaussian with means given by their exact values and standard deviations of 10% of their exact values.

Figures 8.9a-c present the estimated temperature fields at $t = 900$ s, obtained with Liu and West's algorithm, by using 100, 250 and 500 particles. Such comparison of the effects of the number of particles on the solution was made with Liu and West's filter because it accounts for uncertainties on the model parameters, as well as on the evolution models. As the number of particles increased, the estimated temperature field tended to the exact one. Therefore, $N = 500$ particles were used for the results presented in Figures 8.10a and b, which show the temperature field in the regions obtained with the SIR and ASIR filters, respectively. Figures 8.11a-c present the transient temperature

TABLE 8.4

Results for Laser Heating

Filter	Number of Particles	RMS Error Mean (°C)	RMS Error Standard Deviation (°C)	CPU Time
SIR	$N = 100$	0.118	0.009	19 min 48 s
	$N = 250$	0.076	0.005	46 min 27 s
ASIR	$N = 100$	0.162	0.021	37 min 49 s
	$N = 250$	0.135	0.026	92 min 50 s
LW	$N = 100$	0.172	0.029	32 h 19 min
	$N = 250$	0.150	0.032	81 h 02 min

Source: From Ref. [205].

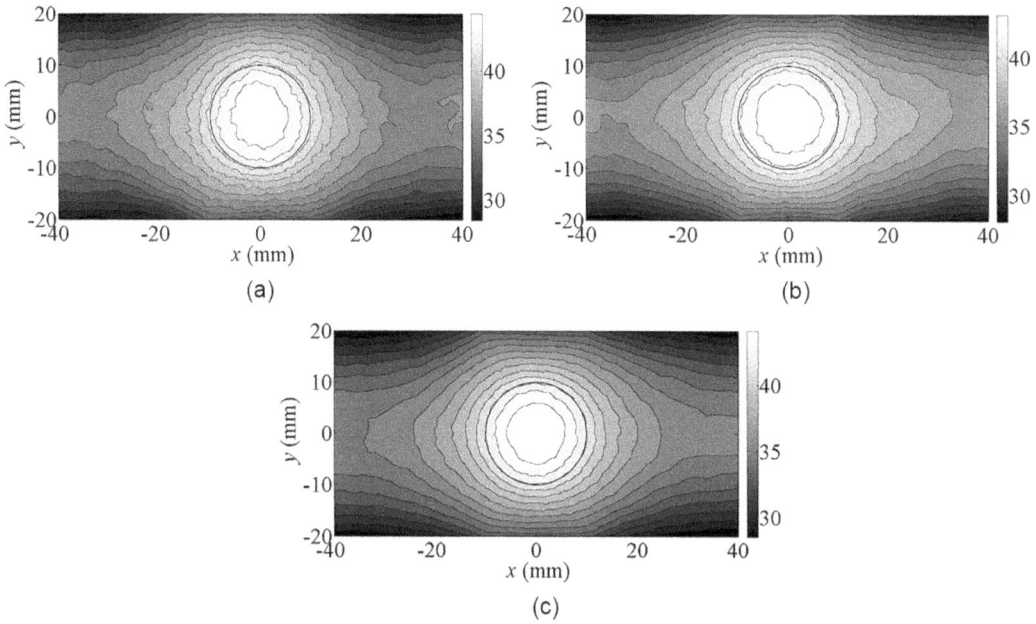

FIGURE 8.9 Estimated temperature field at $t = 900$ s obtained with Liu & West's algorithm: (a) $N = 100$ particles, (b) $N = 250$ particles, (c) $N = 500$ particles. (From Ref. [205].)

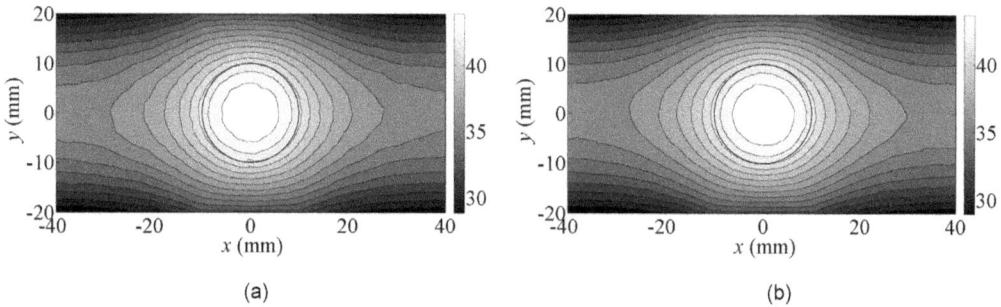

FIGURE 8.10 Estimated temperature fields at $t = 900$ s obtained by using $N = 500$ particles with: (a) SIR; (b) ASIR. (From Ref. [205].)

variations obtained with the three filters, for $N = 500$ particles, at the measurement position. Such as for the case involving the laser heating, these figures show an excellent agreement between estimated and exact temperatures, with relatively small confidence intervals for the three algorithms.

The experimental validation of the state estimation solution with the algorithm of Liu and West was performed with laser heating in the near-infrared range of a cylindrical phantom in [210]. A tumor was simulated inside a plastic phantom by a disk that contained Fe_2O_3 nanoparticles, which were used to promote a localized absorption of the near-infrared laser beam. The state estimation problem was solved using transient temperature measurements available at a single position within the phantom.

Figures 8.12a and b present the comparisons of the estimated and measured transient temperature variations for laser powers of 156.6 and 220 mW, respectively. These measurements were used for the solution of the state estimation problem. The associated 99% confidence intervals of the estimated and measured temperature variations are also included in these figures. It can be noticed in Figures 8.12a and b that the temperature variations estimated with the particle filter matched the measurements used for the solution of the state estimation problem, for both laser powers, at the graph scale. Note also the larger temperature variations observed for the larger power.

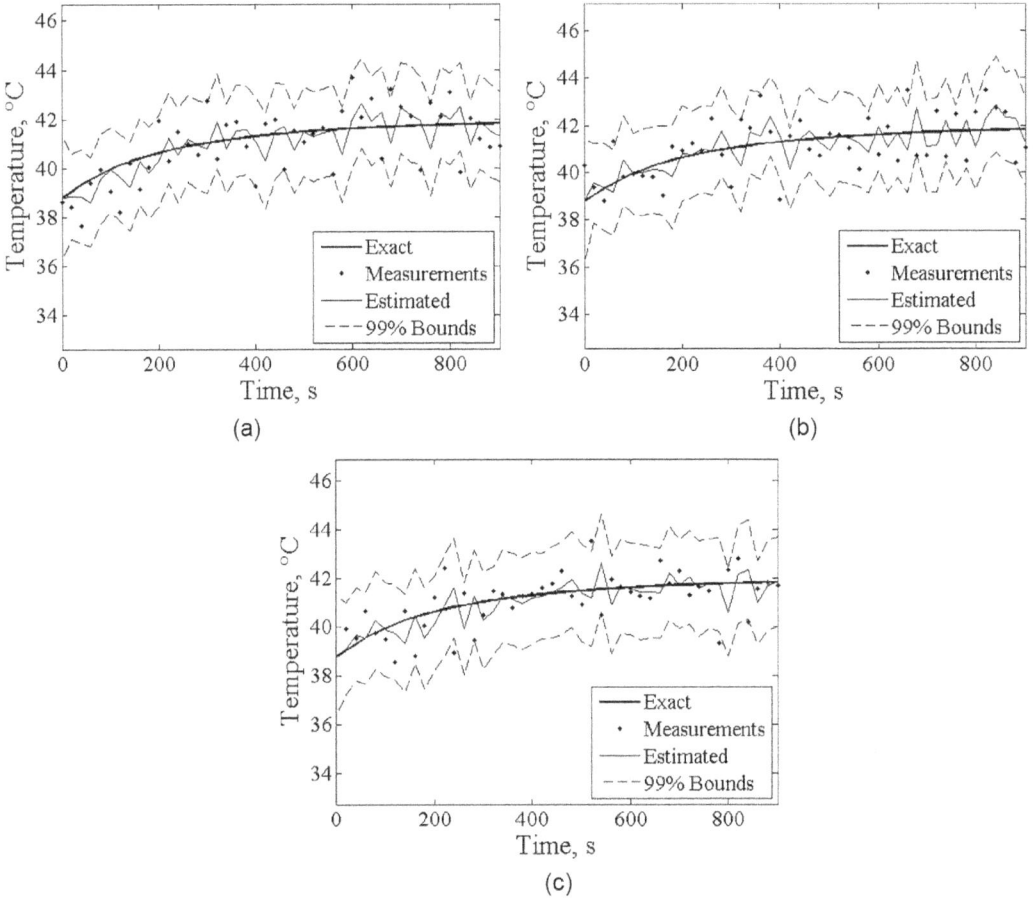

FIGURE 8.11 Estimated temperatures at the measurement position obtained by using $N = 500$ particles with the algorithms: (a) SIR, (b) ASIR, (c) Liu & West. (From Ref. [205].)

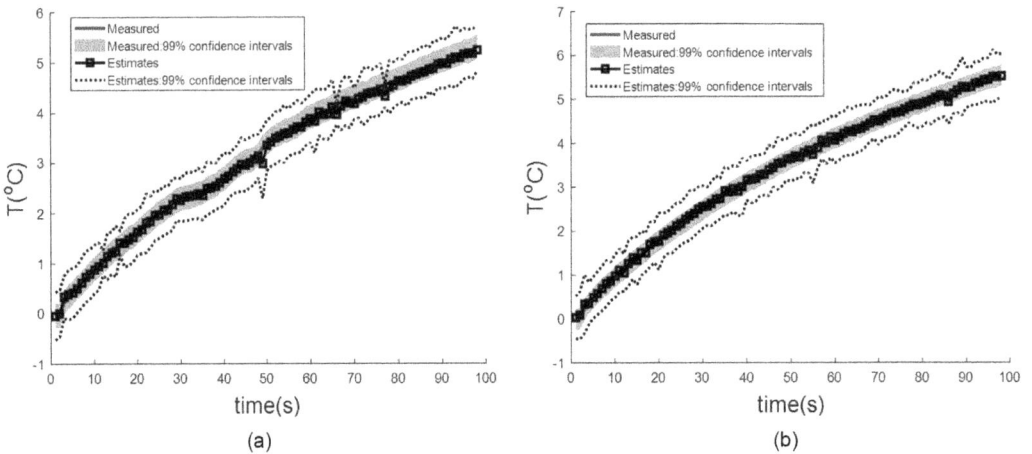

FIGURE 8.12 Comparison of the measured and estimated transient temperature variations: (a) laser power of 156.6 mW; (b) laser power of 220 mW. (From Ref. [210].)

As a validation of the solution of the state estimation problem, the estimated temperature variations were compared to other measurements not used for the solution of the state estimation problem. Figures 8.13a-d present a comparison of the estimated radial temperature variations with the temperature measurements at the boundary exposed to the laser radiation, at selected times ($t = 60$ s and $t = 90$ s), for two laser powers (156.6 and 220 mW). The measurements shown in Figure 8.13 were taken with an infrared camera. The associated 99% confidence intervals of the estimated and measured temperatures are also included in Figure 8.13. This figure shows that the temperature variations estimated with the proposed methodology were in excellent agreement with the measurements. The measurements and the estimated temperatures assumed a Gaussian profile, due to the laser collimator used in the experiments and to the heat transfer process. Note in Figure 8.13 the larger temperature variations as time and laser power increase. This figure shows that the surface temperature spatial and transient variations were correctly estimated through the solution of the state estimation problem with Liu and West's version of the particle filter.

Besides the estimation of the state variables of interest for this problem, that is, the temperature and fluence rate fields, Liu and West's algorithm of the particle filter allowed for simultaneous estimation of the non-dynamic model parameters. Figure 8.14 presents the estimated 99% confidence intervals for the model parameters, including the irradiances, for both laser powers. The means of the Gaussian priors used for each parameter are also presented in this figure. Figure 8.14 shows that the confidence intervals of the model parameters were sequentially reduced and tended to the mean values used for the priors, as the information provided by the measurements was taken into account in the solution of the state estimation problem. Moreover, the confidence intervals for the physical

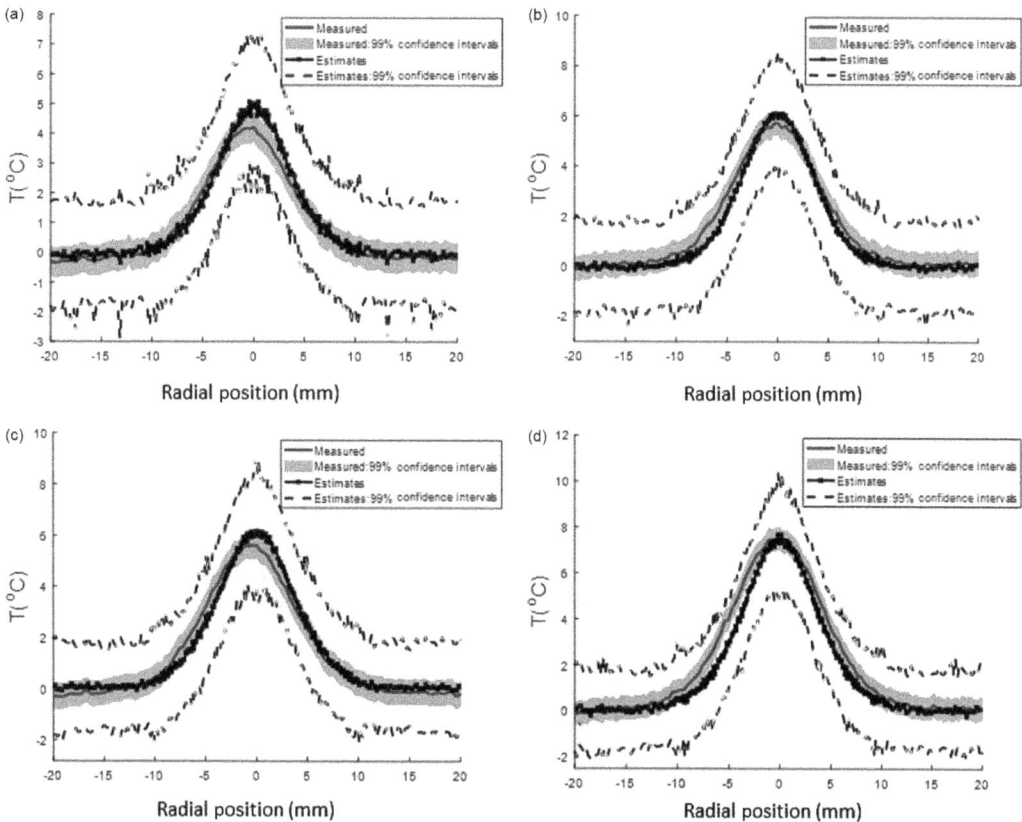

FIGURE 8.13 Comparison of the estimated radial temperature variation with the measurements obtained with an infrared camera: (a) $t = 60$ s and 156.6 mW; (b) $t = 60$ s and 220 mW; (c) $t = 90$ s and 156.6 mW; (d) $t = 90$ s and 220 mW. (From Ref. [210].)

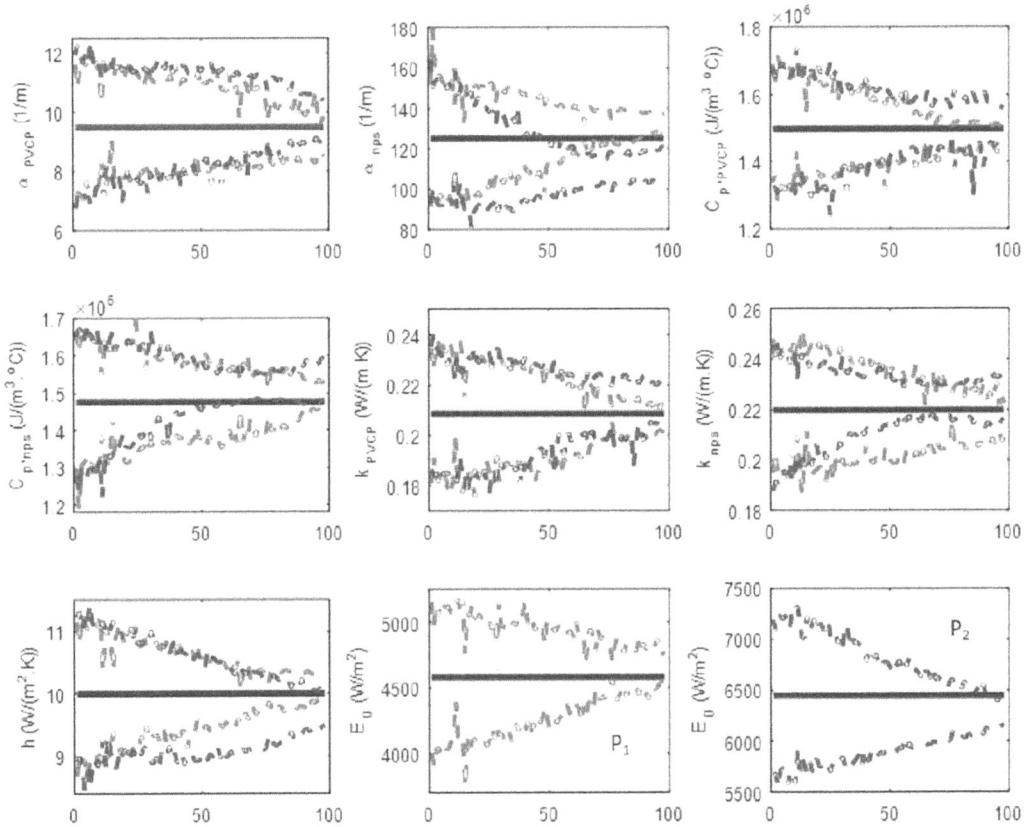

FIGURE 8.14 Confidence intervals (99%) of the model parameters sequentially estimated with the particle filter. (From Ref. [210].)

properties, which were estimated with the two different laser powers, were quite similar because these quantities are not functions of the incident irradiance. On the other hand, the confidence intervals of the irradiances tended toward the two different prior means, as these parameters were sequentially estimated for the two laser powers.

PROBLEMS

8.1 Consider a slab of thickness L, initially at the uniform temperature T_0, which is subjected to a uniform heat flux $q(t)$ over one of its surfaces. The other surface exchanges heat by convection and linearized radiation with a medium at a temperature T_∞ with a heat transfer coefficient h. Temperature gradients are neglected inside the slab and a lumped formulation is used. The formulation for this problem is thus given by:

$$\frac{d\theta(t)}{dt} + m\theta(t) = \frac{mq(t)}{h} \quad \text{for} \quad t > 0$$

$$\theta = \theta_0 \quad \text{for} \quad t = 0$$

where

$$\theta(t) = T(t) - T_\infty; \quad \theta_0 = T_0 - T_\infty; \quad m = \frac{h}{\rho c L}$$

and ρ is the density and c is the specific heat of the homogeneous material of the slab.

Let the heat flux $q(t) = q_0$ be constant and deterministically known. Use the SIR algorithm of the particle filter method to estimate the transient temperature in a slab made of aluminum ($\rho = 2707 \, \text{kg/m}^3$, $c = 896 \, \text{J/kg K}$), with thickness $L = 0.03 \, \text{m}$, for $q_0 = 8000 \, \text{W/m}^2$, $T_\infty = 20°\text{C}$, $h = 50 \, \text{W/m}^2$ K and $T_0 = 50°\text{C}$. Measurements of the transient temperature of the slab are assumed available. These measurements contain additive, uncorrelated, Gaussian errors, with zero mean and a constant standard deviation σ_z. The errors in the state evolution model are also supposed to be additive, uncorrelated, Gaussian, with zero mean and a constant standard deviation σ_θ. Examine the effects of σ_z and σ_θ on the inverse problem solution.

8.2 Repeat Problem 8.1 but now consider the heat flux $q(t) = q_0 f(t)$, with known q_0 but unknown time variation $f(t)$. Use a Gaussian random walk model for $f(t)$.

8.3 Repeat Problems 8.1 and 8.2 but solve the state estimation problem with the ASIR algorithm of the particle filter method.

8.4 Repeat Problems 8.1 and 8.2 but solve the state estimation problem with the Liu and West algorithm of the particle filter method by considering Gaussian uncertainties for the model parameters.

Appendix
Approximate Bayesian Computation

The subject of Approximate Bayesian Computation (ABC) has been delayed to an appendix in this book because the likelihood has been assumed as a Gaussian distribution in the previous chapters. In fact, errors in temperature measurements can be fairly modeled as Gaussian, as illustrated by Figure 1.6. Besides that, for most of the applications, temperature measurements can be easily performed with quite reliable and inexpensive apparatuses that are commercially available.

On the other hand, there are many situations in practice where the likelihood is not exactly known and cannot be approximated in terms of analytical distributions. There might also be situations in which the calculation of the actual likelihood demands large computational times and resources, thus requiring model reduction or data compression, such as shown by examples in the previous chapters. While the Approximation Error Model (AEM) approach was quite effective for the examples presented in this book, the Gaussian model for the total (measurement and modeling) errors cannot be arbitrarily used in general. It is also important to note that the data compression scheme used in Section 6.4 still resulted in an analytical likelihood given by a Gaussian distribution, that is, of the same kind of the original likelihood based on the actual measured data. However, other data compression techniques, like the ones quite popular nowadays based on artificial intelligence, might not result on an analytical likelihood.

ABC can deal with measurement/total errors that may not be appropriately modeled in terms of analytical distributions or the computation of the likelihood function is very time consuming [A.1-A.9]. In this appendix, we present a robust and powerful ABC algorithm, which can be used for simultaneous model selection and model calibration (estimation of the model parameters) [A.2-A.4]. This algorithm is then illustrated by an inverse bioheat transfer problem [A.10].

A.1 SIMULTANEOUS MODEL SELECTION AND MODEL CALIBRATION WITH APPROXIMATE BAYESIAN COMPUTATION

The algorithm of Toni et al. [A.2], which is an extension of the Sequential Monte Carlo algorithm of Sisson et al. [A.1], is presented here for simultaneous model selection and model calibration. The objective is to obtain the combined posterior distribution $\pi(\mathbf{P},\mathbf{M}|\mathbf{Y})$, where $\mathbf{M}$ is a vector of integer numbers that index the models that are considered in the analysis, $\mathbf{P}$ is the vector of model parameters and $\mathbf{Y}$ is the vector of measurements.

In this algorithm [A.2], a model is initially selected from its prior $\pi(\mathbf{M})$, and then a sample for the parameters of this model is obtained from their prior distributions. This process of selecting the model and sampling the parameters is repeated until a pre-specified number of samples (particles), N_s, are accepted. A user-selected distance function, $d(\mathbf{Y},\mathbf{T})$, between measurements ($\mathbf{Y}$) and estimated dependent variables ($\mathbf{T}$), is used as the acceptance criterion instead of the likelihood, with a specified tolerance (ε), that is, $d(\mathbf{Y},\mathbf{T}) \leq \varepsilon$. The set of accepted particles is named population. Weights are calculated for each particle and normalized according to each model. New populations of size N_s are then generated by sampling the model from its prior and sampling the model parameters from their estimated distributions at the previous population. This process is repeated until the tolerance for the last population, ε_{N_P}, is satisfied, where N_P indicates the number of populations. Tolerances are usually set large for the first populations, in order to avoid lack of convergence when the sampling is dominated by the priors. The information provided by the measurements becomes more significant as the populations advance. Consequently, a larger number of particles are selected

TABLE A.1

ABC Algorithm of Toni et al. [A.2]

1. Define the number of populations, N_P, the size of each population (number of particles), N_s, and the tolerances $\varepsilon_1, \varepsilon_2, \ldots, \varepsilon_{N_P}$ for each iteration (population) used for selecting the model and its parameters. Also, specify the distance function $d(\mathbf{Y}, \mathbf{T})$ that substitutes the likelihood function. Set the population indicator $n = 0$.

2. Set the particle indicator $i = 1$, where each particle represents, at each iteration, a model and its parameters.

3. Sample the model M^* from the prior distribution for the models, $\pi(\mathbf{M})$.

4. If $n = 0$, sample the candidate parameters $\mathbf{P}^{**}$ from the prior distribution for the parameters of model M^*, that is, $\pi(\mathbf{P} \mid M^*)$. Otherwise, sample $\mathbf{P}^*$ from the parameters of model M^* in the previous population, $\mathbf{P}_{n-1}^i(M^*)$, with weights $w_{n-1}^i(M^*)$, and perturb this particle to obtain $\mathbf{P}^{**} \approx K_n(\mathbf{P}^*, \mathbf{P}^{**})$, where K_n is a perturbation kernel.

5. If $\pi(\mathbf{P}^{**}) = 0$, return to step 3. Otherwise, simulate from the direct problem (operator f) a candidate set of observable variables with model M^* and parameter $\mathbf{P}^{**}$, that is, $\mathbf{T} = f(\mathbf{P}^{**}, M^*)$.

6. If $d(\mathbf{Y}, \mathbf{T}) > \varepsilon_n$, return to step 3. Otherwise, set $M_n^i = M^*$, $\mathbf{P}_n^i(M^*) = \mathbf{P}^{**}$ and calculate the particle weight:

$$w_n^i(M^*) = \begin{cases} 1 & \text{if } n = 0 \\ \dfrac{\pi(\mathbf{P}_n^i \mid M^*)}{\displaystyle\sum_{k=1}^{N_s}\left\{ w_{n-1}^k(M^*) K_n\left[\mathbf{P}_{n-1}^k(M^*), \mathbf{P}_n^i(M^*)\right]\right\}} & \text{if } n > 0 \end{cases}$$

7. If $i < N_s$, where N_s is the number of particles, set $i = i + 1$ and go to step 3.

8. Normalize the weights.

9. If $n < N_P$, where N_P is the number of iterations (populations), set $n = n + 1$ and go to step 2. Otherwise, terminate the iterations.

for the correct model and the samples for the parameters better reflect the real posterior distribution. Convergence tolerances are then gradually reduced for accurate model selection and parameter estimation as the populations advance.

The ABC algorithm of Toni et al. [A.2] is summarized in Table A.1.

A.2 AN APPLICATION OF APPROXIMATE BAYESIAN COMPUTATION IN BIOHEAT TRANSFER

The physical problem considered here involved the laser heating of a biological tissue composed of 65% wt of water and 35% wt of protein [A.10]. The tissue consisted of a cylindrical geometry, with 3 mm of radius and 3 mm of height. Initially, the tissue was assumed at the uniform body temperature, T_b. A laser heating was then applied at the surface of the tissue at $z = H$, while the boundaries at $r = R$ and $z = 0$ remained at the body temperature, T_b. The boundary at $z = H$ also exchanged heat by convection and linearized radiation with the surrounding environment at the temperature T_∞, with a heat transfer coefficient h_∞. Figure A.1 illustrates the axisymmetric physical problem when the domain contains only the original homogeneous tissue (a) and when part of the tissue has been thermally damaged due to the laser heating (b).

The mathematical formulation for this physical problem is given by:

$$\rho(r,z)\frac{\partial h(r,z,t)}{\partial t} = \nabla \cdot \left[k(r,z)\nabla T(r,z,t)\right] + Q(r,z) \quad 0 < r < R, \quad 0 < z < H, \quad t > 0 \qquad \text{(A.2.1a)}$$

$$T(r,z,t) = T_b \quad 0 < r < R, \quad 0 < z < H, \quad t = 0 \qquad \text{(A.2.1b)}$$

$$\frac{\partial T(r,z,t)}{\partial r} = 0 \quad r = 0, \quad 0 < z < H, \quad t > 0 \qquad \text{(A.2.1c)}$$

$$T(r,z,t) = T_b \quad r = R, \quad 0 < z < H, \quad t > 0 \qquad \text{(A.2.1d)}$$

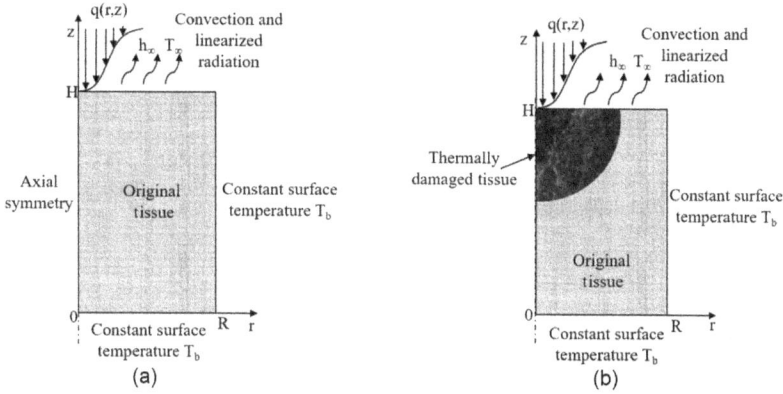

FIGURE A.1 Laser heating of a biological tissue. (a) Original tissue; (b) Tissue after thermal damage. (From Ref. [A.10].)

$$T(r,z,t) = T_b \quad 0 < r < R, \quad z = 0, \quad t > 0 \tag{A.2.1e}$$

$$-k(r,z)\frac{\partial T(r,z,t)}{\partial z} = h_\infty(T - T_\infty) \quad 0 < r < R, \quad z = H, \quad t > 0 \tag{A.2.1f}$$

where $h(r,z,t)$ is the specific enthalpy.

The source term in equation (A.2.1a) includes the energy supplied by blood perfusion, tissue metabolism and the heating laser, that is:

$$Q(r,z) = Q_b(r,z,\Omega) + Q_m(\Omega) + Q_l(r,z) \tag{A.2.2}$$

The blood perfusion source term is:

$$Q_b(r,z,\Omega) = \rho_b c_b \omega_b(\Omega)\left[T_b - T(r,z)\right] \tag{A.2.3}$$

where T_b represents the arterial blood temperature, ρ_b the density and c_b the specific heat of the blood. The perfusion coefficient ω_b was considered as a function of the tissue thermal damage, Ω:

$$\omega_b = \begin{cases} \left(1 + 25\Omega - 260\Omega^2\right)\omega_0 & \Omega \leq 0.1 \\ (1-\Omega)\omega_0 & 0.1 < \Omega \leq 1 \\ 0 & \Omega > 1 \end{cases} \tag{A.2.4}$$

where ω_0 is the blood perfusion of the healthy tissue. Values of thermal damage up to $\Omega = 0.1$ increased the perfusion due to vasodilation of the heated tissue, while values of thermal damage in $0.1 < \Omega \leq 1$ decreased the blood perfusion due to blood clotting. The value of Ω equal to unity ($\Omega = 1$) was associated with blood coagulation and tissue necrosis, when perfusion was then ceased.

The term $Q_m(\Omega)$ in equation (A.2.2) corresponds to the metabolic heat generation, which was considered uniform in the original homogeneous tissue. On the other hand, the metabolic heat generation was null when the tissue became necrotic ($\Omega = 1$) due to the thermal damage. The energy provided by the laser was modeled with Beer-Lambert's law due to the small size of the region [A-10].

Two different models were used here to predict the thermal damage resulting from the laser heating, namely, an Arrhenius model (Model 1) and a two-state model (Model 2) [A.10]. The Arrhenius model considers that the necrosis of the tissue occurs when $\Omega = 1$. The Arrhenius formulation is given by equation (A.2.5) and relates the damage of the tissue to the temperature (T) and time (t) of heating. The parameters of Model 1 are the universal gas constant (R_u), activation energy, E_a, and frequency factor, which is expressed in power form, that is, 10^κ [A.10].

$$\Omega = \int_0^t 10^\kappa \exp\left(-\frac{E_a}{R_u T}\right) dt' \tag{A.2.5}$$

The two-state model (Model 2) is given by Ref. [A.10]:

$$S(t,T) = \frac{1}{1 + \exp\left[-\left(\frac{\gamma}{T} - \beta - \alpha t\right)\right]} \tag{A.2.6}$$

The tissue thermal damage in the two-state model is given by Ref. [A.10]:

$$\Omega = \ln\left\{\frac{S(0)}{S(t)}\right\} \tag{A.2.7}$$

These two competing models were considered to represent the synthetic experimental data. For the results presented below, synthetic transient measurements of the necrosis front at the heated surface were assumed available with a frequency of 100 Hz. The synthetic measurements were Gaussian, with means given by the solution of the direct problem and standard deviations of 1% of these mean values. In order to avoid an inverse crime, the inverse problem was solved on a mesh much coarser than that used to generate the synthetic measurements. Uniform transition kernels were used to generate the particles for the parameters in step 4 of the algorithm shown in Table A.1, and the competing Arrhenius and two-state models were assumed as equally probable to represent the measured data. Eleven populations of this algorithm, with 1000 particles each, were used for the solution of the inverse problem. These numbers of populations and particles, as well as the tolerances for the convergence of each population, were selected based on numerical experiments. The function used in step 6 of the algorithm of Toni et al. [A.2] (see Table A.1), in order to verify the agreement between the measurements **Y** and the estimated responses **T**, was the Euclidean distance between these two vectors. The convergence tolerances were gradually reduced as the populations advanced, and the tolerance for the final population was set based on Morozov's discrepancy principle. The following vector of tolerances $\varepsilon = [4.2 \times 10^{-3}, 1.1 \times 10^{-3}, 8.5 \times 10^{-4}, 6.8 \times 10^{-4}, 4.7 \times 10^{-4}, 3.8 \times 10^{-4}, 3.0 \times 10^{-4}, 2.1 \times 10^{-4}, 1.3 \times 10^{-4}, 8.5 \times 10^{-5}, 6.4 \times 10^{-5}]$ m was used as the convergence criterion.

We present the results for a case where the synthetic measurements were obtained with Model 2 (two-state model). Figure A.2 shows the numbers of particles selected for each model as the

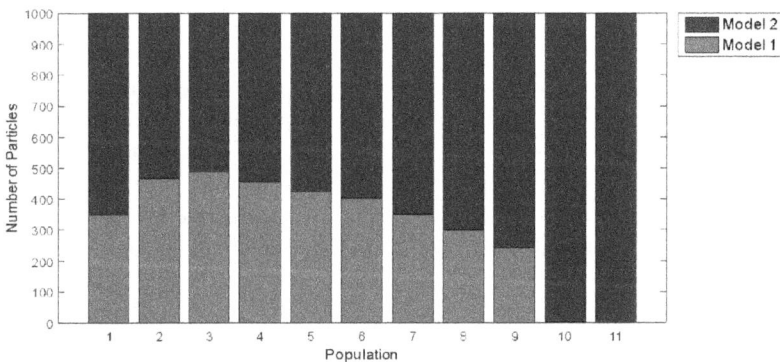

FIGURE A.2 Number of accepted particles at each population—measurements generated with Model 2. (From Ref. [A.10].)

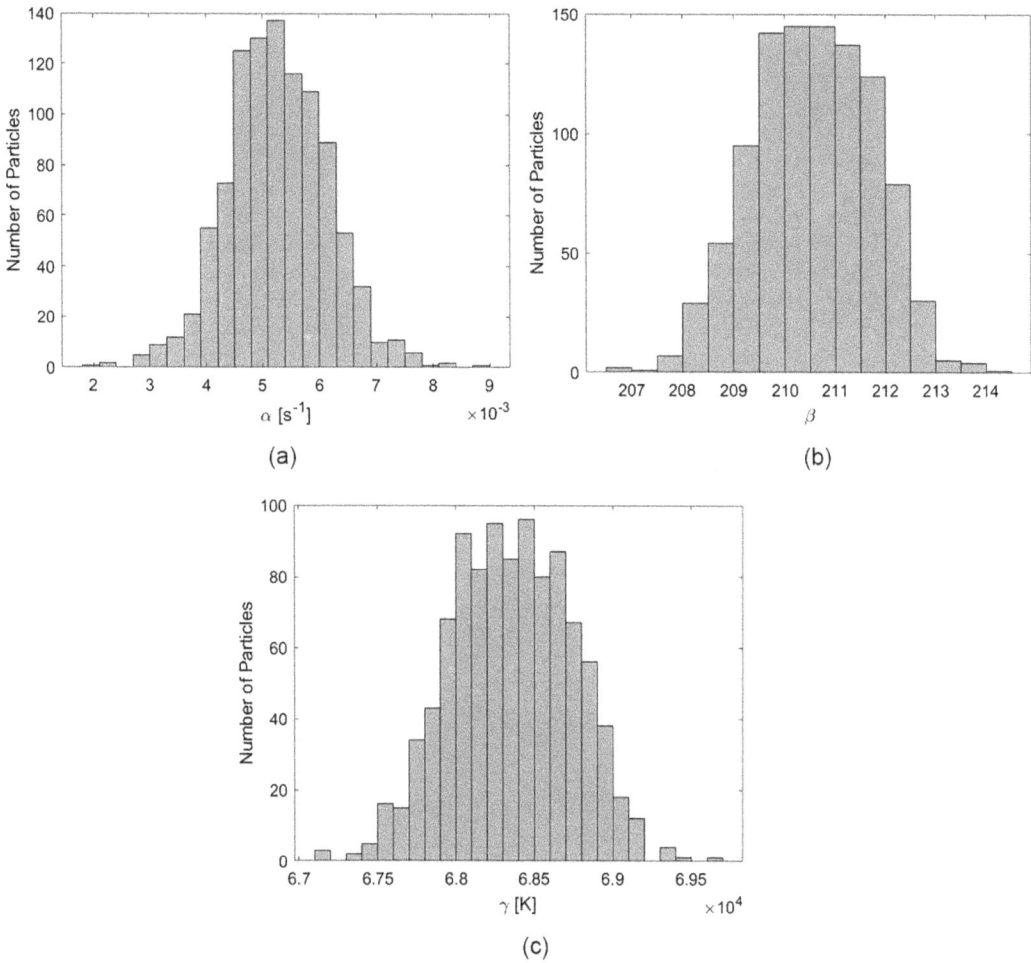

FIGURE A.3   Histograms for the parameters of Model 2—measurements generated with Model 2: (a) α, (b) β, (c) γ. (From Ref. [A.10].)

populations advanced. The algorithm of Toni et al. [A.2] was capable of selecting the correct model in few populations, as shown by Figure A.2. Note in this figure that Model 2 was correctly the only one selected at populations 10 and 11. The histograms of the model parameters at the final population, presented by Figure A.3, exhibited Gaussian behaviors, centered at mean values quite close to the exact parameter values used to generate the simulated measurements, and with small variances.

REFERENCES

A.1. Sisson, S. A., Fan, Y., Tanaka, M. M., Sequential Monte Carlo without likelihoods, *Proceedings of the National Academy of Sciences*, **104**(6), 1760–1765, 2007.

A.2. Toni, T., Welch, D., Strelkowa, N., Ipsen, A., Stumpf, M. P., Approximate Bayesian computation scheme for parameter inference and model selection in dynamical systems, *Journal of the Royal Society Interface*, **6**(31), 187–202, 2009.

A.3. Toni, T., Stumpf, M. P., Parameter inference and model selection in signaling pathway models, in *Computational biology*, Edited by Fenyö, D., Humana Press, Totowa, NJ, pp. 283–295, 2010.

A.4. Toni, T., Stumpf, M. P., Simulation-based model selection for dynamical systems in systems and population biology, *Bioinformatics*, **26**(1), 104–110, 2010.

A.5. Del Moral, P., Doucet, A., Jasra, A. An adaptive sequential Monte Carlo method for approximate Bayesian computation, *Statistics and Computing*, **22**(5), 1009–1020, 2012.

A.6. Beaumont, M. A., Zhang, W., Balding, D. J., Approximate Bayesian computation in population genetics, *Genetics*, **162**(4), 2025–2035, 2002.

A.7. Marjoram, P., Molitor, J., Plagnol, V., Tavaré, S., Markov chain Monte Carlo without likelihoods, *Proceedings of the National Academy of Sciences*, **100**(26), 15324–15328, 2003.

A.8. Wegmann, D., Leuenberger, C., Excoffier, L., Efficient approximate Bayesian computation coupled with Markov chain Monte Carlo without likelihood, *Genetics*, **182**(4), 1207–1218, 2009.

A.9. Costa, J. M., Orlande, H. R. B., Lione, V. O., Lima, A. G., Cardoso, T. C., Varón, L. A., Simultaneous model selection and model calibration for the proliferation of tumor and normal cells during in vitro chemotherapy experiments, *Journal of Computational Biology*, **25**(12), 1285–1300, 2018.

A.10. Loiola, B. R., Orlande, H. R. B., Dulikravich, G. S., Approximate Bayesian computation applied to the identification of thermal damage of biological tissues due to laser irradiation, *International Journal of Thermal Sciences*, **151**, 106243, 2020.

References

1. Hadamard, J., *Lectures on Cauchy's problem in linear differential equations*, Yale University Press, New Haven, CT, 1923.
2. Alifanov, O. M., Determination of heat loads from a solution of the nonlinear inverse problem, *High Temperature,* **15**(3), 498–504, 1977.
3. Tikhonov, A. N., Arsenin, V. Y., *Solution of ill-posed problems*, Winston & Sons, Washington, DC, 1977.
4. Beck, J. V., Arnold, K. J., *Parameter estimation in engineering and science*, Wiley Interscience, New York, 1977.
5. Beck, J. V., Blackwell, B., St. Clair, C. R., *Inverse heat conduction: Ill-posed problems*, Wiley Interscience, New York, 1985.
6. Alifanov, O. M., *Inverse heat transfer problems*, Springer-Verlag, New York, 1994.
7. Alifanov, O. M., Artyukhin, E., Rumyantsev, A., *Extreme methods for solving ill-posed problems with applications to inverse heat transfer problems*, Begell House, New York, 1995.
8. Woodbury, K., *Inverse engineering handbook*, CRC Press, Boca Raton, FL, 2002.
9. Sabatier, P. C., *Applied inverse problems*, Springer Verlag, Hamburg, 1978.
10. Morozov, V. A., *Methods for solving incorrectly posed problems*, Springer Verlag, New York, 1984.
11. Murio, D. A., *The mollification method and the numerical solution of ill-posed problems*, Wiley Interscience, New York, 1993.
12. Trujillo, D. M., Busby, H. R., *Practical inverse analysis in engineering*, CRC Press, Boca Raton, FL, 1997.
13. Hensel, E., *Inverse theory and applications for engineers*, Prentice Hall, Upper Saddle River, NJ, 1991.
14. Kurpisz, K., Nowak, A. J., *Inverse thermal problems*, WIT Press, Southampton, UK, 1995.
15. Vogel, C., *Computational methods for inverse problems*, SIAM, New York, 2002.
16. Yagola A. G., Kochikov, I. V., Kuramshina, G. M., Pentin, Y. A., *Inverse problems of vibrational spectroscopy,* VSP, Netherlands, 1999.
17. Kaipio, J., Somersalo, E., *Statistical and computational inverse problems*, Applied Mathematical Sciences 160, Springer-Verlag, New York, 2004.
18. Calvetti, D., Somersalo, E., *Introduction to Bayesian scientific computing*, Springer, New York, 2007.
19. Tan, S., Fox, C., Nicholls, G., *Inverse problems,* Course notes for Physics 707, University of Auckland, Auckland, 2006.
20. Tarantola, A., *Inverse problem theory*, Elsevier, New York, 1987.
21. Bertero, M., Boccacci, P. *Introduction to inverse problems in imaging*, CRC Press, Boca Raton, FL, 1998.
22. Orlande, H. R. B., Fudym, F., Maillet, D., Cotta, R., *Thermal measurements and inverse techniques*, CRC Press, Boca Raton, FL, 2011.
23. Toivanen, J. M., Tarvainen, T., Huttunen, J. M. J., Savolainen, T., Pulkkinen, A., Orlande, H. R. B., Kaipio, J. P., Kolehmainen, V., Thermal tomography utilizing truncated Fourier series approximation of the heat diffusion equation, *International Journal of Heat and Mass Transfer*, **108**, 860–867, 2017.
24. Toivanen, J. M., Tarvainen, T., Huttunen, J. M. J., Savolainen, T., Orlande, H. R. B., Kaipio, J. P., Kolehmainen, V., 3D Thermal tomography with experimental measurement data, *International Journal of Heat and Mass Transfer*, **78**, 1126–1134, 2014.
25. Toivanen, J. M., Kolehmainen, V., Tarvainen, T., Orlande, H. R. B., Kaipio, J. P., Simultaneous estimation of spatially distributed thermal conductivity, heat capacity and surface heat transfer coefficient in thermal tomography, *International Journal of Heat and Mass Transfer*, **55**, 7958–7968, 2012.
26. Tikhonov, A. N., Solution of incorrectly formulated problems and the regularization method, *Soviet Mathematics: Doklady*, **4**(4), 1035–1038, 1963.
27. Tikhonov, A. N., Regularization of incorrectly posed problems, *Soviet Mathematics: Doklady*, **4**(6), 1624–1627, 1963.
28. Tikhonov, A. N., Inverse problems in heat conduction, *Journal of Engineering Physics*, **29**(1), 816–820, 1975.
29. Alifanov, O. M., Solution of an inverse problem of heat-conduction by iterative methods, *Journal of Engineering Physics*, **26**(4), 471–476, 1974.

30. Artyukhin, E. A., Nenarokomov, A. V., Coefficient inverse heat conduction problem, *Journal of Engineering Physics*, **53**, 1085–1090, 1988.

31. Alifanov, O. M., Kerov, N. V., Determination of external thermal load parameters by solving the two-dimensional inverse heat-conduction problem, *Journal of Engineering Physics*, **41**(4), 1049–1053, 1981.

32. Alifanov, O. M., Klibanov, M. V., Uniqueness conditions and method of solution of the coefficient inverse problem of thermal conductivity, *Journal of Engineering Physics*, **48**(6), 730–735, 1985.

33. Alifanov, O. M., Mikhailov, V. V., Solution of the nonlinear inverse thermal conductivity problem by the iteration method, *Journal of Engineering Physics*, **35**(6), 1501–1506, 1978.

34. Alifanov, O. M., Mikhailov, V. V., Determining thermal loads from the data of temperature measurements in a solid, *High Temperature*, **21**(5), 724–730, 1983.

35. Alifanov, O. M., Mikhailov, V. V., Solution of the overdetermined inverse problem of thermal conductivity involving inaccurate data, *High Temperature*, **23**(1), 112–117, 1985.

36. Alifanov, O. M., Rumyantsev, S. V., One method of solving incorrectly stated problems, *Journal of Engineering Physics*, **34**(2), 223–226, 1978.

37. Alifanov, O. M., Rumyantsev, S. V., On the stability of iterative methods for the solution of linear ill-posed problems, *Soviet Mathematics: Doklady*, **20**(5) 1133–1136, 1979.

38. Alifanov, O. M., Rumyantsev, S. V., Regularizing gradient algorithms for inverse thermal-conduction problems, *Journal of Engineering Physics*, **39**(2), 858–861, 1980.

39. Alifanov, O. M., Tryanin, A. P., Determination of the coefficient of internal heat exchange and the effective thermal conductivity of a porous solid on the basis of a nonstationary experiment, *Journal of Engineering Physics*, **48**(3), 356–365, 1985.

40. Artyukhin, E. A., Optimum planning of experiments in the identification of heat transfer processes, *Journal of Engineering Physics*, **56**, 256–259, 1989.

41. Artyukhin, E. A., Reconstruction of the thermal conductivity coefficient from the solution of the nonlinear inverse problem, *Journal of Engineering Physics*, **41**(4), 1054–1058, 1981.

42. Alifanov, O. M., Nenarokomov, A. V., Three-dimensional boundary inverse heat conduction problem for regular coordinate systems, *Inverse Problems in Engineering*, **7**(4), 335–362, 1999.

43. Artyukhin, E. A., Rumyantsev, S. V., Descent steps in gradient methods of solution of inverse heat conduction problems, *Journal of Engineering Physics*, **39**, 865–868, 1981.

44. Beck, J. V., Calculation of surface heat flux from an internal temperature history, *ASME Paper 62-HT-46*, 1962.

45. Giedt, W. H., The determination of transient temperatures and heat transfer at a gas-metal interface applied to a 40-mm gun barrel, *Jet Propulsion*, **25**, 158–162, 1955.

46. Stolz, G., Numerical solutions to an inverse problem of heat conduction for simple shapes, *ASME Journal of Heat Transfer*, **82**, 20–26, 1960.

47. Calderón, A. P., On an inverse boundary value problem, in *Seminar on numerical analysis and its applications to continuum physics,* Edited by Meyer, W. H., Raupp, M. A., Sociedade Brasileira de Matematica, Rio de Janeiro, Brazil, pp. 65–73, 1980, or *Journal of Computational and Applied Mathematics*, **25**(2–3), pp. 133–138, 2006.

48. Kaipio, K., Fox, C., The Bayesian framework for inverse problems in heat transfer, *Heat Transfer Engineering*, **32**, 718–753, 2011.

49. Orlande, H. R. B., Inverse problems in heat transfer: New trends on solution methodologies and applications, *Journal of Heat Transfer*, **134**, 031011, 2012.

50. ASME V&V 20–2009, Standard for verification and validation in computational fluid dynamics and heat transfer, ASME, 2009.

51. Beck, J. V., Sequential methods in parameter estimation, *Third International Conference on Inverse Problems in Engineering, Tutorial Session*, Port Ludlow, WA, 1999.

52. Maybeck, P., Stochastic models, estimation and control, Academic Press, New York, 1979.

53. Arulampalam, S., Maskell, S., Gordon, N., Clapp, T., A tutorial on particle filters for on-line non-linear/non-Gaussian Bayesian tracking, *IEEE Transactions on Signal Processing*, **50**, 174–188, 2001.

54. Ristic, B., Arulampalam, S., Gordon, N., *Beyond the Kalman filter*, Artech House, Boston, MA, 2004.

55. Doucet, A., Freitas, N., Gordon, N., *Sequential Monte Carlo methods in practice*, Springer, New York, 2001.

56. Özisik, M. N., *Heat conduction* (2nd Edition), Wiley, Hoboken, NJ, 1994.

57. Beck, J. V., Criteria for comparison of methods of solution of the inverse heat conduction problems, *Nuclear Engineering and Design*, **53**, 11–22, 1979.

58. Colaço, M. J., Orlande, H. R. B., Dulikravich, G. S., Inverse and optimization problems in heat transfer, *Journal of the Brazilian Society of Mechanical Sciences and Engineering*, XXVIII, 1–24, 2006.

59. Gamerman, D., Lopes, H. F., *Markov Chain Monte Carlo: Stochastic simulation for Bayesian inference* (2nd Edition), Chapman & Hall/CRC Press, Boca Raton, FL, 2006.
60. Brooks, S., Gelman, A., Jones, G., Meng, X., *Handbook of Markov Chain Monte Carlo*, Chapman & Hall/CRC Press, Boca Raton, FL, 2011.
61. Fonseca, H. M., Orlande, H. R. B., Fudym, O., Sepúlveda, F., A statistical inversion approach for local thermal diffusivity and heat flux simultaneous estimation, *Quantitative InfraRed Thermography*, **19**, 170–189, 2014.
62. Hansen, P. C., The L-curve and its use in the numerical treatment of inverse problems, in *Computational inverse problems in electrocardiology*, Edited by Johnston, P., WIT Press, Southampton, pp. 119–142, 2001.
63. Lee, P. M., *Bayesian statistics*, Oxford University Press, London, 2004.
64. Bayes, T., An essay towards solving a problem in the doctrine of chances, by the late Rv. Mr. Bayes, F. R. S. Communicated by Mr. Price in a Letter to John Cannon, A. M. R. F. S., *Philosophical Transactions* **53**, 370–418, 1763
65. McGrayne, S. B., *The theory that would not die: How Bayes' rule cracked the enigma code, hunted down Russian submarines, and emerged triumphant from two centuries of controversy*, Yale University Press, Devon, 2011.
66. Wang, J., Zabaras, N., A Bayesian inference approach to the inverse heat conduction problem, *International Journal of Heat and Mass Transfer*, **47**, 3927–3941, 2004.
67. Winkler, R., *An introduction to Bayesian inference and decision*, Probabilistic Publishing, Gainsville, 2003.
68. Levenberg, K., A method for the solution of certain non-linear problems in least-squares, *Quarterly of Applied Mathematics*, **2**, 164–168, 1944.
69. Marquardt, D. W., An algorithm for least squares estimation of nonlinear parameters, *Journal of the Society for Industrial and Applied Mathematics*, **11**, 431–441, 1963.
70. Farebrother, R. W., *Linear least squares computations*, Marcel Dekker, New York, 1988.
71. Bard, Y. B., *Nonlinear parameter estimation*, Academic Press, New York, 1974.
72. Moore, D. S., McCabe, G. P., *Introduction to the practice of statistics* (2nd Edition), W.H. Freeman and Co., New York, 1993.
73. Dennis, J., Schnabel, R., *Numerical methods for unconstrained optimization and nonlinear equations*, Prentice Hall, Upper Saddle River, NJ, 1983.
74. Moré, J. J., The Levenberg-Marquardt algorithm: Implementation and theory, in *Numerical analysis, lecture notes in mathematics* (Vol. 630), Edited by Watson, G. A., Springer-Verlag, Berlin, pp. 105–116, 1977.
75. Pierre, D. A., *Optimization theory with applications*, Courier Corporation, Chelmsford, MA, 1986.
76. Fletcher, R., Reeves C. M., Function minimization by conjugate gradients, *Computer Journal*, **7**, 149–154, 1964.
77. Polak, E., *Computational methods in optimization*, Academic Press, New York, 1985.
78. Hestenes, M. R., Stiefel, E., Methods of conjugate gradients for solving linear systems, *Journal of Research of the National Bureau of Standards*, **49**, 409–436, 1952.
79. Powell, M. J. D., Restart procedures for the conjugate gradient method, *Mathematical Programming*, **12**, 241–254, 1977.
80. Boiger, R., Leitao, A., Svaiter, B. F., Range-relaxed criteria for choosing the Lagrange multipliers in nonstationary iterated Tikhonov method, *IMA Journal of Numerical Analysis*, **40**(1), 606–627, 2020.
81. Machado, M. P., Margotti, F., Leitão, A., On the choice of Lagrange multipliers in the iterated Tikhonov method for linear ill-posed equations in Banach spaces, *Inverse Problems in Science and Engineering*, 1–31, 2019.
82. Mejias, M. M., Orlande, H. R. B., Ozisik, M. N., A comparison of different parameter estimation techniques for the identification of thermal conductivity components of orthotropic solids, In *Third International Conference on Inverse Problems in Engineering*, Port Ludlow, WA, 13–18, 1999.
83. Press, W. H., Flannery, B. F., Teukolsky S. A., Wetterling W. T., *Numerical recipes*, Cambridge University Press, New York, 1989.
84. Daniel, J. W., *The approximate minimization of functionals*, Prentice-Hall Inc., Englewood Cliffs, NJ, 1971.
85. Taktak, R., Design and validation of optimal experiments for estimating thermal properties of composite materials, Ph.D. Thesis, Department of Mechanical Engineering, Michigan State University, 1992.
86. Özişik, M. N., Orlande, H. R. B., Colaço, M. J., Cotta, R. M., *Finite difference methods in heat transfer*. CRC Press, Boca Raton, FL, 2017.

87. Fadale, T. D., Nenarokomov, A. V., Emery, A. F., Two approaches to optimal sensor locations, *Journal of Heat Transfer*, **117**, 373–379, 1995.

88. Mejias, M. M., Orlande, H. R. B., Ozisik, M. N., On the choice of boundary conditions for the estimation of the thermal conductivity components of orthotropic solids. In 2000 *National Heat Transfer Conference*, Pittsburgh, PA, 2000.

89. Taktak, R., Beck, J. V., Scott, E. P., Optimal experimental design for estimating thermal properties of composite materials, *International Journal of Heat Mass Transfer*, **36**(12), 2977–2986, 1993.

90. Mikhailov, M.D, Ozisik, M. N., *Unified analysis and solutions of heat and mass diffusion*, John Wiley & Sons, New York, 1984.

91. Dowding, K. J., Beck, J. V., Blackwell, B., Estimation of directional-dependent thermal properties in a carbon-carbon composite, *International Journal of Heat Mass Transfer*, **39**(15), 3157–3164, 1996.

92. Jarny, Y., Özisik, M. N., Bardon J. P., A general optimization method using adjoint equation for solving multidimensional inverse heat conduction, *International Journal Heat and Mass Transfer*, **34**, 2911–2919, 1991.

93. Bokar, J., Özisik, M. N., An inverse problem for the estimation of radiation temperature source term in a sphere, *Inverse of Problems in Engineering*, **1**, 191–205, 1995.

94. Orlande, H. R. B., Wellele, O., Ruperti Jr, N. J., Colaco, M. J., Delmas, A., Identification of the thermophysical properties of semi-transparent materials, In *International Heat Transfer Conference 13*, Begell House Inc, 2006.

95. ASTM Standard E1461-01, *Standard test method for thermal diffusivity by the flash method*, ASTM, West Conshohocken, PA, 2001.

96. Lazard, M., André, S., Maillet, D., Diffusivity measurement of semi-transparent media: Model of the coupled transient heat transfer and experiments on glass, silica glass and zinc selenide, *International Journal of Heat Mass Transfer*, **47**, 477–487, 2004.

97. Wellele, O., Orlande, H. R. B., Ruperti Jr, N., Colaço, M. J., Delmas, A., Coupled conduction–radiation in semi-transparent materials at high temperatures, *Journal of Physics and Chemistry of Solids*, **67**(9–10), 2230–2240, 2006.

98. Ozisik, M. N., *Radiative transfer and interactions with conduction and convection*, Werbel & Peck, New York, 1973.

99. Metropolis, N., Rosenbluth, A., Rosenbluth, M., Teller, A., Teller, E., Equation of state calculation by fast computing machines, *Journal of Chemical Physics*, **21**, 1087–1092, 1953.

100. Hastings, W. K., Monte Carlo sampling methods using Markov Chains and their applications, *Biometrika*, **57**, 97–109, 1970.

101. Haario, H., Saksman, E., Tamminen, J., An adaptive Metropolis algorithm, *Bernoulli*, **7**, 223–242, 2001.

102. Cui, T., Bayesian calibration of geothermal reservoir models via Markov Chain Monte Carlo, Ph.D. Thesis, The University of Auckland, 2010.

103. Orlande, H. R. B., Colaço, M. J., Dulikravich, G. S., Bayesian estimation of the thermal conductivity components of orthotropic solids. In *Fifth National Congress of Mechanical Engineering*, Brazil, 2008.

104. Mejias, M. M., Orlande, H. R. B., Ozisik, M. N., Effects of the heating process and body dimensions on the estimation of the thermal conductivity components of orthotropic solids, *Inverse Problems in Engineering*, **11**(1), 75–89, 2003.

105. Nagasaka, Y., Nagashima, A., Simultaneous measurement of the thermal conductivity and the thermal diffusivity of liquids by the transient hot-wire method, *Review of Scientific Instruments*, **52**(2), 229–232, 1981.

106. Assael, M. J., Dix, M., Gialou, K., Vozar, L., Wakeham, W. A., Application of the transient hot-wire technique to the measurement of the thermal conductivity of solids, *International Journal of Thermophysics*, **23**(3), 615–633, 2002.

107. Blackwell, J., A transient-flow method for determination of thermal constants of insulating materials in bulk part I-Theory, *Journal of Applied Physics*, **25**(2), 137–144, 1954.

108. Thomson, N. H., Orlande, H. R. B., Computation of sensitivity coefficients and estimation of thermophysical properties with the line heat source method, In *III European Conference on Computational Mechanics*, 478–478, Springer, Dordrecht, 2006.

109. Banaszkiewicz, M., Seiferlin, K., Spohn, T., Kargl, G., Kömle, N., A new method for the determination of thermal conductivity and thermal diffusivity from linear heat source measurements, *Review of Scientific Instruments*, **68**(11), 4184–4190, 1997.

110. Sassi, M. B., Dos Santos, C. A., Da Silva, Z. E., Gurgel, J. M., Junior, J., Heat conduction models for the transient hot wire technique, *High Temperatures-High Pressures*, **38**(2), 97–117, 2009.

111. Lamien, B., Orlande, H. R. B., Simultaneous estimation of thermal conductivity and volumetric heat capacity of viscous liquids with the line heat source probe via Bayesian inference, *High Temperatures-High Pressures*, **42**(3), 151–174, 2013.

112. Qiu, T. Q., Tien, C. L., Femtosecond laser heating of multi-layer metals-I. Analysis, *International Journal of Heat and Mass Transfer*, **37**(17), 2789–2797, 1994.

113. Ozisik, M. N., Tzou, D. Y., On the wave theory in heat conduction, *ASME Journal of Heat Transfer*, **116**, 526–535, 1994.

114. Tzou, D. Y., Chiffelle, R. J., Ozisik, M. N., The lattice temperature in the microscopic two-step model, *ASME Journal of Heat Transfer*, **116**, 1034–1038, 1994.

115. Qiu, T. Q., Tien, C. L., Size effects on nonequilibrium laser heating of metal films, *ASME Journal of Heat Transfer*, **115**, 842–847, 1993.

116. Orlande, H. R. B., Özisik, M. N., Tzou, D. Y., Inverse analysis for estimating the electron-phonon coupling factor in thin metal films, *Journal of Applied Physics*, **78**(3), 1843–1849, 1995.

117. Tzou, D. Y., *Macro- To Micro-Scale heat transfer: The lagging behavior*, CRC Press, Boca Raton, FL, 1996.

118. Battaglia, J.-L., Maillet, D., Modeling in heat transfer, Chapter 1 in *Thermal measurements and inverse techniques*, Edited by Orlande, H. R. B., Fudym, O., Maillet, D., Cotta, R. M., CRC Press, Boca Raton, FL, 2011.

119. Nóbrega, P. H. A., Orlande, H. R. B., Battaglia, J.-L., Bayesian estimation of thermophysical parameters of thin metal films heated by fast laser pulses, *International Communications in Heat and Mass Transfer*, **38**(9), 1172–1177, 2011.

120. Geweke, J., Evaluating the Accuracy of sampling-based approaches to the calculation of posterior moments, Vol. 196. Federal Reserve Bank of Minneapolis, Research Department, Minneapolis, MN, 1991.

121. Gelman, A., Rubin, D., Inference from iterative simulation using multiple sequences, *Statistical Science*, **7**, 457–472, 1992.

122. Gelman, A., Shirley, K., Inference from simulations and monitoring convergence, Chapter 6 in *Handbook of Markov Chain Monte Carlo*, Edited by Brooks, S., Gelman, A., Jones, G., Meng, X., CRC Press, Boca Raton, FL, pp. 163–174, 2011.

123. Roberts, G. O., Gelman, A., Gilks, W. R., Weak convergence and optimal scaling of random walk Metropolis algorithms, *The Annals of Applied Probability*, **7**(1), 110–120, 1997.

124. Rosenthal, J., Optimal proposal distribution and adaptive MCMC, Chapter 4 in *Handbook of Markov Chain Monte Carlo*, Edited by Brooks, S., Gelman, A., Jones, G., Meng, X., CRC Press, Boca Raton, FL, pp. 93–112, 2011.

125. Orlande, H. R. B., Colaço, M. J., Dulikravich, G. S., Approximation of the likelihood function in the Bayesian technique for the solution of inverse problems, *Inverse Problems in Science and Engineering*, **16**(6), 677–692, 2008.

126. Christen, J., Fox, C., Markov chain Monte Carlo using an approximation, *Journal of Computational and Graphical Statistics*, **14**(4), 795–810, 2005.

127. Nissinen, A., Modelling errors in electrical impedance tomography, Dissertation in Forestry and Natural Sciences, University of Eastern Finland, 2011.

128. Nissinen, A., Heikkinen, L., Kaipio, J., The Bayesian approximation error approach for electrical impedance tomography: Experimental results, *Measurement Science and Technology*, **19**, 015501, 2008.

129. Nissinen, A., Heikkinen, L., Kolehmainen, V., Kaipio, J., Compensation of errors due to discretization, domain truncation and unknown contact impedances in electrical impedance tomography, *Measurement Science and Technology*, **20**, 105504, 2009.

130. Nissinen, A., Kolehmainen, V., Kaipio, J., Compensation of modeling errors due to unknown boundary domain in electrical impedance tomography, *IEEE Transactions on Medical Imaging*, **30**, 231–242, 2011.

131. Nissinen, A., Kolehmainen, V., Kaipio, J., Reconstruction of domain boundary and conductivity in electrical impedance tomography using the approximation error approach, *International Journal for Uncertainty Quantification*, **1**, 203–222, 2011.

132. Lamien, B., Orlande, H. R. B., Approximation error model to account for convective effects in liquids characterized by the line heat source probe, In *Fourth Inverse Problems, Design and Optimization Symposium*, Albi, France, 2013.

133. Kardestunner, H., Horrie, D. (eds.), *Finite element handbook*, McGraw Hill, New York, 1987.

134. Lebedev, L. P., Vorovich, I. I., Gladwell, G. M., *Functional analysis: Applications in mechanics and inverse problems* (Vol. 41), Springer Science & Business Media, Berlin, Germany, 2012.

135. Le Masson, P., Loulou, T., Artioukhine, E., Estimations of a 2D convection heat transfer coefficient during a metallurgical "Jominy end-quench" test: Comparison between two methods and experimental validation, *Inverse Problems in Science and Engineering*, **12**(6), 595–617, 2004.

136. Machado, H. A., Orlande, H. R. B., Inverse analysis of estimating the timewise and spacewise variation of the wall heat flux in a parallel plate channel, *International Journal of Numerical Methods for Heat and Fluid Flow*, **7**, 696–710, 1997.

137. Rodrigues, F. A., Orlande, H. R. B., Dulikravich, G. S., Simultaneous estimation of spatially-dependent diffusion coefficient and source-term in a nonlinear 1D diffusion problem, *Mathematics and Computers in Simulation*, **66**, 409–424, 2004.

138. Saker, L. F., Orlande, H. R. B., Huang, C., Kanevce, G., Kanevce, L., Simultaneous estimation of the spacewise and timewise variations of mass and heat transfer coefficients in drying, *Inverse Problems in Science and Engineering*, **15**, 137–150, 2007.

139. Luikov, A. V., *Heat and mass transfer in capillary-porous bodies*, Pergamon Press, Oxford, 1966.

140. Tick, J., Pulkkinen, A., Tarvainen, T., Image reconstruction with uncertainty quantification in photoacoustic tomography, *The Journal of the Acoustical Society of America*, **139**(4), 1951–1961, 2016.

141. Chung, J., Saibaba, A. K., Brown, M., Westman, E., Efficient generalized Golub–Kahan based methods for dynamic inverse problems, *Inverse Problems*, **34**(2), 024005, 2018.

142. Song, H. R., Fuentes, M., Ghosh, S., A comparative study of Gaussian geostatistical models and Gaussian Markov random field models, *Journal of Multivariate Analysis*, **99**(8), 1681–1697, 2008.

143. Williams, C. K., Rasmussen, C. E., *Gaussian processes for machine learning*, Cambridge, MA, MIT Press, 2006.

144. Orlande, H. R. B., Lutaif, N. A., Gontijo, J. A. R., Estimation of the kidney metabolic heat generation rate, *International Journal for Numerical Methods in Biomedical Engineering*, **35**(9), e3224, 2019.

145. Alaeian, M., Orlande, H. R. B., Machado, J. C., Temperature estimation of inflamed bowel by the photoacoustic inverse approach, *International Journal for Numerical Methods in Biomedical Engineering*, **36**(3), e3300, 2020.

146. Beard, P., Biomedical photoacoustic imaging, *Interface Focus*, **1**(4), 602–631, 2011.

147. Ripoll, J., Ntziachristos, V., Quantitative point source photoacoustic inversion formulas for scattering and absorbing media, *Physical Review E*, **71**(3), 031912, 2005.

148. Yin, L., Wang, Q., Zhang, Q., Jiang, H., Tomographic imaging of absolute optical absorption coefficient in turbid media using combined photoacoustic and diffusing light measurements, *Optics Letters*, **32**(17), 2556–2558, 2007.

149. Alaeian, M., Orlande, H. R. B., Inverse photoacoustic technique for parameter and temperature estimation in tissues, *Heat Transfer Engineering*, **38**(18), 1573–1594, 2017.

150. Alaeian, M., Orlande, H. R. B., Lamien, B., Application of the photoacoustic technique for temperature measurements during hyperthermia, *Inverse Problems in Science and Engineering*, **27**, 1–21, 2018.

151. Ke, H., Tai, S., Wang, L. V., Photoacoustic thermography of tissue, *Journal of Biomedical Optics*, **19**(2), 026003, 2014.

152. Pramanik, M., Wang, L. V., Thermoacoustic and photoacoustic sensing of temperature, *Journal of Biomedical Optics*, **14**(5), 054024, 2009.

153. Welch, A. J., Van Gemert, M. J. (eds.), *Optical-thermal response of laser-irradiated tissue* (Vol. 2), Springer, New York, 2011.

154. Yao, D. K., Zhang, C., Maslov, K. I., Wang, L. V., Photoacoustic measurement of the Gruneisen parameter of tissue, *Journal of Biomedical Optics*, **19**(1), 017007, 2014.

155. Cox, B. T., Kara, S., Arridge, S. R., Beard, P. C., k-space propagation models for acoustically heterogeneous media: Application to biomedical photoacoustics, *The Journal of the Acoustical Society of America*, **121**(6), 3453–3464, 2007.

156. Yuan, X., Borup, D., Wiskin, J., Berggren, M., Johnson, S. A., Simulation of acoustic wave propagation in dispersive media with relaxation losses by using FDTD method with PML absorbing boundary condition, *IEEE Transactions on Ultrasonics, Ferroelectrics and Frequency Control*, **46**(1), 14–23, 1999.

157. Abreu, L. A., Orlande, H. R. B., Colaço, M. J., Kaipio, J., Kolehmainen, V., Pacheco, C. C., Cotta, R. M., Detection of contact failures with the Markov chain Monte Carlo method by using integral transformed measurement, *International Journal of Thermal Sciences*, **132**, 486–497, 2018.

158. Emery, A. F., Generating simulated responses for stochastic systems using polynomial chaos and wick products, In *Proceedings of Inverse Problems, Design and Optimization Symposium (IPDO-2004)*, Rio de Janeiro, 29–36, 2004.

159. Bialecki, R. A., Kassab, A. J., Fic, A., Proper orthogonal decomposition and modal analysis for acceleration of transient FEM thermal analysis, *International Journal for Numerical Methods in Engineering*, **62**, 774–797, 2005.

160. Bialecki, R. A., Kassab, A. J., Ostrowski, Z., Application of the proper orthogonal decomposition in steady state inverse problems, in *Inverse problems in engineering mechanics IV*, Edited by Tanaka, M., Elsevier, Amsterdam, pp. 3–12, 2003.

161. Fic, A., Bialecki, R. A., Kassab, A. J., Solving transient nonlinear heat conduction problems by proper orthogonal decomposition and the finite-element method, *Numerical Heat Transfer, Part B: Fundamentals*, **48**, 103–124, 2005.

162. Ostrowski, Z., *Application of proper orthogonal decomposition to the solution of inverse problems*, Silesian University of Technology, Gliwice, Poland, 2006.

163. Petit, D., Girault, M., Model Reduction in inverse heat transfer problems, In *Proceedings of Inverse Problems, Design and Optimization Symposium (IPDO-2004)*, Rio de Janeiro, 37–54, 2004.

164. Berger, J., Orlande, H. R. B., Mendes, N., Proper generalized decomposition model reduction in the Bayesian framework for solving inverse heat transfer problems, *Inverse Problems in Science and Engineering*, **25**(2), 260–278, 2017.

165. Cotta, R. M., Hybrid numerical-analytical approach to nonlinear diffusion problems, *Numerical Heat Transfer, Part B- Fundamentals*, **127**, 217–226, 1990.

166. Cotta, R. M., *Integral transforms in computational heat and fluid flow*, CRC Press, Boca Raton, FL, 1993.

167. Cotta, R. M., Benchmark results in computational heat and fluid flow: The integral transform method, *International Journal of Heat and Mass Transfer*, **37**, 381–394, 1984.

168. Cotta, R. M., Mikhailov, M. D., *Heat conduction: Lumped analysis, integral transforms, symbolic computation*, Wiley-Interscience, New York, 1997.

169. Cotta, R. M., *The integral transform method in thermal and fluids sciences and engineering*, Begell House, New York, 1998.

170. Abreu, L. A. S., Orlande, H. R. B., Naveira-Cotta, C. P., Quaresma, J. N. N., Cotta, R. M., Kaipio, J., Kolehmainen, V., Identification of contact failures in multi-layered composites, *ASME 2011 International Design Engineering Technical Conference and Computers and Information in Engineering Conference*, Washington, DC, 479–487, 2011.

171. Abreu, L. A. S., Orlande, H. R. B., Kaipio, J., Kolehmainen, V., Cotta, R. M., Quaresma, J. N. N., Identification of contact failures in multilayered composites with the Markov Chain Monte Carlo method, *Journal of Heat Transfer*, **136**, 101302, 2014.

172. Orlande, H. R. B., Dulikravich, G. S., Neumayer, M., Watzenig, D., Colaço, M. J., Accelerated Bayesian inference for the estimation of spatially varying heat flux in a heat conduction problem, *Numerical Heat Transfer, Part A: Applications*, **65**(1), 1–25, 2014.

173. Fonseca, H., Fudym, O., Orlande, H. R. B., Batsale, J. C., Nodal predictive error model and Bayesian approach for thermal diffusivity and heat source mapping, *Comptes Rendus Mécanique*, **338**, 434–449, 2010.

174. Fonseca, H., Orlande, H. R. B., Fudym, O., Estimation of position-dependent transient heat source with the Kalman filter, *Inverse Problems in Science and Engineering*, **20**, 1079–1099, 2012.

175. Naveira-Cotta, C., Orlande, H. R. B., Cotta, R., Combining integral transforms and Bayesian inference in the simultaneous identification of variable thermal conductivity and thermal capacity in heterogeneous media, *ASME Journal of Heat Transfer*, **133**, 111301–111311, 2011.

176. Naveira-Cotta, C., Orlande, H. R. B., Cotta, R., Integral transforms and Bayesian inference in the identification of variable thermal conductivity in two-phase dispersed systems, *Numerical Heat Transfer: Part B, Fundamentals*, **57**, 173–202, 2010.

177. Naveira-Cotta, C., Cotta, R., Orlande, H. R. B., Inverse analysis with integral transformed temperature fields: Identification of thermophysical properties in heterogeneous media, *International Journal of Heat and Mass Transfer*, **54**, 150–1519, 2011.

178. Knupp, D., Naveira-Cotta, C., Cotta, R., Orlande, H. R. B., Experimental identification of thermophysical properties in heterogeneous materials with integral transformation of temperature measurements from infrared thermography, *Experimental Heat Transfer*, **26**, 1–25, 2013.

179. Knupp, D. C., Naveira-Cotta, C. P., Ayres, J. V., Cotta, R. M., Orlande, H. R. B. Theoretical-experimental analysis of heat transfer in nonhomogeneous solids via improved lumped formulation, integral transforms and infrared thermography, *International Journal of Thermal Sciences*, **62**, 71–84, 2012.

180. Kaipio, J., Duncan S., Seppanen, A., Somersalo, E., Voutilainen, A, State estimation for process imaging, in *Handbook of process imaging for automatic control*, Edited by Scott, D., McCann, H., CRC Press, Boca Raton, FL, pp. 207–235, 2005.

181. Kalman, R., A new approach to linear filtering and prediction problems, *ASME Journal of Basic Engineering*, **82**, 35–45, 1960.

182. Sorenson, H., Least-squares estimation: From Gauss to Kalman, *IEEE Spectrum*, **7**, 63–68, 1970.

183. Andrieu, C., Doucet, A., Robert, C. P., Computational advances for and from Bayesian analysis, *Statistical Science*, **19**(1), 118–127, 2004.

184. Andrieu, C., Doucet, A., Singh, S. S., Tadic, V. B., Particle methods for change detection, system identification, and control, *Proceedings of the IEEE*, **92**(3), 423–438, 2004.

185. Carpenter, J., Clifford, P., Fearnhead, P., An improved particle filter for non-linear problems, *IEEE Proceedings of Part F: Radar and Sonar Navigation*, **146**, 2–7, 1999.

186. Doucet, A., Godsill, S., Andrieu, C., On sequential Monte Carlo sampling methods for Bayesian filtering, *Statistics and Computing*, **10**, 197–208, 2000.

187. Johansen, A. M., Doucet, A., A note on auxiliary particle filters, *Statistics and Probability Letters*, **78**(12), 1498–1504, 2008.

188. Liu, J., Chen, R., Sequential Monte Carlo methods for dynamical systems, *Journal of American Statistical Association*, **93**, 1032–1044, 1998.

189. Del Moral, P., Doucet, A., Jasra, A., Sequential Monte Carlo samplers, *Journal of the Royal Statistical Society: Series B (Statistical Methodology)*, 68(3), 411–436, 2006.

190. Del Moral, P., Jasra, A., Sequential Monte Carlo for Bayesian computation, in *Bayesian statistics*, Edited by Bernardo, J. M., Bayarri, M. J., Berger, J. O., Dawid, A. P., Heckerman, D., Smith, A. F. M., West, M., Oxford University, London, pp. 1–34, 2007.

191. Orlande, H. R. B., Colaco, M. J., Dulikravich, G. S., Vianna, F., Da Silva, W., Fonseca, H., Fudym, O., State estimation problems in heat transfer, *International Journal for Uncertainty Quantification*, **2**(3), 239–258, 2012.

192. Vianna, F., Orlande, H. R. B., Dulikravich, G. S., Optimal heating control to prevent solid deposits in pipelines, In *Proceedings of 5th European Conference on Computational Fluid Dynamics: ECCOMAS CFD*, Lisbon, 2010.

193. Vianna, F., Orlande, H. R. B., Dulikravich, G. S., Estimation of the temperature field in pipeline by using the Kalman filter, In *Proceedings of 2nd International Congress of Serbian Society of Mechanics (IConSSM 2009)*, 1–5 June, Palić, Serbia, 2009.

194. Pacheco, C. C., Orlande, H. R. B., Colaço, M. J., Dulikravich, G. S., Estimation of a location-and time-dependent high-magnitude heat flux in a heat conduction problem using the Kalman filter and the approximation error model, *Numerical Heat Transfer, Part A: Applications*, **68**(11), 1198–1219, 2015.

195. Simon, D., *Optimal state estimation: Kalman, H Infinity, and nonlinear approaches*, John Wiley & Sons, Inc., Hoboken, NJ, 2006.

196. Pacheco, C. C., Orlande, H. R. B., Colaço, M. J., Dulikravich, G. S. Real-time identification of a high-magnitude boundary heat flux on a plate, *Inverse Problems in Science and Engineering*, **24**(9), 1661–1679, 2016.

197. Pacheco, C., Orlande, H. R. B., Colaco, M., Dulikravich, G. S. State estimation problems in PRF-shift magnetic resonance thermometry, *International Journal of Numerical Methods for Heat and Fluid Flow*, **28**(2), 315–335, 2018.

198. Pacheco, C. C., Orlande, H. R. B., Colaço, M. J., Dulikravich, G. S., Varón, L. A., Lamien, B., Real-time temperature estimation with enhanced spatial resolution during MR-guided hyperthermia therapy, *Numerical Heat Transfer, Part A: Applications*, **77**(8), 782–806, 2020.

199. Da Silva, W. B., Rochoux, M. C., Orlande, H. R. B., Colaço, M. J., Fudym, O., El Hafi, M., Cuenot, B., Ricci, S., Application of particle filters to regional-scale wildfire spread, *High Temperatures High Pressures*, **43**, 415–440, 2014.

200. Liu, J., West, M., Combined parameter and state estimation in simulation based filtering, in *Sequential Monte Carlo Methods in Practice*, Edited by Doucet, A., Freitas, N., Gordon, N., Springer, New York, pp. 197–217, 2001.

201. Sheinson, D. M., Niemi, J., Meiring, W., Comparison of the performance of particle filter algorithms applied to tracking of a disease epidemic, *Mathematical Biosciences*, Elsevier Inc., **255**, 21–32, 2014.

202. West, M., Approximating posterior distributions by mixture, *Journal of the Royal Statistical Society B*, **55**, 409–422, 1993.

203. Rios, M. P., Lopes, H. F., The extended Liu and West filter: Parameter learning in Markov switching stochastic volatility models, in *State-space models: Applications in economics and finance, statistics and econometrics for finance 1*, Edited by Zeng, Y. and Wu, S., Springer, New York, pp. 23–62, 2013.

204. Rothermel, R. C., A mathematical model for predicting fire spread in wildland fuels, Research Paper INT-115, US Department of Agriculture Forest Service, 1972.

205. Lamien, B., Varon, L. A. B., Orlande, H. R. B., Elicabe, G. E., State estimation in bioheat transfer: A comparison of particle filter algorithms, *International Journal of Numerical Methods for Heat and Fluid Flow*, **27**(3), 615–638, 2017.

206. Lamien, B., Orlande, H. R. B., Eliçabe, G. E. Inverse problem in the hyperthermia therapy of cancer with laser heating and plasmonic nanoparticles, *Inverse Problems in Science and Engineering*, **25**(4), 608–631, 2017.

207. Lamien, B., Orlande, H. R. B., Eliçabe, G., Particle filter and approximation error model for state estimation in hyperthermia, *Journal of Heat Transfer*, **139**(1), 12001, 2017.

208. Varon, L. A. B., Orlande, H. R. B., Elicabe, G. E., Estimation of state variables in the hyperthermia therapy of cancer with heating imposed by radiofrequency electromagnetic waves, *International Journal of Thermal Sciences*, **98**, 228–236, 2015.

209. Varon, L. A. B., Orlande, H. R. B., Eliçabe, G. E., Combined parameter and state estimation in the radio frequency hyperthermia treatment of cancer, *Numerical Heat Transfer, Part A: Applications*, **70**(6), 581–594, 2016.

210. Lamien, B., Orlande, H. R. B., Antonio Bermeo Varón, L., Leite Queiroga Basto, R., Enrique Eliçabe, G., Silva dos Santos, D., Cotta, R., Estimation of the temperature field in laser-induced hyperthermia experiments with a phantom, *International Journal of Hyperthermia*, **35**(1), 279–290, 2018.

211. Pennes, H. H., Analysis of tissue and arterial blood temperatures in the resting human forearm, *Journal of Applied Physiology*, **1**(2), 93–122, 1948.

Index

Note: **Bold** page numbers refer to tables and *italic* page numbers refer to figures

For Product Safety Concerns and Information please contact our EU
representative GPSR@taylorandfrancis.com
Taylor & Francis Verlag GmbH, Kaufingerstraße 24, 80331 München, Germany

www.ingramcontent.com/pod-product-compliance
Lightning Source LLC
Chambersburg PA
CBHW061343210326
41598CB00035B/5866

9 780367 725266